Herbal Radiomodulators

Applications in Medicine, Homeland Defence and Space

This book is dedicated with love to

My Dear Mother

Mrs. Shanti Arora

[*My Mother, truly as Her name signified, was an embodiment of Peace. She lived a short, nonetheless eventful, saintly life. She prayed for the welfare of all living beings and taught me the true meaning of the Sanskrit Prayer (given below). She continues to reside in my heart and will be there till eternity*!]

सर्वे भवन्तु सुखिनः
सर्वे सन्तु निरामयाः।
सर्वे भद्राणि पश्यन्तु
मा कश्चिद् दुःखभाग् भवेत् ।।

ॐ शान्तिः शान्तिः शान्तिः

Sarve bhavantu sukhinah
Sarve santu nirāmayāh |
Sarve bhadrāni paśyantu
Mā kaścit dukha bhāgbhavet ||

Om Shanti, Shanti, Shanti

May all be happy
May all be free from disease, pain and suffering
May all see good in everyone
May all be free from suffering

Om Peace! Peace! Peace!

Herbal Radiomodulators

Applications in Medicine, Homeland Defence and Space

Editor

Rajesh Arora, PhD

Scientist
Defence Research and Development Service
Institute of Nuclear Medicine and Allied Sciences
(Defence Research and Development Organization)
Delhi, India

www.cabi.org

CABI is a trading name of CAB International

CABI Head Office
Nosworthy Way
Wallingford
Oxfordshire OX10 8DE
UK

Tel: +44 (0)1491 832111
Fax: +44 (0)1491 833508
E-mail: cabi@cabi.org
Website: www.cabi.org

CABI North American Office
875 Massachusetts Avenue
7th Floor
Cambridge, MA 02139
USA

Tel: +1 617 395 4056
Fax: +1 617 354 6875
E-mail: cabi-nao@cabi.org

A catalogue record for this book is available from the British Library, London, UK.
Library of Congress Cataloging-in-Publication Data

Herbal radiomodulators : applications in medicine, homeland defence, and space / edited by Rajesh Arora.
p. ; cm.
Includes bibliographical references and index.
ISBN 978-1-84593-395-1 (alk. paper)
1. Radiation-protective agents. 2. Herbs--Therapeutic use. 3.
2. Radiotherapy. I. Arora, Rajesh. II. Title.
[DNLM: 1. Radiation-Protective Agents--pharmacology. 2. Radiation-Protective Agents--therapeutic use. 3. Phytotherapy. 4. Plant Preparations--therapeutic use. 5. Plants, Medicinal. 6. Radiation Injuries--drug therapy. WN 650 H534 2008]

RM849.H47 2008
615'.321--dc22

2008005764

ISBN-13: 978 1 84593 395 1

Printed and bound in the UK from copy supplied by the author and Ms. Preeti Arora by Biddles Ltd, Kings Lynn.

Contents

Contributors

Alexandru Almasan, Department of Cancer Biology, Cleveland Clinic, Cleveland, Ohio 44195, USA

Rajesh Arora, Division of Radiation Biology and Radioprotection, Institute of Nuclear Medicine and Allied Sciences, Brig SK Mazumdar Road, Delhi-110054, India

Debashish Banerjee, Bio-organic Division, Bhabha Atomic Research Centre (BARC), Trombay, Mumbai-400085, India

Ivan Bašić, Department of Animal Physiology, University of Zagreb, Faculty of Science, Rooseveltov trg 6 HR-10000, Zagreb, Croatia

F. Bayıroğlu, Department of Physiology, Faculty of Veterinary Medicine, Yüzüncü Yıl University, Van, Turkey

Vesna Benković, Department of Animal Physiology, University of Zagreb, Faculty of Science, Rooseveltov trg 6 HR-10000, Zagreb, Croatia

Carmia Borek, Department of Public Health and Family Medicine, Nutrition Infectious Disease Unit, Tufts University School of Medicine, Boston, MA 02111, USA

M.E. Büyükokuroğlu, Department of Chemistry (Biochemistry Division), Faculty of Science and Arts, Afyon Kocatepe University, Afyonkarahisar, Turkey

Subrata Chattopadhyay, Bio-organic Division, Bhabha Atomic Research Centre (BARC), Trombay, Mumbai-400085, India

William Couldwell, Dept of Neurosurgery, University of Utah, 175 N Medical Drive, Salt Lake City, UT 84132, USA

Peter A. Crooks, College of Pharmacy, University of Kentucky, Lexington, KY 40536, USA

T.P.A. Devasagayam, Bhabha Atomic Research Centre (BARC), Trombay, Mumbai-400085, India

Andrei V. Gudkov, Dept of Cell Stress Biology, Roswell Park Cancer Institute, Buffalo, NY 14227, USA

Damodar Gupta, Dept of Cell Stress Biology, Roswell Park Cancer Institute, Buffalo, NY 14227, USA

N. M. Abdel-Hamid, College of Pharmacy, Minia University, Egypt

Seyed Jalal Hosseinimehr, Dept of Medicinal Chemistry, Faculty of Pharmacy, Mazandaran University of Medical Sciences, Sari, Iran

Ganesh C. Jagetia, Department of Zoology, Mizoram University, Tanhril, Aizawl-796009, Mizoram, India

Wendy Maddocks-Jennings, MJ Health, 16 Milverton Avenue, Palmerston North 4410, New Zealand

Gordana Joksic, Vinca Institute of Nuclear Sciences, 11001 Belgrade, Serbia and Montenegro

James Joseph, Human Nutrition Research Center on Aging, USDA-ARS, Tufts University, Boston, MA 02111, USA

Jisha Joy, Amala Cancer Research Center, Trichur-680555, Kerala, India
Jayashree P. Kamat, Bhabha Atomic Research Centre (BARC), Trombay, Mumbai, India
Anica Horvat-Knežević, Department of Animal Physiology, University of Zagreb, Faculty of Science, Rooseveltov trg 6 HR-10000, Zagreb, Croatia
M. Koç, Department of Radiation Oncology, Faculty of Medicine, Ataturk University, Erzurum, Turkey
R. Kumar, Division of Radiation Biology and Radioprotection, Institute of Nuclear Medicine and Allied Sciences, Brig SK Mazumdar Road, Delhi-110054, India
Michael R. Landauer, Armed Forces Radiobiology Research Institute AFRRI- Bldg. 42, 8901 Wisconsin Avenue, Bethesda, MD 20889-5603, USA
Andreja Leskovac, Vinca Institute of Nuclear Sciences, Belgrade, Serbia
Roger M. Macklis, Department of Radiation Oncology, Cleveland Clinic, Cleveland, Ohio 44195, USA
Lucien C. Manchester, The University of Texas Health Science Center, Department of Cellular and Structural Biology, San Antonio, 7703 Floyd Curl Drive-MC 7762, San Antonio, Texas 78229-3900, USA
D.K. Maurya, Bhabha Atomic Research Centre (BARC), Trombay, Mumbai 400085, India
Rajiv R. Mohan, Mason Eye Institute and College of Veterinary Medicine, University of Missouri-Columbia, Columbia, MO 65212, USA
Ralph Moss, Cancer Decisions, Pennsylvania, USA
Cemek Mustafa, Department of Chemistry (Biochemistry Division), Faculty of Science and Arts, Afyon Kocatepe University, Afyonkarahisar, Turkey
C.K.K. Nair, Amala Cancer Research Center, Trichur-680555, Kerala, India
Toba Niazi, Department of Neurosurgery, University of Utah, 175 N Medical Drive, Salt Lake City, UT 84132, USA
Nada Oršolić, Department of Animal Physiology, University of Zagreb, Faculty of Science, Rooseveltov trg 6 HR-10000, Zagreb, Croatia
Sandra Petrović, Vinca Institute of Nuclear Sciences, Belgrade, Serbia
Bernard M. Rabin, University of Maryland, Baltimore County, USA
Russel J. Reiter, The University of Texas Health Science Center, Department of Cellular and Structural Biology, San Antonio, 7703 Floyd Curl Drive-MC 7762, San Antonio, Texas 78229-3900, USA
Abdullah Sakarcan, Division of Pediatric Nephrology, USC School of Medicine, University of South Carolina, Columbia SC, USA
Goksel Sener, Department of Pharmacology, School of Pharmacy, Marmara University, Haydarpasa 34668, Istanbul, Turkey
A. Sharma, Division of Radiation Biology and Radioprotection, Institute of Nuclear Medicine and Allied Sciences, Brig SK Mazumdar Road, Delhi-110054, India
Barbara Shukitt-Hale, Human Nutrition Research Center on Aging, USDA-ARS, Tufts University, Boston, MA 02111, USA

Hiroshi Tamura, The University of Texas Health Science Center, Department of Cellular and Structural Biology, San Antonio, 7703 Floyd Curl Drive-MC 7762, San Antonio, Texas 78229-3900, USA

Dun-Xian Tan, The University of Texas Health Science Center, Department of Cellular and Structural Biology, San Antonio, 7703 Floyd Curl Drive-MC 7762, San Antonio, Texas 78229-3900, USA

M. Pilar Terron, The University of Texas Health Science Center, Department of Cellular and Structural Biology, San Antonio, Texas, USA

Rakshamani Tripathi, V.B.S. Purvanchal University, Jaunpur, India

R.P. Tripathi, Institute of Nuclear Medicine and Allied Sciences, Brig SK Mazumdar Road, Delhi-110054, India

P. Uma Devi, Pathirissery Mana, ARA-B-35A, Plavilakonam, Thachottukavu, Peyad P.O., Thiruvananthapuram-695573, Kerala, India

David R. Worthen, College of Pharmacy, University of Kentucky, Lexington, KY 40536 USA

Berrak C. Yegen, Department of Pharmacology, School of Pharmacy, Marmara University, Haydarpasa 34668, Istanbul, Turkey

Preface

Nature surprisingly unfolds problems that test the ingenuity of man! The mysterious radiation has been around ever since the Universe came into being. However, over the last few centuries it has been exploited by humans for useful applications. Though seemingly useful, radiation acts like a double-edged sword and its indiscriminate use can even have life threatening consequences. While radiation exposures to life forms can occur in a variety of natural settings, human intervention has substantially increased their probability.

Strategic and tactical wars of the future involve the risk of nuclear weapon usage, while the use of radiological/nuclear material by organized terror groups also poses a legitimate threat. Such incidents could result in significant loss of human lives, besides socio-economic losses and irreparable damage to the environment. From a military standpoint, keeping the man behind the machine fit for defensive operational requirements, often in entirely new terrains and across conventional borders in diverse and extremely demanding conditions, is a major challenge. In a nuclear theatre of operations, the task becomes even more arduous and protecting soldiers, paramilitary personnel and emergency responders against the deleterious radiation effects assumes paramount importance.

While the human race is trying to handle problems created by itself, there are other problems too that require attention. Radiation emanating from the Universe is a big impediment in man's quest to understand the Universe better. As man embarks on space missions to explore the untouched, almost infinite regions, several mysteries are likely to unfold–but how will man handle the deleterious effects of radiation?

Radiotherapy is an important modality for treating cancer and protecting normal tissue, while differentially sensitizing tumours is another challenge that confronts researchers/ clinicians.

The aforementioned scenarios have necessitated the development of radiomodulators for clinical and field applications. Radioprotective drugs hold immense promise for saving precious human lives in a radiation environment and are likely to find application in scenarios like wars, terrorist incidents, reactor accidents and spillages, space explorations etc. as alluded to earlier, while radiosensitizers can substantially improve the results of radiotherapy. The use of currently available synthetic radiomodulators is fraught with toxicity, and hence the need to discover and develop new, more effective, yet less toxic radiomodulatory drugs from natural sources.

The flora of several endemic regions of the world have evolved amidst highly stressful environmental conditions over millions of years, and consequently have developed an array of mechanisms to cope with the diverse stresses, often by synthesizing a repertoire of secondary metabolites. Several of these secondary metabolites exhibit diverse pharmacological properties and have been effectively utilized for radiomodulation.

The present book is an attempt to present a global perspective on the recent developments in the area of radioprotection and radiosensitization by drugs of herbal origin. The book is organized into six major sections that cover the entire gamut of radiomodulation.

Section I provides an overview of the various facets of radiomodulation, discusses some key issues and highlights the plausible application of herbal radiomodulators (radioprotectors and radiosensitizers) in diverse areas e.g., in the clinic, homeland defence, military and space. This section besides providing a bird's-eye view of radiomodulation by herbals suggests key areas where attention needs to be focused in future.

Section II, comprising seven chapters, discusses the use of phytomedicine for radioprotection and medical management of radiation injuries. The radioprotective effects of herbs from different regions of the world, including India, Europe, Egypt and New Zealand etc., have been covered in this section. In addition, the role of melatonin, which is found in several fruits and edible plants, in rendering protection against ionizing radiation has also been discussed.

In recent years, dietary compounds and nutraceuticals have received increasing attention since they can obviate the need for taking additional drugs that often cause severe side effects. Section III provides an insight on the use of dietary ingredients/nutraceuticals in rendering biological as well as behavioural radioprotection. Five chapters of this section are exclusively devoted to this topic. The potential use of such agents for protecting military personnel, medical first responders and astronauts is tremendous. Radiation-induced behavioural perturbations can severely impair the operational efficiency of soldiers/ first responders hampering crucial rescue missions during nuclear emergencies. The availability of effective herbal radioprotectors should greatly improve the operational efficiency of soldiers operating in hostile conditions and radiation environments, besides helping civilian populations cope better post-irradiation.

Elucidation of the mode of action of radioprotective compounds is of paramount importance since it could help in designing highly effective radioprotective drugs in future. Section IV comprises four chapters that are exclusively devoted to mechanistic studies aimed at elucidating the mechanism of action of herbal radiomodulators, particularly the role of herbal drugs in protecting mitochondria, membrane proteins and intrinsic antioxidants.

Cancer is a leading cause of morbidity and mortality around the globe and radiosensitizers can modulate the radioresponse. Herbal radiosensitizers that act differentially can help effectively tackle the scourge of cancer, thereby benefiting patients. Four chapters of Section V specifically deal with the utilization of natural radiomodulatory plant products for improving cancer therapy.

The last section (Section VI) focuses on the enigmatic connection between the use of antioxidants and radiotherapy and discusses the pros and cons of their use during radiotherapy.

Advances in the area of herbal radiomodulators are taking place rapidly and we are not far from the day when over-the-counter drugs from plants (possibly from the many that have been discussed in this book) will be available as radioprotectors for emergency use.

Radiomodulation by natural products is an upcoming field of research where more attention needs to be focused. I hope that the book will instil enthusiasm and interest in the specialists to undertake research in this area for the benefit of mankind. The simple presentation should appeal equally to one and all, who would like to protect themselves from the harmful effects of radiation in the event of a radiation emergency.

Human life is precious and cannot be equated in terms of money! Therefore, if it should be possible as a result of the efforts of this book to save even a single human life or alleviate some suffering resulting due to radiation exposures in any part of the world, the purpose of the book would be served.

15th January, 2008

Rajesh Arora
(rajesharoradr@rediffmail.com)

The views expressed here and elsewhere in the book are my own and not those of the Government.

Foreword

Nuclear radiation, ever since its discovery by Wilhelm Conrad Roentgen in 1895, has been used in the agriculture, industry, medical and power sectors for a wide variety of useful applications. However, every technology has its advantages and disadvantages that go hand in hand. The nuclear era in which we live today is both a bane and a boon. It has offered immense benefits to mankind and at the same time has posed innumerable problems and challenges that need to be confronted. With increasing use of nuclear energy, there is an increased risk of radiation exposure to various life forms, and protecting humans against the harmful effects of radiation is a major challenge that needs an urgent solution. Development of effective radiation countermeasure agents for protecting humans against both planned and unplanned radiation exposures assumes importance in this context.

Advances in radiation sciences, particularly the understanding of radiation effects on biological systems, have paved the way towards development of radiomodulators and radiorecovery agents that can be effectively utilized to achieve protection against the deleterious effects of ionizing radiation. Research endeavours with synthetic radioprotectors in the past have met with little success primarily due to toxicity-related problems. Therefore, worldwide attention in recent years has turned towards developing radiation countermeasure agents and radiosensitizers from natural products for diverse applications. Herbals have shown immense promise as effective radiomodulators and the future appears quite bright.

Herbal drugs have been in use in several traditional systems of medicine for several hundreds and even thousands of years for treating various human ailments all over the world since they offer holistic treatment. In view of their practical usage often without any side effects in the clinic, they are considered highly acceptable, effective and non-toxic. As per World Health Organization estimates, almost 80% of the world's population still depends on herbal drugs for solving their medical problems since they are veritably reliable alternatives to the costly modern medicine, which the masses in most developing nations can't afford. In addition, herbals provide a wonderful platform for the synthesis of new chemical entities and development of novel drugs. These factors have contributed to the renewal of interest in herbals.

The Institute of Nuclear Medicine and Allied Sciences, Delhi, a constituent laboratory of DRDO, has done pioneering work in the area of herbal radioprotectors and it is only appropriate that Dr. Rajesh Arora has made efforts to compile this book on a most topical subject of universal relevance. I am very hopeful that the book would be of general interest to civil populations at large, and particularly useful to those who come into contact with radiation, including rescue agencies, emergency first responders, military personnel, scientists, physicians, academicians, space scientists and the drug industry fraternity.

I congratulate the editor for compiling such a timely book presenting the state-of-the-art in the area of radiomodulation.

Dr. W. Selvamurthy
Distinguished Scientist and Chief Controller Research and Development
Ministry of Defence (Government of India)
Defence Research and Development Organization, New Delhi, India

Foreword

Natural products have been a source of a wide variety of drugs that are used today in modern medicine. The current resurgence of interest in medicinal and aromatic plants is a global phenomenon. More and more countries, including the advanced nations, are looking at herbals with a keen eye on finding new chemical entities for treating diseases that afflict humans. Today a number of drugs that find use in western medicine are of plant origin, or synthesized taking cues from nature. A plethora of plants has been used for mitigation and treatment of free radical-mediated ailments like cancer, diabetes, neurodegenerative disorders, etc. Since free radicals are implicated in radiation-mediated damage, it's no wonder then that the nuclear effects of ionizing radiation can be modified by appropriate use of herbal drugs. To this end, research has been carried out in India and abroad, and a number of plants have been found to render effective radioprotection. Several plant-derived molecules have also been found to radiosensitize tumours, benefiting cancer patients. Although a wealth of information has been generated in recent years on radiomodulation by herbal drugs, it has not been appropriately documented so far.

Since radiation has an effect on almost everyone on this planet in one way or the other, this unique book, which focuses on radiomodulators of plant origin mainly of benefit to humans, will be of interest to a wide range of audience. I hope that the book will be a welcome collection for botanists, herbalists, radiation biologists, pharmacists, chemists, the pharma industry and the general public at large since it provides information generated from the latest research done the world over on how to protect oneself from the harmful effects of ionizing radiation.

I congratulate the editor for publication of this very useful book.

Dr. R.C. Sawhney
Director and Secretary Life Science Research Board
Government of India, Ministry of Defence
Defence Research and Development Organization
New Delhi, India

Foreword

In today's changing global scenario, it has become imperative to develop non-toxic, yet effective, radiomodulatory agents for applications in medicine, defence, homeland security and space. Low intensity conflicts and terror threats further necessitate the development of radiation countermeasure agents, and their inclusion in emergency stockpiles is every nation's dream. However, there are hardly any effective agents available and the synthetic drugs that are currently available are toxic at their effective doses. Therefore, in recent years there has been a paradigm shift towards herbal radioprotectors. These agents can be effectively used for mitigating the harmful effects of both ionizing and non-ionizing radiation, and their potential as radiation countermeasure agents is immense. Nonetheless, the development of radioprotectors, mitigators and radiorecovery agents is a highly challenging and demanding task requiring concerted efforts of dedicated teams from diverse disciplines.

The Institute of Nuclear Medicine and Allied Sciences (INMAS) has been engaged in frontline research in several areas of biomedical sciences for over five decades, and is poised as a research centre of excellence in radiation sciences with clinical and research expertise amalgamated together under one umbrella. Researchers at the institute have strived to utilize the potential of nuclear energy for peaceful purposes and the vision of Brigadier S.K. Mazumdar, the founder Director, who is considered the father of Indian Nuclear Medicine, has led to fruition. The institute has contributed significantly in the area of herbal radioprotectors and several high-altitude region Himalayan plant species have yielded promising results. It is, therefore, apposite that Dr. Rajesh Arora took up the Herculean task to compile this book.

Biological radioprotection has been transformed into a thrust area worldwide and path-breaking research is being done on radioprotectors of natural origin. Currently studies are mainly focussing on development of rapid high-throughput bioassays, toxicity and efficacy evaluation of natural radioprotectors as well as purified and isolated phytomolecules in higher animal models in both pre-irradiation and post-irradiation settings, elucidation of intricate mechanisms operating at the cellular/molecular level, phytochemical characterization and formulation development for human applications.

Despite rapid advances the world over in the area of radiomodulation, there has been a paucity of a comprehensive book devoted to this important subject area. This book reviews and presents up-to-date information in the area of radiomodulation by herbal drugs and also provides a forward-looking vision.

I congratulate the editor for his yeoman service and am confident that this interesting book would serve as a useful reference book for coming years greatly benefiting the armed forces and society.

Dr. Rajendra P. Tripathi
Director
Institute of Nuclear Medicine and Allied Sciences
Defence Research and Development Organization
Delhi, India

Acknowledgements

I would like to express my deepest sense of gratitude to Shri M. Natarajan, Scientific Adviser to Defence Minister, Secretary and Director General Research and Development (DRDO), Ministry of Defence, Government of India, for providing inspiration and support in manifold ways.

I am extremely thankful to Dr. W. Selvamurthy, Distinguished Scientist and Chief Controller Research and Development (Life Sciences and Human Resource), DRDO Headquarters, New Delhi, India, for whole-hearted support and taking keen interest in the progress of the book.

I do not have words to express my thanks to Dr. R.P. Tripathi, Director, INMAS, Delhi, India, for his unstinting support and guidance.

I would like to convey my heartfelt thanks to Dr. Rajeev Varshney, who extended his helping hand with the affection of an elder brother. I am also grateful to Dr. R.C. Sawhney, Director (Life Sciences) and Mr. V. Biswas of our corporate HQ for their kind cooperation.

I would be failing in our duty if I do not thank my family members, who were often neglected due to my commitment. My father Mr. Gopal Kumar Arora, a retired Defence Personnel with rich pragmatic experience of nearly four decades in the Indian Air Force, kept on infusing hope in me with his sincere advice and constructive criticism. My wife, Mrs. Preeti Arora, helped in multifarious ways and provided the much needed emotional support required during such endeavours. I have incorporated several of their inputs in the book. My two kids, Tanmay and Geetansh, who are presently too young to understand and know what the book is about, saw me busy at home for nearly two years and kept asking me when the book was going to be published. After all, I was snatching their precious quality time by not being with them. These little angels with their divine innocence kept me going!

I am thankful to my family members, particularly Ms. Dolly and Mr. H. Singh, Ms. Nisha and Mr. Dipak Naz, Mr. Vijay, Shree and Kamal Arora, Ms. B. Kaur and Mr. B.K. Sethi for moral support.

My teachers were a constant source of inspiration and persuaded me to put in my best. They have blessed me in several ways!

I wish to place on record thanks to all the contributors, who submitted their manuscripts within a short time and cooperated wholeheartedly. It was a pleasant experience working with the contributors.

I am grateful to CABI Publishing House for publishing the book in record time. Special thanks go to Ms. Rachel Cutts, Assistant Editor, CABI Publishing House, Wallingford, UK, for her constant help. I thank Ms. Fiona Harrison, Production Editor, CABI for taking care of the minute details during the final stages of publication of the book. I am extremely grateful to Mr. Nigel Ferrar, Publisher, CABI, who helped whole-heartedly in the publication of this book.

I thank Mr. D.C. Tiwari, Neeraj and Ravi for their help in photography/software.

I know not how to thank my constant companion, God, who has been there with me in the journey of my life in happy and despondent moments, in failures and successes and perhaps much more!

15th January, 2008 Rajesh Arora

About the Editor

Dr. Rajesh Arora is a senior scientist at the Institute of Nuclear Medicine and Allied Sciences, Delhi, India. Dr. Arora has made significant contributions in the area of novel radioprotective drug design and development. His pioneering work on the development of radiation countermeasure agents from herbals has laid a strong foundation for research in the area of radioprotection worldwide.

Dr. Arora is a recipient of a number of prestigious fellowships and awards, including two gold medals for his professional accomplishments. Dr. Arora has been awarded the DRDO Laboratory Technology Group Award twice (in 2001 and 2008) for his noteworthy contributions to the herbal radioprotective drug development programme of the Defence Research and Development Organization, Ministry of Defence, Government of India. Dr. Arora has also been cited in the Who's Who of the World, USA, Who's Who in Science and Engineering, USA and International Biography, UK.

Dr. Arora has authored over 100 scientific articles in refereed journals, books and conference proceedings and has more than one dozen patents to his credit. He has also published four books that have been well received by the scientific fraternity. Dr. Arora is a life member of several national and international professional societies and is on the panel for reviewing papers submitted to reputed peer-reviewed national and international journals. He is a recognized Ph.D guide of Delhi University, Jamia Hamdard University and a paper setter/Ph.D thesis examiner of several Universities. He has supervised more than two dozen students for their B.Tech/M.Tech/M.Sc and Ph.D degrees.

Dr. Arora received a B.Sc degree from the University of Jodhpur, India, M.Sc and M. Phil degrees from the University of Meerut, India, a Ph.D degree jointly from the Central Institute of Medicinal and Aromatic Plants (CSIR), Lucknow, India and University of Meerut, India. He also holds post-graduate diplomas in Human Resource Development and Marketing Management and an MBA degree in Human Resource Management. He was professionally trained at the Central Drug Research Institute, Lucknow, India, Sanjay Gandhi Post-Graduate Institute of Medical Sciences, Lucknow, India, Defence Institute of Advanced Technology, Pune, Defence Institute of Physiology and Allied Sciences, Delhi and Institute of Technology Management, Mussorie. In addition to his research, Dr. Arora also coordinates Technical and Human Resource Development activities at his Institute for leveraging the crucial human capital, indispensable for enhancing frontline areas of science and technology. E-mail: rajesharoradr@rediffmail.com

SECTION I

HERBALS FOR RADIOMODULATION (RADIOPROTECTION/RADIOSENSITIZATION): AN OVERVIEW

Chapter 1

Radiomodulatory Compounds of Herbal Origin for New Frontiers in Medicine, Homeland Security, Management of Radiological Incidents and Space Applications

Rajesh Arora, R. Kumar, A. Sharma and R.P. Tripathi

Introduction

The existence of radiation is perhaps as old as the Universe itself. In our galaxy, radiation emanating from the Sun and lightning strikes nearly 4.5 billion years ago formed the basis for the evolution of life on Earth. Life evolved circa 2.5 billion years later, and radiation aided in shaping life via the synthesis of biological molecules, frequently moulding the evolutionary processes. Although humans evolved much later on Earth, they soon discovered radiation and developed ways to harness the tremendous energy of the atom. By virtue of their ingenuity, they have exercised a considerable degree of control over radiation and have utilized it for a variety of purposes, often turning it into a friend from a foe, and vice versa.

Types of Radiation

There are two types of radiation viz., ionizing and non-ionizing. Ionizing radiation refers to electromagnetic radiation, X-ray or γ-rays or particulate radiation (neutrons and α-particles) that possess sufficient energy to induce ionization of atoms or molecules (i.e., is capable of removing electrons from their outermost orbit). Ionizing radiation is differentiated on the basis of their linear energy transfer (LET). LET refers to the amount of energy deposited by a particular type of radiation per unit of path length. Low LET radiation is sparsely ionizing, whereas high LET radiation is densely ionizing and more efficient in producing biological damage (Prasad, 1999; Bump and Malaker, 1998). Non-ionizing radiation on the other hand refers to any type of electromagnetic radiation that does not carry enough energy per quantum to ionize atoms or molecules. Examples include near UV, visible light, infrared, microwave/radiowaves and radiofrequency radiation etc.

Radiation: The Double-edged Sword

Radiation affects the biological system in a variety of ways. On the one hand, as alluded to earlier, it has helped life forms evolve and on the other it manifests deleterious effects on life forms. This paradoxical situation poses several questions before us, which at times are difficult to address. The power of the atom is being harnessed for solving our energy needs, improving agriculture and industry and as an invaluable tool in modern medicine. However, the use of nuclear energy with good intent (for peaceful purposes) also poses risks.

Radiation-induced Biological Effects

Protecting biological systems against the initial transfer of radiation energy, which is likely to occur in less than a billionth of a second, is virtually next to impossible as the transfer of energy is extremely rapid. Since high LET radiation primarily causes direct damage, the resulting injury is difficult to modify. Therefore, most endeavours have focused on modifying radiation injury resulting due to low-LET radiation. Exposure of biological systems to ionizing radiation results in the formation of reactive oxygen species (ROS) and reactive nitrogen species (RNS). These reactive species inflict damage to the various bio-macromolecules like DNA, lipids and proteins present in the cell (Prasad, 1999). The effects of radiation at cellular level are depicted in Fig. 1.1.

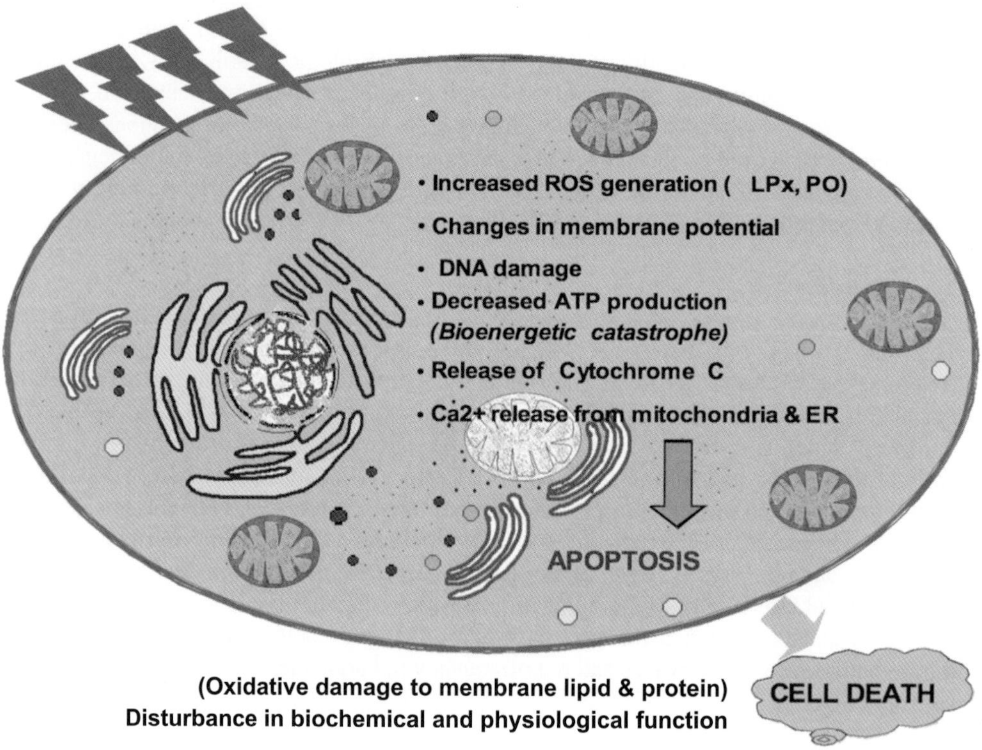

Fig. 1.1. The effect of radiation on the cell.

The health effects resulting from exposure to ionizing radiation are classified into two categories: stochastic (probabilistic) and deterministic. The former often take several years to develop (e.g., cancer appearing several years after exposure), while deterministic effects (e.g., cataract induction, haematologic deficiencies, damage to skin and fertility impairment) at high radiation doses are manifested with certainty above a threshold. Individuals exposed to radiation may develop radiation sickness or exhibit signs of radiation injury.

Radiomodulators: Definition and Types

Radiomodulators are agents that modify the radiation response in biological systems and include both chemical and natural compounds. Radiomodulators are of two types: radioprotectors and radiosensitizers.

Radioprotectors and their classification

"A radioprotective agent (radioprotector) is a chemical or a drug that reduces the damaging effects of radiation, when administered to living organisms (Arora *et al.*, 2006a,b; Sharma and Arora, 2006)". From a clinical perspective, Urtasun (1998) considers "radioprotectors as compounds that protect against radiation damage to targeted normal cells, but do not provide similar protection to tumour cells".

Thomas M. Seed of the Catholic University of America, USA has defined radioprotectors as "any medicinal agent or device applied prior to or during radiation exposure that actively prevents or limits injury, whether that injury be at the molecular, cellular, tissue or organ system level (Seed, 2005)".

World-over researchers have classified radioprotectors differently based on their mechanism of action, utility, route and time of administration or sheer convenience. An attempt has been made to list the various classifications.

Initially radioprotectors were classified into two main categories: i) agents administered pre-irradiation i.e., agents that protect against the deleterious effects of radiation when administered prior to irradiation, and ii) agents administered post-irradiation i.e., agents that are effective when administered after the organism has been irradiated and still exhibit efficacy. These are primarily radiorecovery agents, as they stimulate recovery from radiation damage.

However, presently not all researchers agree with this categorization. Currently, only agents administered pre-irradiation are considered in the category of radioprotectors.

M.V. Vasin of the Institute of Aviation and Space Medicine, Moscow, Russia, in 1999 classified prophylactic antiradiation drugs (radioprotectors) on the basis of their mode of action into the following categories: drugs having short-term and long-term action; drugs that stimulate radioresistance; drugs that suppress symptoms of primary radiation reaction; drugs that detoxify early; and drugs that act by absorption or elimination of radionuclides from an organism (Vasin, 1999).

Nair and co-workers (2001) classified radioprotective agents into three categories viz., radioprotectors, adaptogens and absorbents. They further opine that radioprotectors include compounds like antioxidants and others that possess sulfhydryl groups, while adaptogens are non-toxic stimulators of radioresistance. The last category comprises absorbents that protect against internal radiation injuries resulting due to ingestion of radionuclides. Such agents prevent the incorporation and absorption of radionuclides like ^{90}Sr, ^{137}Cs, ^{239}Pu, ^{131}I etc.

In recent years an altogether different kind of categorization of radiation countermeasure (RCM) agents has been proposed, which is now most widely accepted amongst radiation biologists (Stone *et al.*, 2004). According to this classification, radioprotectors are prophylactic agents that are administered prior to irradiation. The term 'mitigator' refers to agents that are administered during irradiation or immediately after radiation exposure, but before the manifestation of radiation injury; while the term 'radiation therapeutics' is used for agents that are administered after the manifestation of clinical symptoms (Stone *et al.*, 2004).

Herbal Radioprotectors

Herbal radioprotectors can broadly be defined as "extracts, fractionated extracts, isolated bioactive constituents or bioactivity-modulating molecules derived from natural (including algae, fungi, bryophytes, pteridophytes, gymnosperms and angiosperms) or genetically modified plant sources or their artificially generated combinations that render protection against the deleterious effects of ionizing radiation when administered before irradiation" (Arora and Goel, 2000; Arora, 2007a).

Radioprotective Drug Development Initiatives

Research and development on radioprotective drugs started nearly fifty years ago. Patt (1949) reported the protective role of cysteine against X-radiation. Radioprotective drugs were developed by several nations, including the United States, former USSR, Germany, France and China etc. primarily during the 'Cold War' era. Amongst the various drugs developed and tested, the sulfhydryls were found to be the most promising (Bump and Malaker, 1998; Nair *et al.*, 2001; Coleman *et al.*, 2003; Arora *et al.*, 2005a; Arora, 2007b). Over 5000 compounds were synthesized and screened by the 1980s at the Walter Reed Army Institute of Research, USA (Weiss and Simic, 1988; Giambarresi and Jacobs, 1987; Sweeney, 1979). The discovery of amifostine (WR-2721 or Ethiophos) marked the arrival of the clinically useful drug, which is currently considered the 'Gold Standard' in radioprotection.

Though amifostine has been approved by the Food and Drug Administration (FDA) for use in the clinic, the United States Army has not approved it for protection against radiation-induced lethality in a military setting since it induces nausea and vomiting, has poor bioavailability when administered via oral route, is ineffective in post-administration scenarios and exhibits a narrow therapeutic window (Capizzi and Oster, 1995). Several phosphorothioates and thiosulfonates (e.g., WR-159243, WR-3302, WR-1551 and WR-2926) have also not been very successful due to their toxicity (Capizzi and Oster, 1995).

A number of drugs of both synthetic and biological origin have been tested for radioprotection in the last few decades in various parts of the world (Weiss and Landauer, 2003; Arora *et al.*, 2005a,b). These molecular drugs mainly comprise antioxidants, cytoprotective agents, angiotensin I-converting enzyme (ACE) inhibitors, metalloelements, non-steroidal anti-inflammatory drugs (NSAIDS), immunomodulators, sulfhydryl compounds, lipopolysaccharides, prostaglandins and DNA binding ligands (Bump and Malaker, 1998). Combined modalities have also been evaluated with some degree of success. Pre-administration of phosphorothioates with post-administration of immunomodulatory agents has been reported to synergistically reduce radiation damage (Weiss *et al.*, 1990; Neta, 1988).

Haemopoietic growth factors and cytokines e.g., interleukin, colony stimulating factors, stem cell factor and anti-apoptotic cytokine combinations have been evaluated for radioprotection and radiorecovery (Neta, 1988; Singh and Yadav, 2005). Several recombinant DNA technology products e.g., interferon, interleukin, GM-CSF have been tested for their radioprotective efficacy, particularly for the treatment of radiation injury victims. Advances in stem cell technology, particularly for replenishing the haemopoietic, gastro-intestinal and central nervous system, would only find practical application in coming years.

Radiosensitizers

Radiosensitizers, like radioprotectors, have also been defined differently by researchers. Some definitions are presented:

"Radiosensitizers are compounds that when combined with radiation will achieve greater tumour inactivation than would have been expected from the additive effect of each modality (Urtasun, 1998)".

"Radiosensitizers are agents that do not have a therapeutic effect of their own, but act to enhance the therapeutic effect of radiation (Bump *et al.*, 2003)".

Necessity of Biological Radioprotection

Both planned and unplanned radiation exposures pose a threat to humans and can occur in a variety of settings (Fig. 1.2).

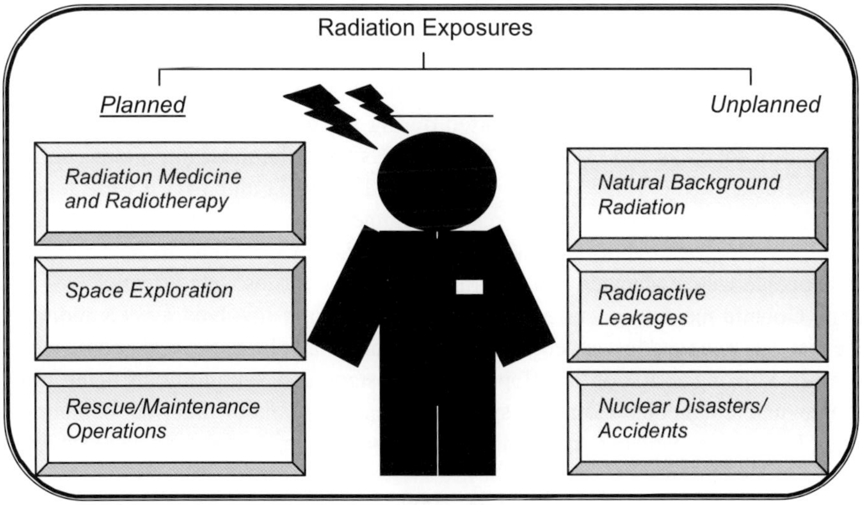

Fig. 1.2. Planned and unplanned radiation exposures.

Radiation exposures can occur without any prior warning in the following settings: radiography, nuclear medicine, radiotherapy, radiological imaging, radionuclide production, biomedical research, military, public domain, transportation, nuclear reactors, space flights etc. There is, therefore, a need to protect humans against planned and unplanned exposures.

Why Herbals as Radiomodulators?

The synthetic radioprotectors and radiosensitizers currently available are toxic at their effective doses, and consequently natural products, particularly herbals, are receiving attention (for review see Arora *et al.*, 2005a, 2006a,b; Arora, 2007a; Garg *et al.*, 2005). Medicinal plants with radioprotective properties contain a plethora of bioactive compounds that exhibit diverse properties e.g., pro- or antioxidative, immunostimulative, cell proliferative, anti-inflammatory and antimicrobial etc. and usually act in a dose-dependent manner (see Chapters 2, 4, 5, 6, 7, 14–17 of this book). In view of their long history of safety in terms of traditional usage in humans for several years, holistic mode of action, countering of the toxic effects of certain constituents, synergism, novel source for new drugs etc. herbals have been considered worthy as radiomodulators.

Applications of Radioprotectors in the Management of Radiation Incidents/Radiological Emergencies

Several radiation incidents have been reported the world over and some are discussed: the Techay-Iset river system (former USSR) was contaminated by radioactive waste (uranium-fission products) in the 1950s released from a radiation production facility. Over 28,000 people were exposed and nearly 1000 people manifested chronic radiation syndrome.

The Three Mile Island, USA, incident occurred in March 1979 in the United States due to a relief-valve fault in the 850-megawatt reactor that caused more than 70% core meltdown. Nearly 10 million Ci of ^{131}Xe, 40,000Ci of ^{85}Kr and 20Ci of ^{131}I was released resulting in an average dose of ca. 2mRem to the local population.

The Chernobyl incident occurred in April 1986, giving a glimpse of what a full-scale radiation disaster would look like. External irradiation resulted due to deposited radioactive materials (primarily 137Cs in the long term) and the dietary ingestion of radionuclides (131I in milk and leafy vegetables were reported during the first month and 134Cs and 137Cs were found in foods later). 131I, 134Cs and 137Cs were the most important contributors to the total dose, while 95Zr, 103Ru, 106Ru, 132Te, 140Ba and 141Ce were reported in air or deposition. Two human deaths resulted immediately due to trauma, and 28 deaths from radiation injuries, while 115 cases of acute radiation syndrome were recorded. Over 116,500 people were evacuated and several European countries were affected.

The Goiania incident occurred in Brazil in 1987, and involved a ^{137}Cs radiotherapy device that was removed by scavengers from an abandoned clinic. Several people who were exposed to ionizing radiation developed clinical signs and symptoms of acute radiation syndrome and cutaneous radiation syndrome, external and/or internal contamination. Several places were also contaminated and more than 0.1 million people were screened for external contamination.

The criticality incident occurred in 1999 at the Tokai-mura uranium conversion facility and the three workers who were severely exposed to a mixed neutron/gamma field with estimated doses in the range of 3–24.5 Gy/Eq died with multiple organ failure within 211 days. People staying in the immediate vicinity (350 m) of the facility were evacuated and more than 300,000 people within a 10 km radius were advised to remain indoors for 18 hrs.

Internal contamination with radiostrontium in Mayak cases was reported. External contamination with ^{210}Po was detected in three patients at the Institute of Biophysics in Moscow.

In 2001, serious accidental radiation exposures occurred in five breast cancer patients undergoing post-operative radiotherapy at the Bialystok Oncology Centre in Poland. The event occurred due to the malfunctioning of the NEPTUN 10P® linear accelerator following a sudden power cut (IAEA, 2004).

The recent Polonium (^{210}P) incident in London, UK, in November 2006 had a great impact on the population and attracted media attention worldwide.

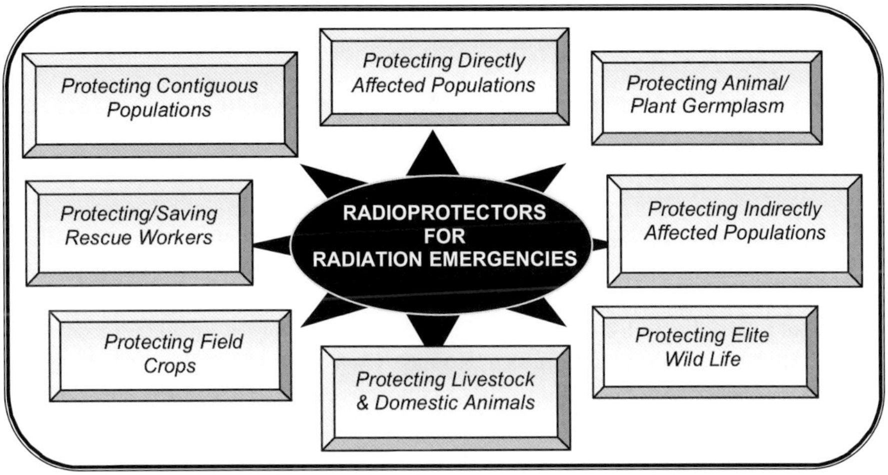

Fig. 1.3. Radioprotectors for the management of radiological incidents.

Tokyo Electric Power Corporation's Kashiwazaki-Kariwa No.3 nuclear power generation unit in Kashiwazaki, Northern Japan, considered as the Achille's heel of nuclear power plants, was affected by an earthquake (6.8 Richter) on July 16, 2007, which sparked a fire in the transformer that led to heavy water leakage. Kashiwazaki-Kariwa plant is the world's largest nuclear power facility with an output capacity of 8.21 million kilowatts.

Radioprotectors can be of immense use in the management of radiological incidents/ emergencies such as those mentioned above (see Fig. 1.3).

A few herbals have been utilized as radiation countermeasure agents therapeutically for managing radiation emergencies. Chernobyl reactor liquidators (emergency workers) were fed a special diet (antioxidant biofactor (AOB)) containing rutin as the major ingredient. AOB reduced clastogenic factors significantly (Emerit *et al.*, 1995, 1997). Food enrichment with beta-carotene was also considered as a feasible strategy for the inhabitants staying near the Chernobyl power station (Weiss and Landauer, 2003). Ben-Amotz *et al.*, (1998) found that children exposed to radiation from the Chernobyl nuclear reactor had increased lipid peroxidation products, which reduced upon supplementation with β-carotene. Nagano and co-workers (2000) studied the effect of diet and bladder cancer incidence in a cohort of atomic-bomb survivors in Japan and found that high consumption of green-yellow vegetables and fruits resulted in protection against bladder cancer. Alginates, water-soluble polysaccharides found in several marine algae, have been shown to effectively remove internally deposited radionuclides in humans (Gong *et al.*, 1991). Such agents hold immense promise in the effective management of radiological emergencies.

Recently, our group has isolated a highly radioresistant bacterium from high radiation environment, which renders significant radioprotection to experimental animals (Kumar *et al.*, 2006). This could be very useful in the management of radiological emergencies, besides applications in other sectors e.g., homeland defence.

There is an increased likelihood of radiological incidents today, and radiation countermeasure agents, including herbals, offer promise since several of these agents are able to repair radiation-induced damage, stimulate the immune system and even chelate the radionuclides bringing about their effective clearance in a short period of time. This area needs more attention and research in coming years.

Applications of Radioprotectors in Homeland Defence

The Arrival of Radiological Terror

In 1996 the Chechen rebels placed a container of caesium-137 in a Moscow park. The radioactive material did not disperse, however, the incident marked the arrival of radiological terror.

In recent times, the high-profile assassination of Alexander Litvinenko, a former KGB (erstwhile Soviet Union's secret service–Komitit Gosudarstvennoy Bezopasnosti) agent in November 2006 utilizing polonium-210, a radioactive element, showed how dangerous radiation can be.

These are only examples of isolated incidents, but what if radioactive materials e.g., cobalt-60, strontium-90, caesium-137 and iridium-192 are used against large populations?

Radiological terror refers to the use of radioactive substances to create damage and mass hysteria. Dr. John Moulder of the Center for Radiation Injury Intervention, Medical College of Wisconsin, USA, defines radiological terrorism as "the use of radioactive material to scare people, to kill people, or to contaminate the environment".

As per IAEA estimates, there are approximately 10,000 radiotherapy units and industrial sources each, and about 300 irradiators the world over. These sources pose a serious threat since radioactive materials could be stolen, despite the precautions in place, and used in the form of a Radiological Dispersal Device (RDD, i.e., dirty bomb) or a Radiological Exposure Device (RED; i.e., deliberately concealed sources) (WHO, 2007). Besides, if deliberate attacks were ever made on nuclear reactors, radiological industries or military units housing nuclear weapons, the consequences would be devastating. Such an attack would result in mass panic, evacuations and saturation of the available medical infrastructure and also contaminate the contiguous areas for months to years, seriously jeopardizing regional/national economy. In addition, radiation emanating from radioactive sources could affect livestock, water reservoirs, regional flora and fauna and food products.

The world dreads a radiological terror attack and several nations (USA, UK, Japan, China, Russia, Germany, Sweden, Finland, Brazil, Argentina etc.) are, therefore, harbouring national stockpiles for radiation emergencies. The contents of available stockpiles include KI, Ca-DTPA, Zn-DTPA, Prussian blue etc. Some countries stockpile some other specific agents, e.g., 2,3-dimercaptopropane-1-sulfonate or generic products and even herbals. These stockpiles are usually maintained in central/regional warehouses, nuclear power plants, hospitals and other strategic locations.

There is an urgent need to develop radioprotectors for use by emergency responders (in most countries they usually comprise professional/volunteer firefighters, emergency public service personnel, police, military personnel (army, navy and air force personnel), medical emergency personnel, paramilitary personnel (civil guards or home guards), coastguards, rescue personnel, Red Cross/volunteer organizations etc.) since in case of a nuclear/radiological exposure they would be the first ones to be exposed in the aftermath of an emergency.

The possibility of a nuclear/radiological attack cannot be ruled out in today's scenario. Kimery (2003) opines that the use of dirty bombs is not a question of if, but when.

Herbal radioprotectors have the potential to fill an urgent need for counter-terrorism applications in homeland defence and the military, as well as protecting emergency and disaster response personnel from radiation hazards. Most herbals (discussed in this book and also many others) render effective radiation protection and can be used for homeland defence e.g., gensitein, *Podophyllum hexandrum*, *Hippophae rhamnoides*, *Tinospora cordifolia*, *Ocimum sanctum*, *Panax ginseng*, *Rhodiola imbricata*, melatonin etc.

Radioprotectors for Defence Applications

For defence applications, radioprotectors are an inevitable necessity (Arora, 2006; Fig. 1.4). With the dismantling of the former USSR (the post Cold War scenario), the probability of large-scale conflict involving nuclear weapons during war has substantially decreased. However, the rapid proliferation of nuclear weapons and indigenous development of nuclear technology by several small, less responsible nations poses a veritable threat. Armed forces personnel and the civil populations of almost all nations pin their hopes on radioprotectants that would help them survive in a nuclear milieu.

Modern day nuclear weapons (1000 kiloton or 1-megaton) are much more powerful than the one that was dropped on Hiroshima and Nagasaki (15 kiloton). Since a war between less responsible nations may involve the use of nuclear weapons, it would be very important to maintain military performance in such a scenario. In the event of a nuclear exposure, defence personnel could be subjected to decrements in performance, and physiological injury that could lead to malfunctioning of the immunological-haematological system resulting in infectious complications (Walker and Conklin, 1987). Most injuries associated with a nuclear weapon detonation will be caused not only by irradiation, but also by the combined effects of radiation, heat and trauma. During combined injury, the individual sub-lethal injuries act synergistically and become lethal.

An Ideal Radioprotector for Military Applications

To date, an ideal radioprotector for military use does not exist, only a wish list does! Giambarresi and Jacobs (1987) initially charted a wish list of field-usable radioprotector nearly 20 years ago, which has been upgraded by Seed (2005) and others. A compilation of the list and also our opinion is reflected in the ensuing section keeping in mind that in today's scenario soldiers operate in diverse environmental and operational conditions, often across borders.

An ideal radioprotector for use in defence should: preferably reduce by a factor of 10 the dose of radiation that produces combat ineffectiveness; offer long-lasting protection (lasting several days); possess no serious side effects and toxicity; be highly protective and effective in a global setting; protect all populations at risk; render protection against early and late effects of different kinds of radiation. In addition, it should be non-performance decrementing; be safe with repeated doses; be easily administrable (self administered as a tablet, capsule, inhaler, skin patch or swab); act in a wide time-window to render protection; be rapid in action following administration; be effective even upon repeated dosing; be chemically stable and possess long shelf-life etc. For a radioprotector to be of military interest, it should be effective when administered post-irradiation, since prior signalling may not always be available, and the soldiers could inadvertently get exposed to radiation.

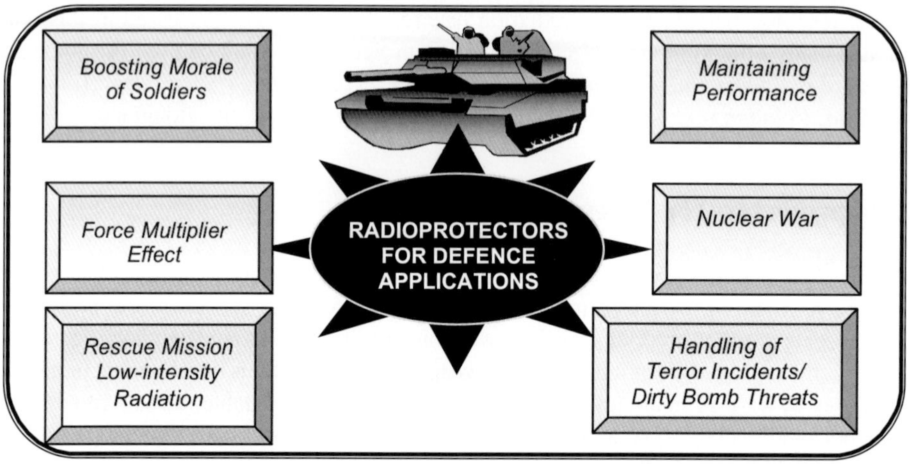

Fig. 1.4. The diverse military applications of radioprotectors.

Of course, the ideal list mentioned above is far from being complete. But in reality, do we have a drug that can be fielded as a radioprotector for defence applications? FDA-approved radioprotective/radiation countermeasure agents currently available (not exclusively for defence applications) include WR-2721, KI, Ca/Zn DTPA, Prussian blue and Granisetron (an antiemetic; 5-hydroxytryptophan inhibitor). A number of other agents not yet approved by the FDA, but under various stages of development/testing, include amifostine-SR (a slow-release formulation), WR-3689, 5-androstenidiol, nitroxides, PGG-Glucan, antiemetics, MnSOD, antioxidants (vitamin E, selenium, N-acetyl cysteine, captopril, Mesna), prostaglandin/COX2 inhibitors, ACE inhibitors, growth factors (G-CSF, GM-CSF, kertainocyte growth factor (KGF), fibroblast growth factor (FGF) etc.), pentoxifylline/Vit E/SOD, pentoxifylline, tempol, stem cell transplants (peripheral blood, bone marrow, umbilical cord blood, liver/CNS Flt3 ligand, IL7) etc.

Is there hope or we are chasing a mirage?

The currently available radioprotectors in the armamentarium of most armies of the world are not suitable for field applications. A number of research laboratories in different parts of the world, often in isolation and in a shroud of secrecy, are engaged in developing radiation countermeasure agents primarily for military use. The spin offs of such research would definitely help in saving precious human lives.

A plethora of herbals and their bioactive constituents have been tested or are under testing as effective radiation countermeasure agents (though not exclusively for military applications) and include *Glycine max* (gensitein), *Podophyllum hexandrum* (mainly aryltetralin lignans and flavonoids), *Hippophae rhamnoides*, *Tinospora cordifolia*, *Ocimum sanctum* and its isolated constituents orientin and vicenin, *Chelidonium majus (Ukrain)*, *Panax ginseng*, *Rhodiola imbricata* etc (Arora and Sharma, 2004; also see Chapters 2, 3, 12–17 of this book).

Radiation can significantly compromise military operations. Even low doses of radiation exposures can lead to decrements in motor and mental function within minutes to hours, while higher doses result in early transient incapacitation. Besides, radiation sickness can result in debilitating psychological effects (e.g., acute anxiety leading to cognitive stress) in military personnel. The development of phobias, general depression and malaise, post-traumatic stress disorder, and psychosomatic complications would only be inevitable consequences of exposure to ionizing radiation in a military setting. Several herbal drugs have been used for rendering behavioural radioprotection. In our experience, *Mentha piperita* and *Zingiber officinalis* have yielded promising results (Sharma *et al.*, 2005, 2006, 2007; Haksar *et al.*, 2006), but many more may be useful in alleviating the symptoms of radiation sickness.

If a nuclear explosion ever occurs, complete destruction will occur within a few kilometres of ground zero. However, the probability of survival increases with distance from ground zero. Three factors are considered sacrosanct in radiation protection philosophy: time (duration of exposure), distance (physical distance between the individual and radiation) and shielding (physical barriers between the individual and radiation). Radiation countermeasure agents should in no way be considered as a panacea since they can only provide limited protection to those individuals who receive low/moderate doses of radiation exposure. It is often said that for the dying, even a ray of hope is enough. Radiation countermeasure agents of herbal origin do offer that hope! All said and done, the greatest hope for mankind is that "a nuclear holocaust never occurs".

Applications of Radioprotectors in Space

Astronauts get exposed to natural radiation in space that can have immediate and long-term health risks (Cucinotta *et al.*, 2001; Seed *et al.*, 2002). As far as our galaxy, the Milky Way, is concerned the three major sources of radiation are: trapped belt radiation (consisting of protons and electrons), galactic cosmic rays (GCRs; constituting of proton and heavy ion particles) and solar particle events (SPEs). The haemopoietic tissue is one of the main targets for both acute radiation effects (e.g., an SPE) and late stochastic risk (caused due to low-dose chronic exposure to space radiation) (Esposito *et al.*, 2001). In addition, astronauts are vulnerable to developing cataracts, cancer and shortening of life due to increased radiation exposures in space.

Developing appropriate countermeasures for prevention of radiation-induced cancers or other disorders, resulting due to long-term manned space missions, is a major challenge before space radiation biologists. A substantial amount of data is available so far as acute human exposure to gamma rays is concerned, however, the biological consequences of low-level exposures to the high-energy charged-particle radiation encountered in space remain to be fully elucidated.

Man has constantly been endeavouring to remain in space for longer durations, so far in orbital stations e.g., the International Space Station, but with manned space missions to nearby planets becoming a reality today, there is an urgent need to develop radiation countermeasure agents that can specifically protect astronauts against the deleterious effects of space radiation.

Extended stay in space poses potential risks to the astronauts. Virtually very few agents are currently available for space applications, but radioprotectants hold immense promise in the area of space (Fig. 1.5) and whatever is available is often utilized as a precautionary measure.

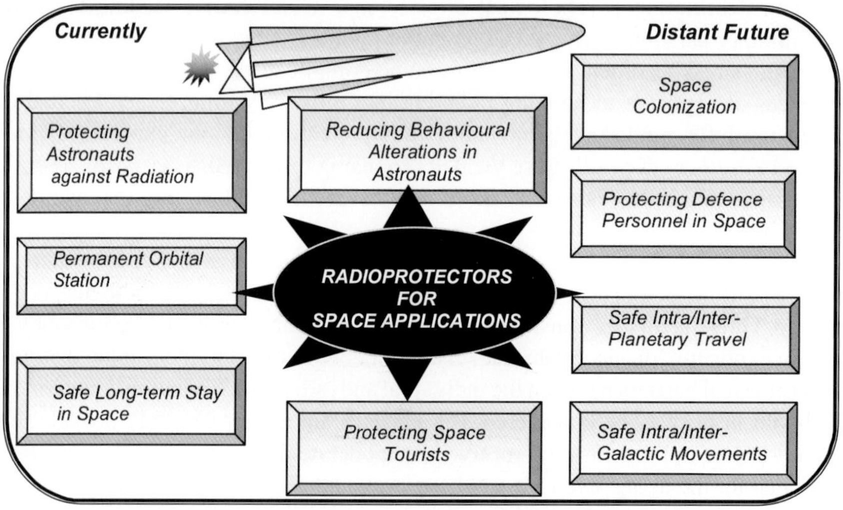

Fig. 1.5. Radioprotectors for possible applications in space.

In fact, amifostine was taken by US astronauts, on their first visit to the moon, to protect them against intense radiation from possible solar flares. Researchers have directed their attention towards developing effective countermeasures against space radiation-induced damage (Kennedy, 2007; Shukitt-Hale *et al.*, 2007).

The promise of herbal agents in rendering protection against space radiation has also been established. Olive oil has been shown to enhance the DNA repair efficacy of space astronauts. Bowman-Birk inhibitor (a soybean-derived serine protease inhibitor) and its enriched extract (BBI concentrate (BBIC)) have been shown to suppress radiation-induced malignant transformation (Kennedy, 1993, 1998). Kennedy *et al.*, (2006) found that BBIC and antioxidants (ascorbic acid, co-enzyme Q10, L-selenomethioninine and vitamin E succinate) were effective in protecting against space radiation-induced cytotoxicity, and BBI, BBIC and antioxidants were effective in protecting against space radiation-induced phenotypic changes associated with transformation of HT Ori-3 cells. These authors opine that since these agents are readily available and have favourable safety profiles, they are potentially useful as countermeasure agents against space radiation-induced biological effects. Blueberry and strawberry extracts have been shown to protect against space radiation (see Chapter 11 of this book).

Several natural radioprotectants that hold promise as countermeasure agents against space radiation include *Ocimum sanctum*, *Podophyllum hexandrum*, *Hippophae rhamnoides*, *Rhodiola sps*, *Ginkgo biloba*, *Chelidonium majus*, *Panax ginseng*, *Spirulina*, *Citrus* sps. (also see Chapters 3, 7, 9, 10–13 of this book). Some of the plant products have in–fact been taken on board space missions by cosmonauts of the former USSR with the aim to reduce radiation exposure and improve overall immunity and performance. *Ganoderma lucidum* (Reishi or Ling Zai) and Indralin have been used by cosmonauts from Russia and Bulgaria. The Cleveland Biolabs under the aegis of a NASA-funded program is evaluating a promising bacterial flagellin (that renders protection in mice against gamma radiation even up to 14 Gy) for possible protection of astronauts during space missions. It is likely that herbals, including several dietary ingredients, may soon find their way to space for protecting astronauts since a manned mission to Mars has been planned by NASA, and radiation from the Sun and the rest of space remains one of the biggest hurdles to this path-breaking mission. Dietary agents possessing radioprotective properties hold immense promise in present and future space missions since there would be no need of taking additional drugs and they do not have any side effects.

Applications of Radiomodulators in Medicine

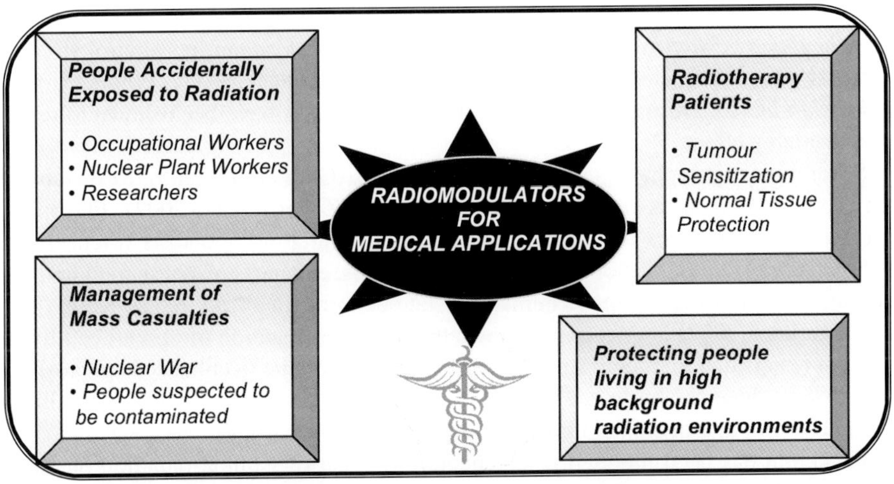

Fig. 1.6. Diverse applications of radiomodulators in medicine.

Radiomodulators can be used for diverse applications in medicine (Fig. 1.6). A number of natural products have the potential of being used in the clinic for radioprotection, as well as radiosensitization of tumour cells.

Herbal Radioprotectors for the Clinic

Some examples of herbals that could be useful in clinical radioprotection are: *Ocimum sanctum*, *Tinospora cordifolia*, *Podophyllum hexandrum*, *Piper betel*, *Andrographis paniculata*, *Mentha piperita*, *Zingiber officinale*, *Panax ginseng*, *Scutellaria baicalensis*, *Viscum album*, *Aloe vera*, *Swertia chirata*, *Ginkgo biloba*, *Spirulina platensis* etc. (Arora *et al.*, 2005a; Puri *et al.*, 2005). Natural products from lower organisms like *Trametes hirsuta*, a novel endophyte isolated from *Nothapodytes foetida*, also hold potential as radioprotectors (Puri *et al.*, 2005; Puri *et al.*, 2006).

A drug from *Ocimum sanctum* is currently undergoing clinical trials for its radioprotective efficacy (Uma Devi, 2006), and many more are in the pipeline. Essential oils like Manuka and Kanuka from New Zealand are useful in alleviating radiation-mediated mucositis (see Chapter 8 of this book).

Herbal Radiosensitizers for the Clinic

Herbals have been reported to act as radiosensitizers under *in vitro* and *in vivo* conditions (Garg *et al.*, 2005; Zoberi *et al.*, 2002; see Chapters 18–22 of this book). Examples of some natural products shown to be useful in radiosensitization include: Withaferin A from *Withania somnifera*, camptothecin from *Camptotheca acuminata*, gossypol from *Gossypium*, L-canavanine from legumes, taxol from *Taxus baccata*, gossypol, Plumbagin from *Plumbago zeylanica*, hypericin from *Hypericum perforatum*, resveratrol from *Vitis vinefera*, *Panax notoginseng*, genistein from soybean and oleandrin etc. Whole extracts of *Azadirachta indica*, *Tinospora cordifolia*, *Alstonia scholaris* have been shown to render a radiosensitizing effect.

Several plant polyphenols radiosensitize tumours (Garg *et al.*, 2005). Resveratrol has been reported to possess radiosensitizing properties (Zoberi *et al.*, 2002). Flavopiridol, a flavone, radiosensitizes malignant glioma cells by initiating apoptosis (Stenzel *et al.*, 2002). Gossypol, a polyphenol found in *Gossypium* sps, enhances response to radiotherapy resulting in tumour regression of human prostate cancer (Xu *et al.*, 2005). Kasten-Pisula and co-workers (2007) have recently shown that radiosensitization of tumour cell lines by Gossypol results due to double-strand break repair and not due to enhanced apoptosis. Soy flavones have been shown to act as radiosensitizers and potentiate tumour cell killing by radiation *in vitro* and in animal tumour model (Raffoul *et al.*, 2007).

Curcumin has been reported to potentiate the effect of gamma radiation on hamster ovary cells (Araujo *et al.*, 1999). Curcumin renders radiosensitizing effects in the prostate cancer cell line by inhibiting the growth of the human prostate PC-3 cells and down regulating the radiation-induced prosurvival factors (Deeb *et al.*, 2003).

Murayama *et al.*, (1987) reported radiosensitization of Chinese hamster V79 cells with oridonin, a Rabdosia diterpenoid, and misonidazole. An enhancement ratio of 1.92 with 0.01 mM oridonin and 1 mM misonidazole was observed by these researchers, who further reported an enhancement ratio of 1.16 and 1.59 respectively in hypoxic cells. A supra-additive effect was observed with combined treatment.

Ellagic acid has been shown to enhance radiation-mediated oxidative stress and consequent cytotoxicity in tumour cells by decreasing antioxidant enzymes like superoxide dismutase (SOD), catalase and glutathione peroxidase (Bhosle *et al.*, 2005).

Intracellular thiols play an important role in rendering resistance to ionizing radiation in cancer cells (Chen *et al.*, 2005a). Since several phytoconstitutents and nature-identical molecules deplete intrinsic antioxidants, they can enhance the radiation effects.

Chen and co-workers (2005b) have shown that Caffeic acid phenethyl ester differentially radiosensitizes colorectal adenocarcinoma (CT26) cells via glutathione depletion and inhibition of NF-KB activity. Caffeine has been shown to reduce the survival of several cell lines following irradiation. Caffeine is a G2 checkpoint inhibitor and releases radiation-induced G2 block (Higuchi *et al.*, 2000). Valenzuela *et al.*, (2000) have also shown the radiosensitizing effect of caffeine. Radiotherapy studies have indicated that caffeine consumption decreases severe late toxicity after radiation to the pelvis (Stelzer *et al.*, 1994).

Even low dosages of hypericin enhance the radiation-induced cell killing in malignant glioma cell lines (Zhang *et al.*, 1996; also see Chapter 20 of this book).

A *Panax notoginseng* extract (PNE) and a purified saponin from the plant (Rb1) has been shown to render differential radiosensitization, with maximum sensitization observed at 10 mg/kg concentration 30 min after injection (Chen *et al.*, 2001).

The camptothecins are known to be efficient radiosensitizers (Rich, 2003). Semisynthetic compounds e.g., 9-amino-camptothecin and 9-nitro-camptotecin have also been reported to be potent radiosensitizers *in vivo*. Kirichenko and Rich (1999) found that under *in vivo* conditions, with the use of 9-AC and irradiation, a DMF of 2.8 is achievable at an isoeffective radiation dose of 28 Gy. The radiosensitizing properties of irinotecan in human lung cancer xenografts and human colon adenocarcinoma (Tamura *et al.*, 1997) and that of Topotecan have been reported in *in vitro* and *in vivo* model systems (Rich, 2003). Chen *et al.*, (1999) have shown a schedule-dependent radiosensitization activity when CPT derivatives like camptostar are combined with radiotherapy.

The taxanes, paclitaxel and docetaxel, promote the assembly of microtubules and induce cell cycle arrest in the most radiosensitive phase (G2/M), thereby promoting radiosensitivity. *In vitro* studies suggest maximum radiosensitization could be achieved by prolonged infusion of paclitaxel (Liebmann *et al.*, 1994). Although a radiosensitizing effect of taxol has been reported, some researchers have shown that the effect is not universal and that radiosensitization is dependent on the origin of cell types. Taxol effects cervical carcinoma cells at other stages of the cell cycle than G2/M and this might be responsible for failure to obtain radiosensitization with cervical carcinoma cells (Geard and Jones, 1994).

Radiosensitization of two pathogenic bacteria, *Escherichia coli* and *Salmonella typhii,* in the presence of thyme and its principal essential oil constituents (carvacrol and thymol) has also been reported (Chiasson *et al.*, 2004).

The cardiac glycosides e.g., digitoxin and ouabain also exhibit radiosensitizing effects. Oleandrin, another cardiac glycoside present in the plant extract Anvirzel, radiosensitized PC-3 human prostate cells with an enhancement factor of 1.32. Susceptibility of PC-3 cells to oleandrin and radiation-induced apoptosis was dependent on activation of caspase-3 (Nasu *et al.*, 2002).

Neem (*Azadirachta indica*) oil enhances the radiosensitivity of cells by interacting with residual damage after X-irradiation, thereby converting sublethal damage or potentially lethal damage into lethal damage, inhibiting double strand break repair and reducing the G(2) phase of the cell cycle in Balb c/3T3 and SCID cells (Kumar *et al.*, 2002).

Several recent studies have documented the radiosensitizing effects of herbals. β-lapachone, an O-napthoquinone with novel chemotherapeutic and radiosensitizing properties, targets cancer cells versus normal cells due to endogenous overexpression of NAD(P)H: quinone oxidoreductase 1(NQ01) (Reinicke, 2007). The radiosensitizing effect of parthenolide, a sesquiterpene lactone found in *Tanacetum parthenium*, has been reported in three prostate cancer cell lines (Sun *et al.*, 2007). Parthenolide renders a radiosensitizing effect by inhibition of the NF-κβ pathway. Kumar *et al.* (2007) found that diospyrin, a napthoquinonoid from *Diospyros montana* Roxb., radiosensitizes MCF-7 breast carcinoma cells by regulating gene expression involved in cell cycle and apoptosis. Lawenda *et al.* (2007) have shown that vitamin E protected normal tissue. However, neither vitamin E nor epigallocatechin gallate exerted a significant radiomodifying effect on the murine breast carcinoma (Mca-IV tumour).

The possibility of discovering many more such compounds from the flora of different regions of the world remains an area worth investigating.

The Multifaceted Mode of Action of Herbal Radioprotectors and Radiosensitizers

Though the precise mechanism by which herbal radioprotectors render radioprotection is far from being completely elucidated (Arora *et al.*, 2005a), some of the mechanisms responsible for radioprotection include: i) augmentation of intrinsic antioxidants like glutathione; ii) metal chelation; iii) enhanced DNA repair and recovery; iv) antiinflammatory response; v) reduced generation of reactive oxygen/nitrogen species; vi) DNA protein cross linking and induction of chromatin compaction; vii) reduced lipid peroxidation and protein oxidation; viii) hypoxia induction; ix) induction of cell cycle arrest; x) increased cell proliferation; xi) increased free radical scavenging; xii) stabilization of cytoplasmic and mitochondrial membrane potential; xiii) modulation of expression of proteins associated with apoptosis etc.

A number of mechanisms are involved in radiosensitization. Some of the commonly reported mechanisms include: i) enhanced generation of ROS/RNS; ii) selective depletion of tumour cell antioxidants and antioxidant enzymes; iii) increased lipid peroxidation depletion of glutathione; iv) elevated levels of lipid peroxidation and DNA damage of tumour cells; v) formation of DNA adducts; vi) inhibition of DNA repair; vi) inhibition of DNA synthesis; vii) induction of cell cycle arrest; viii) induction of apoptosis; ix) depletion of protein kinase C etc.

There is a need to unravel the precise mechanism of action of the various constituents and how they act in combination often rendering a synergistic effect.

Conclusions and Future Directions

Development of radiation countermeasure agents and radiosensitizers is an inevitable necessity in today's scenario. As alluded to earlier in the Chapter, ionizing radiation induces severe damage at various hierarchical levels of organization in living organisms (molecules, cells, tissues, organ, system, whole-body) via both direct and indirect effects, and at high doses can unleash catastrophe affecting a large population. Biological radioprotection is a priority area the world-over due to the increasing risks of radiation exposure (planned and unplanned). However, the available synthetic radiation countermeasure agents are practically ineffective due to their inherent toxicity. The quest for safe, yet non-toxic and effective, RCM agents has been on-going, and in recent years herbal drugs (medicinal, dietary and neutraceutical agents) have shown immense potential. With improved high-throughput screening systems, and the development of appropriate rapid bioassays and molecular imaging techniques, it should be possible to speed the discovery of novel radiomodulating drugs in coming years.

RCM agents of herbal origin are likely to find use in real-life scenarios e.g., rescue missions and dirty bomb incidents. In the immediate future, we can anticipate the availability of effective herbal RCM agents as "first generation radioprotectors/mitigators for emergency responders". The bioactive constituents would form the basis for synthetic derivatization of new, highly effective radiation countermeasure agents. In the future, radiation countermeasure agents will most likely be available as over-the-counter drugs and dietary ingredients/nutraceuticals for diverse prophylactic and therapeutic applications in the realm of medicine, defence, homeland security and space etc. The availability of such highly effective RCM agents for defence and civil applications will instil hope and confidence amongst all those who come in direct contact with radiation. Such agents will drastically change the way we look at radiological incidents/nuclear disasters today, while patients will benefit enormously from the radiosensitizers. The search for radiation countermeasure agents from herbals must continue with increased zeal so that novel agents become available that can simultaneously protect against radiation damage, reduce and repair radiation injury as well as regenerate the damaged cells and tissues. But there is still a long way to go since several intricate mechanisms operate at molecular level and these need to be unravelled.

The concept of "*Vasudheva Kutumbakam*"– the world is one family was propounded in India several thousand years ago by the saints unifying the entire humanity. There is a need for all nations to shed the inhibitions of boundaries and join hands in the endeavour to develop radiation countermeasure agents from the flora that nature has bestowed upon us for the benefit of the whole of mankind!

References

Araujo, M.C., Dias, F.L. and Takahashi, C.S. (1999) Potentiation by turmeric and curcumin of gamma radiation-induced chromosome aberrations in Chinese hamster ovary cells. *Teratogenesis, Carcinogenesis, and Mutagenesis* 19(1), 9–18.

Arora, R. (2006) Phytodefence Molecules for Defence Forces: Utility and Future Prospects. In: Gupta, M., Kumar, I.P. and Arora, R. (eds) *Herbs for Health: Application in Armed Forces.* Institute of Nuclear Medicine and Allied Sciences, Delhi, India, pp. 58–62.

Arora, R. (2007a) Botanicals as Potential Radioprotective and Radiorecovery Agents: Current Status and Emerging Directions for Clinical Applications. *Proceedings of the World Ayurveda Conference*, Pune, India, November 5–12, 2006.

Arora, R. (2007b) Biological Radioprotection: A Phytochemical Perspective. *Proceedings of the Continuing Education Programme on Phytochemistry.* Defence Institute of Physiology and Allied Sciences, Delhi, India, December 10–14, 2007, pp. 19–37.

Arora, R. and Goel, H.C. (2000) Herbal Radioprotectors. *Proceedings of the International Conference on Radiation Biology (Radiobiology 2000)* Organized under the aegis of the Indian Society for Radiation Biology by Regional Cancer Centre (RCC), Thiruvananthapuram, India, from 17–19 February, 2000, pp. 87.

Arora, R. and Sharma, R.K. (2004) Development of Herbal Agents for Biological Radiation Protection. *Proceedings of the Defence Research and Development Organization (DRDO) Life Sciences Scientist's Conference*, Defence Food Research Laboratory, Mysore, India.

Arora, R., Chawla, R., Sagar, R.K., Singh, S., Kumar, R., Sharma, A.K., Singh, S., Prasad, J., Puri, S.C., Gupta, M.L. and Sharma, R.K. (2005b) Low aryl-tetralin lignans containing *Podophyllum hexandrum* extract protects against radiation–mediated lethal damage. *Proceedings of the Conference: Nanobiotechnology: The Road Ahead*, National Botanical Research Institute, Lucknow, India, pp. 29–30.

Arora, R., Chawla, R., Singh, S., Sagar, R.K., Kumar, R., Sharma, A.K., Singh, S., Prasad, J., Sharma, R.K. and Tripathi, R.P. (2006a) Radioprotection by Himalayan High-altitude Region Plants. In: Sharma, R.K. and Arora, R. (eds) *Herbal Drugs: A Twenty First Century Perspective.* Jaypee Brothers Medical Publishers Private Ltd., Delhi, India, pp. 301–325.

Arora, R., Chawla, R., Singh, S., Sagar, R.K., Kumar, R., Sharma, A., Prasad, J., Singh, S., Gurudatta, G.U. and Sharma, R.K. (2006b) Bioprospection for Radioprotective Molecules from Indigenous Flora. In: Govil, J.N., Singh, V.K. and Bhardwaj, R. *Recent Progress in Medicinal Plants.* Studium Press, LLC, USA, Vol. 16, pp. 179–219.

Arora, R., Gupta, D., Chawla, R., Sagar, R., Sharma, A., Kumar, R., Prasad, J., Singh, S., Samanta, N. and Sharma RK. (2005a) Radioprotection by Plant Products: Present status and future Prospects. *Phytotherapy Research* 19(1), 1–22.

Ben-Amotz, A., Yatziv, S., Sela, M., Greenberg, S., Rachmilevich, B., Shwarzman, M. and Weshler, Z. (1998) Effect of natural beta-carotene supplementation in children exposed to radiation from the Chernobyl accident. *Radiation and Environmental Biophysics* 37, 187–193.

Bhosle, S.M., Huilgol, N.G. and Mishra, K.P. (2005) Enhancement of radiation-induced oxidative stress and cytotoxicity in tumour cells by ellagic acid. *Clinica Chimica Acta* 359, 89–100.

Bump, E.A. and Malaker, K. (1998) *Radioprotectors: Chemical, Biological and Clinical Perspective.* CRC Press, Boca Raton, Florida, 431p.

Bump, E.A., Hoffman, S.J. and Foye, W.O. (2003) Radiosensitizers and Radioprotective Agents. In: Abraham, D.J (ed) *Burger's Medicinal Chemistry and Drug Discovery*, Vol 5: Chemotherapeutic Drugs. John Wiley and Sons. Inc., USA, pp. 151–211.

Capizzi, R.L. and Oster, W. (1995) Protection of normal tissue from the cytotoxic effects of chemotherapy and radiation by amifostine: clinical experiences. *European Journal of Cancer* 31, 8–13.

Chen, F.D., Wu, M.C., Wang, H.E., Hwang, J.J., Hong, C.Y., Huang, Y.T., Yen, S.H. and Ou, Y.H. (2001) Sensitization of a tumour, but not normal tissue, to the cytotoxic effect of ionizing radiation using *Panax notoginseng* extract. *American Journal of Chinese Medicine* 29(3/4), 517–524.

Chen, M.F., Keng, P.C., Lin, P.Y., Yang, C.T., Liao, S.K. and Chen, W.C. (2005a) Caffeic acid phenethyl ester decreases pneumonitis after irradiation *in vitro* and *in vivo*. *BMC Cancer* 5, 158.

Chen, M.F., Liao, H.F., Tsai, T.H., Wang, S.Y. and Shiao, M.S. (2005b) Caffeic acid phenethyl ester preferentially sensitizes CT26 colorectal adenocarcinoma to ionizing radiation without affecting bone marrow radioresponse. *International Journal of Radiation Oncology, Biology and Physics* 63, 1252–1261.

Chiasson, F., Borsa, J., Onatottara, B. and Lacoroix, M. (2004) Radiosensitization of *Escherichia coli* and *Salmonella typhi* in ground beef. *Journal of Food Protection* 67(6), 1157–1162.

Coleman, N.E., Blakely, W.F., Fike, J.R., MacVittie, T.J., Metling, N.F., Mitchell, J.B., Moulder, J.E., Preston, J.R., Seed, T.M., Stone, H.B., Tofilon, P.J. and Wong, R.S.L. (2003) Molecular and cellular biology of moderate-dose (1–10Gy) radiation and potential mechanisms of radiation protection. *Radiation Research* 159, 812–834.

Cucinotta, F.A., Schimmerling, W., Wilson, J.W., Peterson, L.E., Badhwar, G.D., Saganti, P.B. and Dicello, J.F. (2001) Space radiation cancer risk projections for exploration missions: uncertainty reduction and mitigation. *NASA Johnson Space Center* (Bulletin-29295), pp. 1–73.

Deeb, D., Yong X, Hao, XU., Xiaohua, J., G.A.O., Janakiraman, N., Chapman R.A. and Gautam, S.C. (2003) Curcumin (diferuloyl-methane) enhances tumor necrosis factor-related apoptosis-inducing lignad-induced apoptosis in LNCaP prostate cancer cells. *Molecular Cancer Therapeutics* 2, 95-103.

Emerit, I., Arutyunyan, R., Oganesian, N., Levy. A., Cernjavski, L., Sarkisian, T., Pogosian, A. and Asrian, K. (1995) Radiation-induced clastogenic factors: anticlastogenic effects of *Ginkgo biloba* extract. *Free Radical Biology and Medicine* 18, 985–991.

Emerit, I., Oganisian, N., Aruyunyian, R., Pogossian, A., Sarkisian, T., Cernjavski L., Levy, A. and Feingold, J. (1997) Oxidative stress-related clastogenic factors in plasma of Chernobyl liquidators: protective effects of antioxidant plant phenols, vitamins and oligoelements. *Mutation Research* 373, 47–54.

Esposito, R.D., Durante, M., Gialanella, G., Grossi, G., Pugliese, M., Scampoli, Jones, T.D. (2001) A model of radiation-induced myelopoiesis in space. *Physica Medica* 17(Supplement 1), 181–182.

Garg, A. K., Buchholz, T. A. and Aggarwal, B.B. (2005) Chemosensitization and radiosensitization of tumours by plant polyphenols. *Antioxidants and Redox Signalling* 7, 1630–1647.

Geard, G.R. and Jones, J.M. (1994) Radiation and taxol effects on synchronized human cervical carcinoma cells. *International Journal of Radiation Oncology, Biology and Physics* 29(3), 565–569.

Giambarresi, L., Jacobs, A.J. (1987) Radioprotectants. In: Conklin, J.J. and Walker, R.I. (eds) *Military Radiobiology*, Academic Press, Inc, London, pp. 265–301.

Gong , Y.F., Huang, Z.J., Qiang, M.Y., Lan, F.X., Bai., G.A., Mao, Y.X., Ma, X.P., and Zhang, F.G. (1991) Suppression of radioactive strontium absorption by sodium alginate in animals and human subjects. *Biomedical and Environmental Sciences* 4, 273–282.

Haksar, A., Sharma, A.K., Chawla, R., Arora, R., Singh, S., Prasad, J., Kumar, R., Tripathi, R.P., Arora, M.P., Islam, F. and Sharma, R.K. (2006) *Zingiber officinale* Rosc. exhibits behavioural radioprotection against radiation–induced conditioned taste aversion in a gender specific manner. *Pharmacology Biochemistry and Behaviour* 84, 179–188.

Higuchi, K., Mitsuhashi, N., Saitoh, J., Maebayashi, K., Sakurai, H., Akimoto, K. and Niibe, H. (2000) Caffeine enhanced radiosensitivity of rat tumour cells with a mutant type p53 by inducing apoptosis in a p53– independent manner. *Cancer Letters* 152, 157–162.

IAEA (2004) Accidental Overexposure of Radiotherapy Patients in Bialystok. International Atomic Energy Agency, Vienna, Austria, pp. 103.

Kasten-Pisula, U., Windhorst, S., Dahm-Dalphi, J., Mayr, G. and Dikomey, E. (2007) Radiosensitization of tumour cell lines by the polyphenol Gossypol results from double-strand break repair and not from enhanced apoptosis. *Radiotherapy and Oncology* 5, 1752–1756.

Kennedy, A.R. (1993) In: Troll, W. and Kennedy, A.R. (eds) *Protease Inhibitor as Cancer Chemopreventive Agents*. Plenum Press, New York, USA, pp. 9–64.

Kennedy, A.R. (1998) Chemopreventive agents: protease inhibitors. *Pharmacology and Therapeutics* 78, 167–209.

Kennedy, A.R. (2007) Countermeasures against space radiation-induced oxidative stress in mice. *Radiation and Environmental Biophysics* 46(2), 201–203.

Kennedy, A.R., Zhou, A., Donahne, J.J. and Ware, J.H. (2006) Protection against adverse biological effects induced by space radiation by the Bowman-Birk inhibitor and antioxidants. *Radiation Research* 166, 327–332.

Kimery, A. (2003) Radiological race. *Homeland Defense Journal* 1, 16–17.

Kirichenko, A.V. and Rich (1999) Radiation enhancement by 9-aminocamptothecin: the effect of fractionation and timing of administration. *International Journal of Radiation Oncology, Biology and Physics* 44(3), 659–664.

Kumar, A., Rao, A.R. and Kimura, H. (2002) Radiosensitizing effects of neem (*Azadirachta indica*) oil. *Phytotherapy Research* 16(1), 74–77.

Kumar, B., Joshi, J., Kumar, A., Pandey, B.N., Hazra, B. and Mishra, K.P. (2007) Radiosensitization by diosypyrin diethylether in MCF-7 breast carcinoma cell line. *Molecular and Cellular Biochemistry* 304(1–2) (online).

Kumar, R., Singh, P.K., Arora, R., Sharma, A.K., Sharma, R.K. and Tripathi, R.P. (2006) A process for isolation of a highly radioresistant bacterium and preparation of a radioprotective extract thereof. Patent application filed with the Indian Patent Office, Delhi, India.

Lawenda, B.D., Smith, D.E., Xu, L., Niemierko, A., Silverstein, J.R., Boucher, Y., Kashiwagi, S., Held, K.D., Jain, R.K., Loeffler, J.S., Eisesnberg, D.M. and Blumberg, J.B. (2007) Do dietary supplements epigallocatechin gallate or vitamin E cause a radiomodifying response on tumours *in vivo*? A pilot study with murine breast cancer. *Journal of Society of Integrative Oncology* 5(1), 11–17.

Liebmann, J, Cook, J.A. and Fisher, J. *et al.*, (1994) *In vitro* studies of Taxol as a radiosensitizer in human tumour cells. *Journal of the National Cancer Institute* 86, 441–446.

Murayama, C., Nagano, Y., Sano, S., Ochiai, M., Fuji, K., Fujita, E. and Mori, T. (1987) Effect of oridonin, a Rabdosia diterpenoids, on radiosensitization with misonidazole. *Experientia* 43(11–12), 1221–1223.

Nagano, J., Kono, S., Preston, D.L., Moriwaki, H., Sharp, G.B., Koyama, K. and Mabuchi, K. (2000) Bladder-cancer incidence in relation to vegetable and fruit consumption: a prospective study of atomic–bomb survivors. *International Journal of Cancer* 86, 132–138.

Nair, C.K.K., Parida, D.K. and Nomura, T. (2001) Radioprotectors in Radiotherapy. *Journal of Radiation Research* 42, 21–37.

Nasu, S., Milas, L., Kawabe, S., Raju, U. and Newman, R. (2002) Enhancement of radiotherapy by oleandrin is a caspase–3 dependent process. *Cancer Letters* 185(2), 145–151.

Neta, R. (1988) Role of cytokines in radiation protection. *Pharmacology and Therapeutics* 39, 261–266.

Patt, H.M., Tyree, E.B., Straube, R.L., Smith, D.E. (1949) Cysteine protection against X-irradiation. *Science* 110, 213–214.

Prasad, K.N. (1999) *Handbook of Radiobiology.* CRC Press, Boca Raton, Florida, USA, p. 344.

Puri, S.C., Nazir, A., Chawla, R., Arora, R., S. Riyaz-ul-Haq, Amna, T., Ahmed, B., Verma, V., Singh, S., Sagar, R.K., Sharma, A.K., Kumar, R., Sharma, R.K. and Qazi, G.N. (2006) The endophytic fungus *Trametes hirsuta* as a novel alternative source of podophyllotoxin and related aryl tetralin lignans. *Journal of Biotechnology* 122, 494–510.

Puri, S.C., Nazir, A., Gupta, V.K., Negi, J.S., Handa, G., Dhar, K.L., Qazi, G.N., Arora, R., Kumar, R., Sharma, A.K., Kumar, P., Prasad, J., Singh, S., Chawla, R., Sagar, R.K. and Sharma, R.K. (2005) A process for preparation of a radioprotective formulation containing aryltetralin lignans. Patent application (NF/235/2005), Indian Patents Office, New Delhi, India.

Raffoul, J.J., Banerjee, S., Singh-Gupta, V., Knoll, Z.E., Fite, A., Zhang, H., Abrams, J., Sarkar, F.H. and Hillman, G.G. (2007) Down-regulation of apurinic/apyrimidinic endonuclease 1/redox factor 1 expression by soy isoflavones enhances prostate cancer radiotherapy *in vitro* and *in vivo*. *Radiation Research* 67(5), 2141–2149.

Reinicke, E.K. (2007) Prodrug development and the role of reactive oxygen species in beta-Lapachone-mediated cell death. PhD Thesis, Case Western Reserve University, USA, pp. 129.

Rich, T.A. (2003) Camptothecin radiation sensitization. In: Choy, H.J. (ed) *Chemoradiation in Cancer Therapy*. Humana Press Inc., Totowa, NJ, USA, pp. 93–103.

Seed, T., Kumar, S., Whitnall, M., Srinivasan, V., Singh, V., Elliott, T., Landauer, M., Miller, A., Chang, C-M., Inla, C., Deen, J., Gehlhaus, M., Jackson, W., Hilyard., Pendergrass, J., Toles, R., Villa, V., Miner, V., Stewart, M., Benjack, J., Danilenko, D. and Farrell, C.K. (2002) New strategies for the prevention of radiation injury: possible implications for countering radiation hazards of long-term space travel. *Journal of Radiation Research* 43, S239–S244.

Seed, T.M. (2005) Radiation protectants: current status and future prospects. *Health Physics* 89(5), 531–545.

Sharma, A., Haksar, A., Chawla, R., Arora, R., Kumar, R., Singh, S., Prasad, J., Gupta, M.L., Lahiri, S.S., Tripathi, R.P., Arora, M.P., Islam, F. and Sharma, R.K. (2007) Behavioural protection by *Mentha spicata* Linn. against radiation-induced conditioned taste aversion: A comparative study with a radioprotective calcium channel-blocking drug. *Indian Journal of Radiation Research* 4(1), 47–56.

Sharma, A., Haksar, A., Chawla, R., Kumar, R., Arora, R., Singh, S., Prasad, J., Islam, F., Arora, M.P. and Sharma, R.K. (2005) *Zingiber officinale* Rosc. modulates gamma radiation–induced conditioned taste aversion. *Pharmacology Biochemistry and Behaviour* 81, 864–870.

Sharma, A.K., Haksar, A., Kumar, R., Arora, R. and Sharma, R.K. (2006) Ginger: Evaluation of its Anti-emetic and Behavioral Radioprotective Properties. In: Govil, J.N., Singh, V.K. and Bhardwaj, R. *Recent Progress in Medicinal Plants*. Studium Press, LLC, USA, Vol.16, 237–264.

Sharma, R.K. and Arora, R. (2006) *Herbal Drugs: A Twenty First Century Perspective*. Jaypee Brothers Medical Publishers Private Ltd., Delhi, India, p. 688.

Shukitt-Hale, B., Carey, A.N., Jenkins, D., Rabin, B.M. and Joseph, J.A. (2007) Beneficial effects of fruit extracts on neuronal function and behaviour in a rodent model of accelerated aging. *Neurobiology of Aging* 28, 1187–1194.

Singh, V.K. and Yadav, V.S. (2005) Role of cytokines and growth factors in radioprotection. *Experimental and Molecular Pathology* 78, 156–169.

Stelzer, K.J., Koh, W.-J., Kurtz, H., Greer, B.E. and Griffin, T.W. (1994) Caffeine consumption is associated with decreased severe late toxicity after radiation to the pelvis. *International Journal of Radiation Oncology Biology and Physics* 30, 355–360.

Stenzel, R.A., Rajesh, D. and Howard, S.P. (2002) Flavopiridol as a radiosensitizer in malignant glimoas. *Proceedings of the American Association for Cancer Research* 43, A975.

Stone, H.B., Moulder, J.E., Coleman, C.N., Ang, K.K., Anscher, M.S., Barcellos-Hoff, M.H., Dynan, W.S., Fike, J.R., Grdina, D.J., Greenberger, J.S., Hauer–Jensen, M., Hill, R.P., Kolesnick, R.N., Macvittie, T.J., Marks, C., McBride, W.H., Metting, N., Pellmar, T., Purucker, M., Robbins, M.E., Schiestl, R.H., Seed, T.M., Tomaszewski, J.E., Travis, E.L., Wallner, P.E., Wolpert, M. and Zaharevitz, D. (2004) Models for evaluating agents intended for the prophylaxis, mitigation and treatment of radiation injuries. *Radiation Research* 162, 711–728.

Sun, Y., St. Clair, D.K., Fang, F., Warren, G.W., Rangnekar, V.M., Crooks, P.A. and St. Clair, W.H. (2007) The radiosensitization effect of parthenolide in prostate cancer cells is mediated by nuclear factor-κβ inhibition and enhanced by the presence of PTEN. *Molecular Cancer Therapeutics.* 6, 2477–2486.

Sweeney, T.R. (1979) A Survey of Compounds from the Antiradiation Drug Development Program of the US Army Medical Research and Development Command, Walter Reed Army Institute of Research, Washington DC, USA.

Tamura, K, Takada, M, Kawase, I. (1997) Enhancement of tumour radio-response by irinotecan in human lung tumour xenografts. *Japanese Journal of Cancer Research* 88, 218–223.

Uma Devi, P. (2006) Development of a radioprotective drugs from *Ocimum sanctum*. In: Sharma, R.K. and Arora, R. (eds) *Herbal Drugs: A Twenty First Century Perspective*. Jaypee Brothers Medical Publishers Pvt. Ltd., Delhi, India, pp. 291–300.

Urtasun, R.C. (1998) Chemical Modifiers of Radiation. In: Lieberl, S.A. and Phillips, T.L. *Text Book*

of Radiation Oncology. WB Saunders Company, Philadelphia, USA, pp. 42–52.
Valenzuela, M.T., Mateos, S., Almodoavar, J.M.R. and McMillan, T.J. (2000) Variation in sensitizing effect of caffeine in human tumour cell lines after γ–irradiation. *Radiotherapy and Oncology* 54, 261–271.
Vasin, M.V. (1999) Classification of radiation protective agents as a basis of modern radiation pharmacology. *Radiatsionnaya Biologiya Radioekologiya* 39, 212–222.
Walker, R.I. and Conklin, J.J. (1987) Military Radiobiology: A Perspective. In: Walker, R.I. and Conklin, J.J. (eds) *Military Radiobiology*, Academic Press Inc., London, UK, pp. 404.
Weiss, J.F. and Landauer, M.R. (2003) Protection against ionizing radiation by antioxidant nutrients and phytochemicals. *Toxicology* 189, 1–20.
Weiss, J.F. and Simic, M.G. (1988) Perspectives in radioprotection. *Pharmacology and Therapeutics* 39, 1–414.
Weiss, J.F., Kumar, K.S., Walden, T.L., Neta, R., Landauer, M.R., Clark, E.P. (1990) Advances in radioprotection through the use of combined agent regimens. *International Journal of Radiation Oncology, Biology and Physics* 57, 709–722.
WHO (2007) *Development of Stockpiles for Radiation Emergencies*. Perez, M. and Carr, Z. (eds) WHO Headquarters, Geneva, Switzerland, pp. 42.
Xu, L, Yang, D., Wang, S., Tang, W., Liu, M., Davis, M., Chen, J., Rae, J.M., Lawrence, T.S. and Lippmann, M.E. (2005) (-)-Gossypol enhances response to radiotherapy and results in tumour regression of human prostate cancer. *Molecular Cancer Therapeutics* 4(2), 197–205.
Zhang, W., Anker, L., Law, R.E., Hinton, D.R., Gopalakrishna, R., Pu, Q., Gundimeda, U., Weiss, M.H. and Couldwell, W.T. (1996). Enhancement of radiosensitivity in human malignant glioma cells by hypericin *in vitro*. *Clinical Cancer Research* 2(5), 843–846.
Zoberi, I., Bradberry, C.M., Curry, H.A., Bisht, K.S., Goswami, P.C. and Roti Roti, J.L. (2002) Radiosensitizing and antiproliferative effects of resveratrol in two human cervical tumour cell lines. *Cancer Letters* 175(2), 165–173.

SECTION II

PHYTOMEDICINE FOR RADIOPROTECTION AND MANAGEMENT OF RADIATION INJURIES

Chapter 2

Indian Medicinal Herbs and Ayurvedic Formulations as Potential Radioprotectors

D.K. Maurya and T.P.A. Devasagayam

Introduction

The effects of radiation exposure on human health have been well documented from the epidemiological studies on the survivors of the Hiroshima and Nagasaki atom bombs. There has been extensive research on radioprotective or anti-radical compounds since World War II on account of the relevance of these compounds in military, clinical and industrial applications. Radiation protection might offer a tactical advantage on the battlefield in the event of nuclear warfare. Radiation protection is also relevant in the present day scenario as there is a global increase in attacks by terrorist groups, and there is a possible threat of nuclear terrorism. Radioprotectors could also reduce the cancer risk to a population exposed to radiation directly or indirectly through industrial and military applications. Hence, understanding the factors governing a cell's response to radiation damage will be of great advantage in utilizing radiation in diagnosis and therapy, as well as in the management of radiation exposure cases, of both intentional and accidental nature.

In spite of more than six decades of research on the development of radioprotectors, or anti-radiation drugs, and the spectacular advances made over the last few decades in the areas of cell and molecular biology, synthetic chemistry and biochemistry, the development of a safe and effective non-toxic radioprotector for human use has remained elusive till today. Several compounds, which have been found very effective in the laboratory studies, have failed in human applications due to toxicity problems or lack of significant protective effects at physiologically relevant doses (Bump and Malakar, 1998; Turner *et al.*, 2002; Weiss and Landauer, 2003; Maurya *et al.*, 2006a; Sharma and Arora, 2006).

Currently Available Strategies/Chemicals

The use of chemicals to protect against the harmful effects of radiation was initiated after World War II, with realization of the need to safeguard humans against atomic weapons. Apart from this, radioprotectors are also essential for protecting normal tissues during radiotherapy of cancer (Nair *et al.*, 2004; see Chapter 1 of this book). In fact, no radioprotective agent is now available, either alone or in combination, which meets all the requisites of an ideal radioprotector (Maisin, 1998). Amifostine is the only one that is currently used for human applications for radioprotecting normal tissues during radiotherapy, even though there are reports about contra-indications in some cases (Tannehill and Mehta, 1996). Though a large number of compounds have been shown to be promising as radioprotectors in laboratory studies, few could pass the transition from bench to bedside. Most of them failed even before reaching the pre-clinical stage due to toxicity and side effects. For clinical application of any compound as a radioprotector or as an adjuvant in radiotherapy, there is a need for absolute certainty about its relative protection factors for tumour and normal tissues accompanied by minimum toxicity to avoid unacceptable clinical risk.

Drawbacks of Chemical/Synthetic Radioprotectors (Molecular Drugs)

Most of the chemical radioprotectors screened after World War II were thiol based. The major drawback of these compounds was their toxic nature (at protective concentrations) and also they were unable to provide post-irradiation protection. The high toxicity of thiol compounds necessitated the search for alternative agents that possess less toxicity and would be highly effective at non-toxic concentrations.

Advantage of Natural Plant Products in Radioprotection

Because of the inherent toxicity of chemicals and synthetic agents at their effective radioprotective concentrations, investigators diverted their attention towards natural products, focussing on their active ingredients. Many natural products and compounds have been investigated in the recent past for their efficacy to protect biological systems against radiation-induced damage (Maurya *et al.,* 2006a; Uma Devi, 2001; Arora and Goel, 2000; Arora and Sharma, 2004; Arora *et al.,* 2003a,b, 2004a,b, 2005a,b,c,d, 2006a,b, 2007a,b, 2008a,b; Arora, 2007, 2008; Chawla *et al.,* 2005, 2006, 2007, 2008; Puri *et al.,* 2005; Kumar *et al.,* 2005a,b; Gupta *et al.,* 2006). An ideal radioprotector should be cheap, should not have toxic implications in a wide dose range, and preferably be orally administered, rapidly absorbed, possess a reasonably good dose-reduction factor and ideally should act through multiple mechanisms (Arora *et al.,* 2005, 2006a,b; Jagetia, 2007a). Since plants and natural products are extensively used in several traditional systems of medicine, screening of radioprotective compounds from them has several advantages because usually they are considered non-toxic and are widely accepted by humans.

Approaches for Screening and Assessing an Ideal Radioprotector

Many short- and long-term assays have been developed for measuring radioprotective properties of natural and synthetic compounds under *in vitro*, *ex vivo* and *in vivo* conditions. Short-term *in vitro* and cell free assays, such as plasmid relaxation assay (Maurya *et al.,* 2006b; Arora *et al.,* 2006a,b), estimation of lipid peroxidation and protein oxidation (Kamat *et al.,* 2000a) provide basic information about the radioprotective effect of the tested compounds. Assays for testing the free radical scavenging also help in screening probable radioprotectors. If a candidate molecule is found to inhibit relaxation of plasmid, peroxidation of lipid and oxidation of protein, along with the free radical scavenging, it can act as a potential radioprotector. The next step is to test these potential molecules using short-term *ex vivo* and *in vivo* assays for their protective ability at cellular level in terms of the DNA damage using comet assay (Maurya *et al.,* 2005a,b, 2006b), apoptosis, antioxidant enzymes, cell survival and frequency of micronuclei (Maurya *et al.,* 2005b). Once a compound gives protection at cellular level, its radioprotective ability should be tested using long-term *in vivo* assays such as clonogenicity assay, survival of mice, and development of various radiation syndromes such as gastrointestinal (GI) using intestinal crypt cell assay, besides hemopoietic and central nervous system (CNS) syndromes.

Development of Radiation-induced Damage and the Mechanism of their Inhibition by Radioprotectors

Ionizing radiation induces damage either directly by hitting target molecules or indirectly by inducing formation of reactive oxygen species in the form of ·OH, ·H, singlet oxygen and peroxyl radicals that follow a cascade of the events leading to DNA damage such as single- or double-strand breaks (DSB), base damage, and DNA-DNA or DNA-protein cross-links, and these lesions cluster as a complex locally multiplying damaged sites (Nair *et al.,* 2001; Jagetia, 2007a).

The radioprotectors can act by various mechanisms (see Fig. 2.1) such as: (i) suppressing the formation of free radicals; (ii) detoxifying the radiation induced reactive species; (iii) inducing the cellular defence mechanisms such as superoxide dismutase (SOD), glutathione etc.; (iv) enhancing DNA repair by triggering one or more cellular DNA repair pathways, and (v) delaying cell division and inducing hypoxia in the tissues (Maurya *et al.*, 2006a).

A detailed diagram describing radiation-induced damages and the mechanism of protection is shown in Fig. 2.1.

Natural Products are Useful for Radioprotection

Many natural antioxidants, whether consumed before or after radiation exposure, are able to confer some level of radioprotection. In addition to achieving beneficial effects from established antioxidants such as vitamins C and E and their derivatives (Sarria and Prasad, 1984; Radner and Kennedy, 1986; Umegaki *et al.*, 1994; Kline *et al.,* 1995; Salvi *et al.,* 2001; Satyamitra *et al.*, 2001; Kumar *et al.,* 2002; Rajagopalan *et al.*, 2002; Jagetia *et al.,* 2004a, 2007c) and folic acid, protection is also conferred by several novel molecules, such as flavonoids, epigallocatechin and other polyphenols. The immune system is protected against radiation by the following natural compounds: polyphenols, lignans, vitamin C, glutamine and arginine, palm carotene, fatty acids, ubiquinone and hydroquinone (Luthra *et al.,* 2008; Arora *et al.,* 2005a). Similarly, the central nervous system is protected by the following components: aged garlic extract and polyphenols. The eye was protected against radiation by vitamin C, fruits and vegetables, as well as by aged garlic extract. Radiation-induced carcinogenesis can be reduced by the following components: zinc, vitamins C and E, selenite, polyphenols, thiols, fatty acids, yellow-green vegetables/fruits, curcumin, niacin and nicotinamide adenine dinucleotide (Singh *et al.,* 1990; Turner *et al.,* 2002; Weiss and Landauer, 2003).

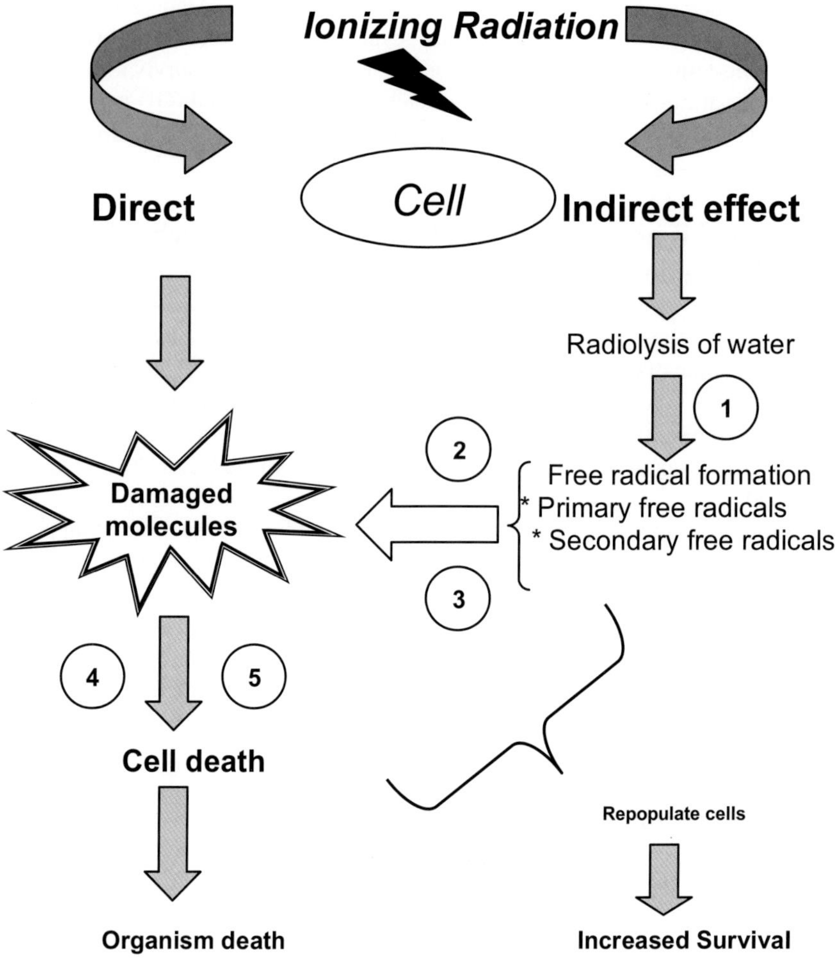

Fig. 2.1. A schematic diagram of radiation-induced damage and the mechanism of its mitigation. (1) represents suppression of free radical formation, (2) detoxification of reactive species, (3) induction of the cellular radioprotectors such as SOD, GSH etc, (4) enhancement of the molecular repair pathways, and (5) delay in cell division and induction of hypoxia in the tissues.

Usefulness of Isolated Natural Phytomolecules in Radioprotection

Several natural dietary ingredients, including vitamins and several isolated compounds, used in India and also other parts of the world show radioprotective effects. For example ascorbic acid, caffeine, chlorophyllin, ferulic acid, glycyrrhizic acid, troxerutin, vanillin, vitamin E and its derivatives like alpha-tocopheryl succinate (α-TS), alpha-tocopherol monoglucoside (TMG) etc.

Caffeine

Caffeine offers radioprotection against oxygen-dependent radiation-induced damage in barley seeds (Kesavan *et al.*, 1973), Chinese hamster ovary cells (Kesavan and Natarajan, 1985), rat liver mitochondria (Kamat *et al.*, 1999, 2000a) and plasmid DNA (Kumar *et al.*, 2001). Caffeine is radioprotective in chromosomes of bone marrow in mice, whether given before or after whole-body gamma irradiation (Farooqi and Kesavan, 1992). Post-treatment with caffeine did not influence radiation-induced damage in mouse L cells (Rauth, 1967), or Chinese hamster cells (Kihlman *et al.*, 1974). Caffeine restored the normal cell cycle, following X-ray-induced arrest in G_2 phase, in one and two-cell mouse embryos (Grinfeld and Jacquet, 1988). Caffeine reduced the concentration of an X-ray-induced protein in melanoma cells (Hughes and Boothman, 1991). Caffeine has also been shown to protect against radiation-induced lethality in mice (George *et al.*, 1999).

Chlorophyllin

Chlorophyllin acts both as an antimutagen (Ong *et al.*, 1986) and as a radioprotector (Morales-Ramirez *et al.*, 1984; Zimmering *et al.*, 1990; Abraham *et al.*, 1994; Morales-Ramirez and Garcia-Rodriquez, 1994; Pimentel *et al.*, 1999; Boloor *et al.*, 2000; Kamat *et al.*, 2000b). Chlorophyllin protects mitochondrial membranes against gamma radiation *in vitro* (Boloor *et al.*, 2000; Kamat *et al.*, 2000b), strand breaks in plasmid DNA (Kumar *et al.*, 1999) and sister chromatid exchange (SCE) in murine bone marrow cells *in vivo* (Morales-Ramirez *et al.*, 1984). It significantly reduces the incidence of micronucleated polychromatic erythrocytes in bone marrow cells upon gamma-ray exposure (Abraham *et al.*, 1994). Chlorophyllin exhibited significant radioprotective activity in *Drosophila melanogaster* (Zimmering *et al.*, 1990). A dose of 100 μg of chlorophyllin/g body weight protected against the induction of sister chromatid exchange in murine bone marrow cells by 1Gy of gamma rays (Morales-Ramirez *et al.*, 1984; Morales-Ramirez and Garcia-Rodriquez, 1994).

Ferulic acid

Ferulic acid is a polyphenol occurring in plant products such as rice, green tea, coffee beans etc. Our laboratory and other groups have shown that ferulic acid offers good radioprotection under various *in vitro*, *ex vivo* and *in vivo* conditions, when applied before radiation exposure (Maurya *et al.*, 2005a). Ferulic acid offers radioprotection by inhibiting damage to DNA (Maurya *et al.*, 2005a) and membrane (Prasad *et al.*, 2006). The possible mechanisms involved include modulation of antioxidant enzyme (Prasad *et al.*, 2006) and the DNA repair process (Maurya *et al.*, 2005a). Administration of ferulic acid 1 h prior to gamma irradiation significantly reduced the DNA damage in mouse blood leukocyte and bone marrow cells. When ferulic acid is given 1 h prior to and/or immediately after gamma radiation exposure it significantly reduces the micronucleated reticulocytes (MNRETs) in mouse blood (Maurya *et al.*, unpublished data). It also renders preferential radioprotection to normal tissue, not to tumour cells under both *ex vivo* and *in vivo* situation of radiation exposure (Maurya and Nair, 2006b).

Glycyrrhizic acid

Glycyrrhizic acid, an active constituent of *Glycyrrhiza glabra* L (Family: Fabaceae), offered protection to plasmid pBR322 DNA from radiation-induced strand breaks with a dose-reduction factor of 2.04 at 2.5 mM concentration. Under *ex vivo* conditions, glycyrrhizic acid protected the cellular DNA of human peripheral blood leukocytes exposed to gamma radiation in a concentration-dependent manner. Intraperitoneal administration of glycyrrhizic acid (4 mg/kg body weight) to mice 1 h prior to radiation exposure protected cellular DNA of peripheral blood leucocytes and bone marrow cells. Pulse radiolysis studies indicated that glycyrrhizic acid offers radioprotection by scavenging free radicals (Gandhi *et al.,* 2004).

Troxerutin

Troxerutin, a derivative of the natural flavonoid rutin extracted from *Sophora japonica* (Japanese Pagoda tree) inhibits lipid peroxidation in membranes of sub-cellular organelles as well as normal tissues of tumour-bearing mice exposed to gamma radiation. Further, it was found that administration of troxerutin resulted in differential protection of DNA in blood leukocytes and bone-marrow cells and not in cells of a tumour in whole-body irradiated tumour-bearing mice (Maurya *et al.,* 2004). Maurya *et al.,* (2005b) have shown that it enhances the process of DNA repair and has concentration-dependent radioprotection to mouse blood and bone marrow cells. Troxerutin significantly inhibited the micronuclei in human peripheral blood lymphocytes and mouse blood reticulocytes.

Vanillin

Vanillin (4-hydroxy-3-methoxybenzaldehyde) is the major component of natural vanilla, which is one of the most widely used and important flavouring materials worldwide. It suppresses the chromosomal aberrations induced by X-rays in V79 cells *in vitro* (Keshava *et al.,* 1998) and in mice *in vivo* (Sasaki *et al.,* 1990). It inhibits lipid peroxidation in rat liver mitochondria and reduces DNA damage in plasmid pBR322 (Kamat *et al.,* 2000c; Kumar *et al.,* 2000). Our recent investigations show that vanillin has the ability to protect against DNA damage in human and mouse peripheral blood leukocyte and splenic lymphocytes, besides enhancing survival in splenic lymphocytes against gamma radiation, and that the possible mechanism may involve scavenging of radicals generated during radiation, apart from modulation of DNA repair observed earlier (Maurya *et al.,* unpublished data).

Indian Herbs and Ayurvedic Formulations for Radioprotection

Indian medicinal plants and Ayurvedic polyherbal preparations have proved to be very effective radioprotectors. A list of radioprotective effects of Indian medicinal plants and polyherbal Ayurvedic formulations is given in Tables 2.1 and 2.2 respectively.

Some individual plants that show a variety of significant protective effects (Table 2.1) are dealt with in detail in the ensuing section:

Indian medicinal plants exhibit potent radioprotective efficacy

Hippophae rhamnoides (Linn.) (Family: Elaeagnaceae)

Hippophae rhamnoides (Sea Buckthorn) has been used in the Indian and Tibetan system of medicine for centuries. The plant has antioxidant, anti-inflammatory, antimicrobial and immunostimulatory properties. Its aqueous extract enhances the survival of Strain 'A' mice when administered 30 min prior to whole-body gamma irradiation (Goel *et al.,* 2000c, 2002). It provides protection to the gastrointestinal system against lethal whole-body gamma irradiation. Administration of a hydroethanolic (50:50 v/v) extract, 30 min before irradiation increased the number of surviving crypts in the jejunum by a factor of 2.02 and villi cellularity by 2.5 fold (Goel *et al.,* 2000c, 2003). Chawla *et al.,* 2007 have recently reported the radioprotective efficacy of fractionated, flavonoid-rich fractions of *Hippophae rhamnoides* berries.

Mentha piperita (Linn.) (Family: Labiatae)

This is an aromatic plant (peppermint) with several reported therapeutic properties, mainly pertaining to treatment of nausea and vomiting (Sharma *et al.,* 2007). The oral administration of Mentha extract (1 g/kg body weight/day) before exposure to gamma radiation was found to be effective in increasing the frequency of radiation-induced endogenous spleen colonies. A significant increase in the weight of the spleen was observed in animals of the Mentha-treated and radiation-exposed group in comparison to the irradiated group on day 10 of post-irradiation. Oral administration of 1 g/kg body weight/day before exposure to gamma radiation protected against radiation-induced chromosomal damage in bone marrow of mice with a dose modifying factor (DMF) value 1.78 (Samarth and Kumar, 2003a). Mentha extract and its oil enhanced the survival of mice (Samarth and Kumar, 2003b), besides improving haematological parameters (Samarth *et al.,* 2004).

Ocimum sanctum (Linn.) (Family: Lamiaceae)

Ocimum sanctum (Tulsi or Indian holy basil) is a medicinal herb widely used in the Ayurvedic system of medicine in India. It is used for treating infections, skin diseases, common cold and cough, malarial fever, besides hepatic disorders. It also possesses antibacterial, anti-inflammatory, antiviral, anticarcinogenic, antioxidant and immunostimulatory activities. Uma Devi and Ganasoundari (1995) reported its radioprotective property for the first time.

Aqueous and alcoholic extracts of leaves possess radioprotective properties, but its aqueous extract (optimum dose was 50 mg/kg body weight; acute LD_{50} was 6 g/kg body weight) was more effective in increasing survival (Uma Devi and Ganasoundari, 1995). Its extract was compared to WR-2721, a standard radioprotector. An intraperitoneal injection of an optimum dose (10mg/kg daily for 5 days) of leaf extract to mice before delivering sub-lethal (2Gy) total-body gamma radiation produced a significantly higher bone marrow stem cell survival than such a pre-treatment with 300mg/kg of WR-2721 (Ganasoundari *et al.,* 1998). Two active components of *O. sanctum*, orientin and vicenin, did not exhibit any systemic toxicity in mice even at a dose of 100mg/kg b.wt and both significantly increased the survival of mice when administered 30min prior to lethal whole-body gamma irradiation. Vicenin gave a DMF value of 1.37, whereas orientin gave 1.30 in the murine system (Uma Devi *et al.,* 1999). These compounds also significantly inhibited the Fenton reaction-induced ·OH generation under *in vitro* conditions (Uma Devi *et al.,* 2000), and protected human lymphocyte chromosomes against radiation (Uma Devi, 2001).

Podophyllum hexandrum (Royle) (Family: Berberidaceae)

Podophyllum hexandrum (Himalayan Mayapple) has been shown to mitigate radiation injuries and especially the haemopoietic syndrome in adult mice. It protects plasmid pBR322 DNA against radiation-induced damage *in vitro.* It enhanced survival of mice and increased levels of liver GST and SOD besides intestinal SOD (Goel *et al.,* 2000a,b; Mittal *et al.,* 2001). It also prevents radiation-induced neuronal damage in post-natal rats exposed *in utero* (Sajikumar and Goel, 2003). Aryltetralin lignans present in Podophyllum, along with flavonoids render a radioprotective effect. *Podophyllum* also protects the haemopoietic system against gamma radiation (Sagar *et al.,* 2006). Aqueous extract of *Podophyllum hexandrum* (RP-1), which has been reported to render more than 82% survival against whole-body lethal (10Gy) gamma irradiation in mice, was further investigated for its immunomodulatory potential. In this study, no significant change could be scored in peritoneal macrophages survival up to the 8th day after whole-body irradiation (Goel *et al.,* 2007).

Recently, whole-body radioprotection by a semi-purified fraction of *Podophyllum hexandrum* has been reported (Lata *et al.,* 2007). The hydro-alcoholic material (REC-2000) extracted from the rhizome offers cytoprotection against gamma radiation and this cytoprotective effect is mediated *via* haemopoietic system stimulation and up-regulation of haeme-oxygenase-1 and the prosurvival multidomain protein Bcl-2 (Arora *et al.,* 2007). The molecular mechanisms involved in radioprotection by *P. hexandrum* have also been reported (Kumar *et al.,* 2007). Recently, a novel quercetin derivative, with radioprotective properties, isolated from *Podophyllum hexandrum* has been characterized by ^{1}H AND ^{13}C NMR (Arora *et al.,* 2008).

Tinospora cordifolia (Miers) (Family: Menispermaceae)

Tinospora cordifolia (Guduchi) is widely used in Ayurvedic medicines. It is known for its immunomodulatory, anti-hepatotoxic, anti-stress and antioxidant properties. A preparation of *T. cordifolia* (RTc) administered i.p. (200mg/kg body weight) to male mice 1hr before whole-body gamma irradiation was evaluated for its radioprotective efficacy, in terms of whole-body survival, spleen colony forming units (CFU), haematological parameters, cell cycle progression and micronuclei induction. Pre-irradiation treatment with RTc rendered 76.3% survival (30 days), compared to 100% mortality in irradiated control and also prevented radiation-induced weight loss (Goel *et al.,* 2001). RTc also restored total lymphocyte counts and increased the S-phase population that was reduced after 2Gy exposures (Goel *et al.,* 2004). Its aqueous extract enhanced the survival of mice against a sub-lethal dose of gamma radiation (Pahadiya and Sharma, 2003). *T. cordifolia* protected Swiss albino mice against radiation injury and helped to regain the weight lost. It has also been shown to reduce radiation-induced damage in the liver cells (Pahadiya and Sharma, 2003; Singh *et al.,* 2004).

Ayurvedic formulations hold immense promise for radioprotection

Some of the polyherbal formulations that show a variety of significant protective effects are given in detail below (also see Table 2.2).

Abana

The active ingredients of Abana include *Terminalia arjuna* (Arjuna), *Centella asiatica* (Gotu-Kola) and *Withania somnifera* (Ashwagandha). Abana protects mouse bone marrow against radiation-induced micronuclei formation (Jagetia and Aruna, 1997). The alcoholic extract of Abana (20mg/kg body weight) provided protection against gastrointestinal (GI) death and enhanced the survival of mice after exposure to gamma radiation. Acute toxicity studies revealed that Abana was non-toxic up to a dose of 1.6g/kg body weight. The LD_{50} dose of Abana was found to be 1.8g/kg body weight. This study demonstrated the ability of Abana as a good radioprotective agent and the optimum protective dose of Abana was 1/90 of its LD_{50} dose (Jagetia *et al.,* 2003a; Baliga *et al.,* 2004).

Cystone

Some of the active ingredients of Cystone are *Rubia cordifolia* (Indian Madder), *Didymocarpus pedicellata* (Shilapushpa), *Saxifraga ligulata* (Pasanavheda), *Cyperus scariosus* (Umbrella's edge), *Achyranthes aspera* (Rough Chaff Tree) and *Tinospora cordifolia* (Guduchi). Cystone regulates the urinary tract function. Treatment of mice with different doses of cystone, consecutively for five days before irradiation, delayed the onset of mortality and reduced the symptoms of radiation sickness (Jagetia and Baliga, 2002a).

Geriforte

Geriforte, a completely natural product with strong antioxidant and anti-aging properties, contains active ingredients like Chyavanprash, *Withania somnifera* (Ashwagandha), *Emblica officinalis* (Indian Gooseberry), *Mucuna urens* (Cow-itch plant) etc. Treatment of mice with different doses of geriforte, consecutively, for five days before irradiation delayed the onset of mortality and reduced radiation sickness. A dose of 10mg/kg body weight ($1/475^{th}$ of the LD_{50}) protected against gastrointestinal (GI) and bone marrow death with a DMF of 1.14 upon gamma radiation exposure (Jagetia and Baliga, 2004b).

Mentat

Mentat is a herbal formulation of several medicinal plants that have been categorized in Ayurveda as "*Medharasayanas*". This formulation includes *Bacopa monniera* (Brahmi), *Centella asiatica* (Gotu-Kola), *Adoxa moschatellina* (Musk root), *Terminalia arjuna* (Arjuna) as some of the active ingredients. Mentat is used to regulate behaviour, improve memory and minimize deficits associated with aging. Mentat administration consecutively for five days before irradiation delayed the onset of mortality and reduced the symptoms of radiation sickness. It also protected against GI syndrome (Jagetia and Baliga, 2003b).

Plants	Model system	Radioprotective effects	References
Acanthopanax senticosus (Family: Araliaceae)	Mouse	Protects against radiation-induced suppression of hemopoiesis.	(Miyanomae and Friendel, 1988)
Aegle marmelos	Mouse, Human	Reduces the symptoms of radiation-induced sickness and increases survival of mice. The possible mechanism is free radical scavenging and elevation of GSH and other antioxidant enzymes. Reduces the frequency of micronuclei in human peripheral blood lymphocytes.	(Jagetia *et al.,* 2003d; Jagetia and Venkatesh, 2005, 2007b; Jagetia *et al.,* 2006a)
Allium sativum (Family: Alliaceae)	Mouse	Protects against gamma-radiation-induced micronuclei (chromosomal damage) *in vivo*.	(Singh *et al.,* 1995, 1996)
Aloe vera (Family: Liliaceae)	Mouse	Protects against radiation-induced injury in intestinal mucosa and skin of Swiss albino mice.	(Gehlot and Saini, 2004a, b)
Citrus aurantium *var. amara (Family: Rutaceae)*	Mouse	Reduces the frequencies of micronucleated polychromatic erythrocytes and normochromatic erythrocytes. Protects mouse bone marrow by a factor of 2.2 against the side effects of gamma radiation.	(Hosseinimehr *et al.,* 2003)
Emblica officinalis (Family: Euphorbiaceae)	Mouse	Pre-treatment inhibits mortality and provides protection against radiation-induced deleterious alteration in intestinal mucosa of mice. It modulates TNF-□ and IL-1□ and prevents radiation-induced gastric damage.	(Singh and Goyal *et al.,* 2004; Bhattacharya *et al.,* 2004; *Jindal* et al., *2004)*
Hippophae rhamnoides (Family: Elaeagnaceae)	Mouse	Aqueous extract enhances the survival of strain 'A' mice when administered 30min prior to whole-body gamma irradiation. It provides protection to the gastrointestinal system against lethal whole-body gamma radiation. Administration of the extract before irradiation increased the number of surviving crypts in the jejunum by a factor of 2.02 and villi cellularity by 2.5 fold.	(Goel *et al.,* 2000, 2002, 2003)

Table 2.1. Radioprotective effect of some medicinal plants.

Table 2.1. contd.

Mentha piperita (Family: Lamiatae)	Mouse	Mentha oil affords protection in terms of survival percentage and haematological parameters in mice. Oral administration before exposure to gamma radiation protects against chromosomal damage in bone marrow of mice with a DMF value of 1.78 and enhances the survival of mice. Protects against radiation-induced conditioned taste aversion.	(Samarth and Kumar, 2003a, 2003b, Samarth *et al.*, 2004; Sharma *et al.*, 2007)
Ocimum sanctum (Family: Lamiaceae)	Mouse	Its extract provides protection to bone marrow and enhances the survival of mice with a DMF of 1.28. Their active ingredient also showed good radioprotective ability.	(Uma Devi and Ganasoundari, 1995; Ganasoundari *et al.*, 1998; Uma Devi *et al.*, 1999; Uma Devi, 2001)
Panax ginseng (Family: Araliaceae)	Mouse	Enhances the jejunal crypt survival and endogenous spleen colony formation and reduces the frequency of radiation induced apoptosis.	(Kim *et al.*, 2001)
Phyllanthus amarus (Family: Euphorbiaceae)	Mouse	Significantly increased the total WBC count, bone marrow cellularity and alpha-esterase activity. It also increases the antioxidant defence enzymes such as CAT, SOD, GST, GPx and GR in both blood and tissue, which were reduced by radiation treatment. It also protects mouse chromosomes against radiation-induced damage.	(Kumar and Kuttan, 2004)
Podophyllum hexandrum (Family: Podophyllaceae)	Mouse, Plasmid	Enhanced liver GST and SOD, intestinal SOD and survival of mice. It protects plasmid pBR322 DNA against radiation-induced damage. Also prevents radiation-induced neuronal damage in postnatal rats exposed in utero. Hydroalcoholic material (REC-2000) extracted from the rhizome offers cytoprotection against gamma radiation. Renders protection to hemopoietic system. Molecular mechanisms reveal modulation of protein expression associated with apoptosis.	(Mittal *et al.*, 2001; Sajikumar and Goel, 2003; Goel *et al.*, 2007; Chawla *et al.*, 2005; Sagar *et al.*, 2006; Lata *et al.*, 2007; Arora *et al.*, 2007; Kumar *et al.*, 2005a,b; 2007)

Table 2.1. contd.

Rubia cordifolia (Family: Rubiaceae)	Plasmid	Protects plasmid pBR322 DNA against strand breaks and microsomal and mitochondrial membranes against lipid peroxidation induced by gamma radiation.	(Shah *et al.,* 2004)
Syzygium cumini (Family: Myrtaceae)	Mouse, Human	Leaf extract reduces radiation-induced micronuclei formation in human peripheral blood lymphocytes and delayed the onset of mortality and reduced the symptoms of radiation sickness in mice.	(Jagetia and Baliga, 2002c, 2003c)
Terminalia chebula (Family: Combretaceae)	Plasmid, Human, Rat	Inhibits gamma-radiation-induced lipid peroxidation in rat liver microsomes and damage to superoxide dismutase enzyme in rat liver mitochondria besides gamma-radiation-induced strand breaks in plasmid pBR322 DNA. It also protects DNA of human peripheral blood leukocytes against gamma-radiation-induced damage.	(Naik *et al.,* 2004; Gandhi and Nair, 2005)
Tinospora cordifolia (Family: Menispermaceae)	Mouse	Protects Swiss albino mice against radiation injury and helps them regain the weight loss. Pre-irradiation treatment with RTc (an extract of *Tinospora cordifolia*) rendered 76.3% survival (30 days), as compared to 100% mortality in irradiated control and prevented radiation-induced weight loss. It restored total lymphocyte counts and increases the S-phase population, which was reduced after 2Gy exposure. Enhanced the survival of mice and modulated macrophage response to radiation.	(Pahadiya and Sharma, 2003; Goel *et al.,* 2001, 2004)
Zingiber officinale (Family: Zingiberaceae)	Mouse	Reduces the severity of radiation sickness and mortality. Protects mice from GI and bone marrow syndromes. Protects against radiation-induced conditioned taste aversion (CTA) in both male and female mice.	(Jagetia *et al.,* 2003e; Sharma *et al.,* 2005, 2006; Haksar *et al.,* 2006)

Table 2.2. Radioprotective effect of some Ayurvedic formulations

Plants	Model system	Radioprotective effects	References
Abana (a polyherbal formulation)	Mouse	Treatment of mice with different doses of abana delayed the onset of mortality and reduced the symptoms of radiation sickness as compared to the irradiated controls. Pre-treatment of mice with abana before irradiation delayed the onset of mortality and reduced the symptoms of radiation sickness. Provided protection against both gastrointestinal and haemopoietic death.	(Jagetia and Aruna, 1997, 2003a; Baliga *et al.*, 2004)
Cystone (an Ayurvedic herbal medicine)	Mouse	Treatment of mice with different doses of cystone, consecutively for five days before irradiation, delayed the onset of mortality and reduced the symptoms of radiation sickness.	(Jagetia and Baliga, 2002a)
Geriforte (Polyherbal formulation)	Mouse	Delayed the onset of mortality and reduces radiation sickness. Protects against GI and bone marrow death with a DMF of 1.14.	(Jagetia and Baliga, 2004b)
Liv 52	Mouse	It significantly reduced the frequency of radiation-induced MPCEs and MNCEs, reduced the symptoms of radiation sickness and increased mouse survival 10 and 30 days after irradiation and increases antioxidant enzymes.	(Jagetia *et al.*, 2006b)
Mentat (a polyherbal formulation)	Mouse	Mentat administration, consecutively for five days before irradiation, delayed the onset of mortality and reduced the symptoms of radiation sickness. It also protects against GI syndrome.	(Jagetia and Baliga, 2003b)
Triphala (an Ayurvedic formulation)	Mouse	Delayed the radiation induced mortality and reduced the symptoms of radiation sickness. Provides protection against gastrointestinal and haematopoetic deaths.	(Jagetia *et al.*, 2002b)
Y Rad A (a herbal formulation)	Mouse	Protects haematological parameters and reduces micronuclei. Inhibits lipid peroxidation and reduced glutathione content in RBC.	(Tripathi *et al.*, 2004)

Triphala

Triphala is a formulation of three herbs viz., *Terminalia chebula*, *Phyllanthus emblica* (or *Emblica officinalis*) and *Terminalia bellerica*. It has been described in the Ayurveda as a "tridoshic rasayana". This is an antioxidant-rich formulation. Triphala and/or its individual plant constituents have antibacterial anti-malarial, anti-fungal, anti-allergic and antiviral activities. It has good radioprotective effect at a dose of 1/28 of its LD_{50} dose.

Triphala enhances the survival of mice and reduces the symptoms of radiation sickness upon exposure to gamma radiation and provides protection to gastrointestinal and haematopoietic deaths (Jagetia *et al.,* 2002b). It was found that radiation-induced mortality was reduced by 60% in mice fed with triphala (1g/kg body weight/day) orally for 7 days prior to whole-body irradiation (WBI) at 7.5Gy, followed by post-irradiation feeding for 7 days (Sandhya *et al.,* 2006).

Conclusion and Future Prospects

Radiation causes damage to different tissues and natural compounds can protect against such damage. Radioprotection by such compounds has potential applications in intentional and accidental radiation exposures. Development of radioprotectors from natural sources will have great value in future because they are easily available, economical and non-toxic. Since people consume them in daily life, there will be no need of clearance from regulatory authorities like the Food and Drug Administration (FDA) for their use in the clinic during various radiation exposure scenarios.

References

Abraham, S.K., Sarma, L. and Kesavan, P.C. (1994) Role of chlorophyllin as an *in vivo* anticlastogen: protection against gamma radiation and chemical clastogens. *Mutation Research* 322, 209–212.

Arora, R. (2007) Botanicals as Potential Radioprotective and Radiorecovery Agents: Current Status and Emerging Directions for Clinical Applications. *Proceedings of the World Ayurveda Conference*, Pune, India, November 5–12, 2006.

Arora, R. (2008) Herbal Drugs for Oncology: Current Status and Future Directions in Cancer Chemoprevention In: *Herbal Drugs: A Cancer Chemopreventive and Therapeutic Perspective.* Arora, R. (ed) Jaypee Brothers Medical Publishers (P) Ltd., Delhi, India (in press).

Arora, R. and Goel, H.C. (2000) Herbal Radioprotectors. *Proceedings of the International Conference on Radiation Biology (Radiobiology 2000)* Organized under the aegis of the Indian Society for Radiation Biology by Regional Cancer Centre (RCC), Thiruvananthapuram, India, from 17–19 February, 2000, pp. 87.

Arora, R. and Sharma, R.K. (2004) Development of Herbal Agents for Biological Radiation Protection. *Proceedings of the Defence Research and Development Organization (DRDO) Life Sciences Scientist's Conference*, Defence Food Research Laboratory, Mysore, India.

Arora, R., Chawla, R., Chaudhary, P., Sharma, R.K., Puri, S.C., Gupta, V.K., Handa, G. and Qazi, G.N. (2004a) Antioxidant activities of fractionated extracts of *Podophyllum hexandrum* Royle. *Proceedings of the International Conference on Chemistry Biology Interface: Synergistic New Frontiers*, University of Delhi, Delhi, India, pp. 25–46.

Arora, R., Chawla, R., Prasad, J., Singh, S., Kumar, R., Sharma, A., Sagar, R., Singh, S. and Sharma, R.K. (2005b) Radioprotective activity of *Rhodiola imbricata*– a high altitude medicinal plant.

Molecular and Cellular Biochemistry 273, 209–223.

Arora, R., Chawla, R., Puri, S.C., Sagar, R.K., Singh, S., Kumar, R., Sharma, A.K., Prasad, J., Singh, S., Kaur, G., Chaudhary, P., Qazi, G.N. and Sharma, R.K. (2005c) Radioprotective activity of low-altitude *Podophyllum hexandrum. Journal of Environmental Pathology, Toxicology and Oncology* 24 (4), 299–314.

Arora, R., Chawla, R., Sagar, R.K., Singh, S., Kumar, R., Sharma, A.K., Singh, S., Prasad, J., Puri, S.C., Gupta, M.L. and Sharma, R.K. (2005d) Low aryl-tetralin lignans containing *Podophyllum hexandrum* extract protects against radiation–mediated lethal damage. *Proceedings of the Conference Nanobiotechnology: the Road Ahead*, National Botanical Research Institute, Lucknow, India, pp. 29–30.

Arora, R., Chawla, R., Singh, S., Sagar, R.K., Kumar, R., Sharma, A.K., Singh, S., Prasad, J., Sharma, R.K. and Tripathi, R.P. (2006b) Radioprotection by Himalayan high–altitude region plants. In: Sharma, R.K. and Arora, R. (eds) *Herbal Drugs: A Twenty First Century Perspective*. Jaypee Brothers Medical Publishers Private Ltd., Delhi, India, pp. 301–325.

Arora, R., Chawla, R., Singh, Shikha., Sagar, R.K., Kumar, R., Sharma, A., Prasad, J., Singh, S., Gurudatta, G.U. and Sharma, R.K. (2006a) Bioprospection for Radioprotective Molecules from Indigenous Flora. In: Govil, J.N., Singh, V.K. and Bhardwaj, R. *Recent Progress in Medicinal Plants*. Studium Press, LLC, USA, Vol. 16, pp. 179–219.

Arora, R., Gupta, D., Chawla, R., Sagar, R., Sharma, A., Kumar, R., Prasad, J., Singh, S., Samanta, N. and Sharma, R.K. (2005a) Radioprotection by plant products: Present status and future prospects. *Phytotherapy Research* 19, 1–22.

Arora, R., Gupta, D., Prasad, J., Sharma, A.K., Singh, S., Sagar, R., Samanta, N., Kumar, R. and Goel, H.C. (2003a) Bioprospection of Himalayan high-altitude region flora for radioprotective biomolecules. *Proceedings of the National Seminar on New Millennium Strategies for Quality, Safety and GMPs for Herbal Drugs/Products*, National Botanical Research Institute, Lucknow, India, p. 51.

Arora, R., Gupta, D., Prasad, J., Singh, S., Sharma, A.K., Sagar, R., Samanta, N., Kumar, R. and Goel, H.C. (2003b) Screening of indigenous medicinal and aromatic plants for their radioprotective potential: Some considerations. *XXIV Annual Conference of Indian Association of Biomedical Scientists*, Laser Science and Technology Centre (LASTEC), Delhi, and Institute of Nuclear Medicine and Allied Sciences, Delhi, India, pp. 86–87.

Arora, R., Gupta, D., Sharma, A.K., Sagar, R.K., Prasad, J., Singh, S., Samanta, N., Kumar, R., Goel, H.C., Sinha, A.K., Chawla, R. and Sharma, R.K. (2004b) Modification of radiation-induced damage in mammalian model systems by natural plant products: implications for radiation protection. *IUPAC Sponsored International Conference on Biodiversity and Natural Products*, University of Delhi, Delhi, India, p. 56.

Arora, R., Puri, S.C., Chawla, R., Sharma, A.K., Kumar, R., Sharma, R.K., Qazi, G.N. and Tripathi R.P. (2008) ^{1}H AND ^{13}C NMR Studies on a novel quercetin derivative (with radioprotective properties) isolated from *Podophyllum hexandrum* Royle. *Special Symposium on Advanced MR Applications and 14th National Magnetic Resonance Society Meeting* (Abstracts and Souvenir), Institute of Nuclear Medicine and Allied Sciences, Delhi, p.154.

Arora, R., Sagar, R., Singh, S., Kumar, R., Sharma, A.K., Prasad, J., Singh, S., Gupta, M., Sharma, R.K., Puri, S.C., Krishna, B., Siddiqui, M.S., Lahiri, S.S., Tripathi, R.P. and Qazi, G.N. (2007c) Cytoprotective effect of *Podophyllum hexandrum* against gamma radiation is mediated via hemopoietic system stimulation and up-regulation of heme-oxygenase-1 and the pro-survival multidomain protein Bcl-2. *Integrative Cancer Therapies* 6(1), 54–65.

Arora, R., Singh, S., Puri S.C. and Sharma, R.K. (2008a) Himalayan Mayapple (*Podophyllum hexandrum* Royle): Traditional Uses, Clinical Indications and Future Prospects. In: Watson, R. R. and Preedy, V.R. *The Encyclopedia of Herbal Medicine in Clinical Practice*, CABI Publishing, Wallingford, Oxon, United Kingdom, p. 26 (in press).

Arora, R., Singh, S., Sagar, R., Chawla, R., Kumar, R., Sharma, A., Gupta, M.L., Lahiri, S.S. and Tripathi, R.P. (2007a) Radioprotection by high-altitude flora and phyto-constituents therefrom: Potential for Armed Forces. *Proceedings of the International Conference on Frontiers of Radiation and Photochemistry*, Mahatma Gandhi University, Kottayam, India, p. 190.

Arora, R., Singh, S., Sagar, R.K., Chawla, R., Kumar, R., Puri, S.C., Singh, S., Prasad, J., Gupta, M.L., Krishna, B., Siddiqui, S.M., Sharma, A.K., Tripathi, R.P., Qazi, G.N. and Sharma, R.K. (2008b) Radiomodulatory and free radical scavenging activity of the fractionated aquo-alcoholic extract of the adaptogenic nutraceutical (*Rhodiola*)– A comparative *in vitro* assessment with ascorbate. *Journal of Herbal Pharmacotherapy* 7(4) (in press).

Arora, R., Singh, S., Sagar, R.K., Kumar, R., Sharma, A., Puri, S.C., Tripathi, R.P. and Qazi, G.N. (2007b) Role of antioxidant and DNA protective activity of a promising fraction of *Podophyllum hexandrum* (REC-2006) in providing protection against γ-radiation, *Proceedings of the International Conference on Frontiers of Radiation and Photochemistry*, Mahatma Gandhi University, Kottayam, India, p.78.

Baliga, M.S., Jagetia, G.C., Venkatesh, P., Reddy, R. and Ulloor, J.N. (2004) Radioprotective effect of abana, a polyherbal drug following total-body irradiation. *British Journal of Radiology* 77, 1027–1035.

Bhattacharya, S., Chatterjee, S., Bauri, A.K., Kamat, J.P., Poduval, T.B., Chattopadhyay, S. and Bandyopadhyay, S.K. (2004) *Emblica officinalis*: A possible preventor of radiation and NSAID induced gastric damage. *Indian Journal of Radiation Research* 1, 33.

Boloor, K.K., Kamat, J.P. and Devasagayam, T.P.A. (2000) Chlorophyllin as a protector of mitochondrial membranes against γ–radiation and photosensitization. *Toxicology* 155, 63–71.

Bump, E.A. and Malaker, K. (1998) *Radioprotectors: Chemical, Biological and Clinical Perspective.* CRC Press, Boca Raton, Florida, USA, pp. 431.

Chawla, R., Arora, R. and Sharma, R.K. (2008) Radioprotective activity of Himalayan Seabuckthorn and Aindri: Role of secondary metabolites. *Proceedings of the National Seminar on Physiological and Biotechnological Approaches to Improve Plant Productivity Centre for Plant Biotechnology,* Chaudhary Charan Singh Haryana Agricultural University, Hissar, India, pp. 93-94.

Chawla, R., Arora, R., Kumar, R., Sharma, A.K., Prasad, J., Singh, S., Sagar, R.K., Chaudhary, P., Shukla, S., Kaur, G., Sharma, R.K., Puri, S.C., Handa, G., Gupta, V.K. and Qazi, G.N. (2005) Antioxidant activities of fractionated extracts of high altitude *Podophyllum hexandrum* in radiation protection. *Molecular and Cellular Biochemistry* 273, 193–208.

Chawla, R., Arora, R., Singh, S, Sagar, R.K., Sharma, R.K., Kumar, R., Sharma, A., Gupta, M.L., Singh, S., Prasad, J., Khan, H.A., Swaroop, A., Sinha, A.K., Gupta, A.K., Tripathi, R.P. and Ahuja, P.S. (2007) Radioprotective and antioxidant activity of fractionated extracts of berries of *Hippophae rhamnoides*. *Journal of Medicinal Food* 10(1), 101–109.

Chawla, R., Arora, R., Singh, S., Sagar, R., Sharma, R.K., Puri, S.C., Khan, H.A., Shakir, M., Shawl, A.S., Kumar, R., Sharma, A., Singh, S., Prasad, J., Tripathi, R.P. and Qazi, G.N. (2006) Variation in aryltetralin lignan content of *Podophyllum hexandrum* influences radioprotective efficacy. *Evidence Based Complementary and Alternative Medicine* 3(4), 503–511.

Farooqi, Z. and Kesavan, P.C. (1992) Radioprotection by caffeine pre- and post-treatment in the bone marrow chromosomes of mice given whole-body γ-irradiation. *Mutation Research* 269, 225–230.

Ganasoundari, A., Uma Devi, P. and Rao, B.S.S. (1998) Enhancement of bone marrow radioprotection and reduction of WR-2721 toxicity by *Ocimum sanctum. Mutation Research* 397, 303–312.

Gandhi, N.M. and Nair, C.K.K. (2005) Radiation protection by *Terminalia chebula*: Some mechanistic aspects. *Molecular and Cellular Biochemistry* 277, 43–48.

Gandhi, N.M., Maurya, D.K., Salvi, V., Kapoor, S., Mukherjee, T. and Nair, C.K.K. (2004) Radioprotection of DNA by glycyrrhizic acid through scavenging free radicals. *Journal of Radiation Research* 45, 461–468.

Gehlot, P. and Saini, M.R. (2004a) Effectiveness of *Aloe vera* leaf extract against low-level exposure to gamma radiation induced injury in intestinal mucosa of Swiss albino mice. *Indian Journal of Radiation Research* 1, 36.

Gehlot, P. and Saini, M.R. (2004b) Radioprotective effects of *Aloe vera* leaf extract on skin of Swiss albino mice after gamma irradiation. *Indian Journal of Radiation Research* 1, 52.

George, K.C., Hebbar, S.A., Kale, S.P. and Kesavan, P.C. (1999) Caffeine protects mice against whole-body lethal dose of gamma irradiation. *Journal of Radiological Protection* 19, 171–176.

Goel, H.C., Arora, R., Prasad, J., Sharma, A.K., Singh, S., Mathew, T.L., Chaurasia, O.P. and Singh, B. (2000b) A process for preparation of a radioprotective herbal extract. Indian Patents Office, Delhi, India (Patent No. 191821).
Goel, H.C., Arora, R., Prasad, J., Sharma, A.K., Singh, S., Mathew, T.L., Chaurasia, O.P. and Singh, B. (2000a) A process for preparation of a radioprotective herbal extract. Indian Patents Office, Delhi, India (Patent No. 192177).
Goel, H.C., Arora, R., Sinha, A.K., Dogra, R., Prasad, J., Singh, S., Sagar, R., Kumar, I.P., Agarwala, P.K., Sharma, A.K., Samanta, N., Gupta, D., Prakash, H., Salin, C.A., Kumar, S. S., Mittal, A. and Pathania, V. (2001) A process for preparation of a radioprotective herbal extract from *Tinospora*. Indian Patents Office, Delhi, India (Patent No. 195062).
Goel, H.C., Arora, R., Sinha, A.K., Joshi, B.P., Prasad, J., Singh, S., Kumar, I.P., Sagar, R., Salin, C.A., Agarwala, P.K., Prakash, H., Gupta, D., Sharma, A.K., Mittal, A., Pathania, V., Mathew, T.L. and Singh, B. (2000c) A process for preparation of a radioprotective herbal extract from *Hippophae rhamnoides*. Indian Patents Office, Delhi, India (Patent No. 192060).
Goel, H.C., Prakash, H., Ali, A. and Bala, M. (2007) *Podophyllum hexandrum* modulates gamma radiation–induced immunosuppression in Balb/c mice: implications in radioprotection. *Molecular and Cellular Biochemistry* 295(1-2), 93–103.
Goel, H.C., Prasad, J., Singh, S., Sagar, R.K., Agrawala, P.K., Bala, M., Sinha, A.K. and Dogra, R. (2004) Radioprotective potential of an herbal extract of *Tinospora cordifolia. Journal of Radiation Research* 45, 61–68.
Goel, H.C., Prasad, J., Singh, S., Sagar, R.K., Kumar, I.P. and Sinha, A.K. (2002) Radioprotection by a herbal preparation of *Hippophae rhamnoides*, RH-3, against whole body lethal irradiation in mice. *Phytomedicine* 9, 15–25.
Goel, H.C., Salin, C.A. and Prakash, H. (2003) Protection of jejunal crypts by RH-3 (a preparation of *Hippophae rhamnoides*) against lethal whole body gamma irradiation. *Phytotherapy Research* 17, 222–226.
Grinfeld, S. and Jacquet, P. (1988) An unusual radiation-induced G2 arrest in the zygote of Balb/c mouse strain. *International Journal of Radiation Biology* 51, 353–363.
Gupta, M., Kumar, I.P. and Arora, R. (eds) (2006) *Herbs for Health: Application in Armed Forces.* Institute of Nuclear Medicine and Allied Sciences, Delhi, India, p. 100.
Haksar, A., Sharma, A.K., Chawla, R., Arora, R., Singh, S., Prasad, J., Kumar, R., Tripathi, R.P., Arora, M.P., Islam, F. and Sharma, R.K. (2006) *Zingiber officinale* Rosc. exhibits behavioural radioprotection against radiation–induced conditioned taste aversion in a gender specific manner. *Pharmacology Biochemistry and Behaviour* 84, 179–188.
Hosseinimehr, S.J., Tavakoli, H., Pourheidari, G., Sobhani, A. and Shafiee, A. (2003) Radioprotective effects of citrus extract against gamma irradiation in mouse bone marrow cells. *Journal of Radiation Research* 44, 237–241.
Hughes, E.N. and Boothman, D.A. (1991) Effect of caffeine on the expression of a major X ray–induced protein in human tumour cells. *Radiation Research* 125, 313–317.
Jagetia, G.C. (2007a) Radioprotective potential of plants and herbs against the effects of ionizing radiation. *Journal of Clinical Biochemistry and Nutrition* 40, 74–81.
Jagetia, G.C. and Aruna, R. (1997) The herbal preparation abana protects against radiation–induced micronuclei in the mouse bone marrow. *Mutation Research* 393, 157–163.
Jagetia, G.C. and Baliga, M.S. (2002a) Cystone, an ayurvedic herbal drug imparts protection to the mice against the lethal effects of gamma radiation: a preliminary study. *Nahrung* 46, 332–336.
Jagetia, G.C. and Baliga, M.S. (2002c) *Syzygium cumuni* (Jamun) reduces the radiation–induced DNA damage in the cultured human peripheral blood lymphocytes: a preliminary study. *Toxicology Letters* 132, 19–25.
Jagetia, G.C. and Baliga, M.S. (2003b) Treatment of mice with a herbal preparation (mentat) protects against radiation–induced mortality. *Phytotherapy Research* 17, 876–881.
Jagetia, G.C. and Baliga, M.S. (2003c) Evaluation of the radioprotective effect of the leaf extract of *Syzygium cumini* (Jamun) in mice exposed to a lethal dose of gamma irradiation. *Nahrung* 47(3), 181–185.

Jagetia, G.C. and Baliga, M.S. (2004b) Evaluation of the radioprotective action of geriforte in mice exposed to different doses of gamma radiation. *American Journal of Chinese Medicine* 32(4), 551–567.

Jagetia, G.C. and Venkatesh, P. (2005) Radioprotection by oral administration of *Aegle marmelos* (L.) Correa *in vivo*. *Journal of Environmental Pathology, Toxicology, and Oncology* 24(4), 315–332.

Jagetia, G.C. and Venkatesh, P. (2007b). Inhibition of radiation–induced clastogenicity by *Aegle marmelos* (L.) correa in mice bone marrow exposed to different doses of gamma radiation. *Human and Experimental Toxicology* 26(2), 111–124.

Jagetia, G.C., Baliga M.S., Malagi, K.J. and Kamath, M.S. (2002b) The evaluation of the radioprotective effect of Triphala (an Ayurvedic rejuvenating drug) in the mice exposed to γ-radiation. *Phytomedicine* 9, 99–108.

Jagetia, G.C., Baliga, M.S., Aruna, R., Rajanikant, G.K. and Jain, V. (2003a) Effect of abana (a herbal preparation) on the radiation–induced mortality in mice. *Journal of Ethnopharmacology* 86, 159–165.

Jagetia, G.C., Baliga, M.S., Venkatesh, P. and Ulloor, J.N. (2003e) Influence of ginger rhizome (*Zingiber officinale* Rosc.) on survival, glutathione and lipid peroxidation in mice after whole-body exposure to gamma radiation. *Radiation Research* 160(5), 584–92

Jagetia, G.C., Ganapathi, N.G., Venkatesh, P., Rao, N. and Baliga, M.S. (2006b) Evaluation of the radioprotective effect of Liv 52 in mice. *Environmental and Molecular Mutagenesis* 47(7), 490–502.

Jagetia, G.C., Rajanikant, G.K. and Mallikarjuna Rao, K.V. (2007c) Ascorbic acid increases healing of excision wounds of mice whole-body exposed to different doses of gamma radiation. *Burns* 33(4), 484–494.

Jagetia, G.C., Rajanikant, G.K., Baliga, M.S., Rao, K.V. and Kumar, V. (2004a) Augmentation of wound healing by ascorbic acid treatment in mice exposed to gamma radiation. *International Journal of Radiation Biology* 80(5), 347–354.

Jagetia, G.C., Venkatesh, P. and Baliga, M.S. (2003d) Evaluation of the radioprotective effect of *Aegle marmelos* (L.) Correa in cultured human peripheral blood lymphocytes exposed to different doses of gamma radiation: a micronucleus. *Mutagenesis* 18(4), 387–393.

Jagetia, G.C., Venkatesh, P., Archana, P., Krishnanand, B.R. and Baliga, M.S. (2006a) Effects of *Aegle marmelos* (L.) Correa on the peripheral blood and small intestine of mice exposed to gamma radiation. *Journal of Environmental Environ Pathology, Toxicology, and Oncology* 25(4), 611–624.

Jindal, A., Sharma, A., Singh, I. and Goyal, P.K. (2004) Radiomodulatory potential of *Emblica officinalis* against enteric lesions in mice. *Indian Journal of Radiation Research* 1, 32.

Kamat, J.P., Boloor, K.K. and Devasagayam, T.P.A. (2000b) Chlorophyllin as an effective antioxidant against membrane damage *in vitro* and *in vivo*. *Biochimica et Biophysica Acta* 1487, 113–127.

Kamat, J.P., Boloor, K.K., Devasagayam, T.P.A. and Kesavan, P.C. (1999) Protection of superoxide dismutase by caffeine in rat liver mitochondria against γ-irradiation. *Current Science* 77, 286–289.

Kamat, J.P., Boloor, K.K., Devasagayam, T.P.A., Jayashree, B. and Kesavan, P.C. (2000a) Differential modification of oxygen–dependent and independent effects of γ-irradiation in rat liver mitochondria by caffeine. *International Journal of Radiation Biology* 76, 1281–1288.

Kamat, J.P., Ghosh, A. and Devasagayam, T.P.A. (2000c) Vanillin as an antioxidant in rat liver mitochondria: Inhibition of protein oxidation and lipid peroxidation induced by photosensitization. *Molecular and Cellular Biochemistry* 209, 47–53.

Kesavan, P.C. and Natarajan, A.T. (1985) Protection and potentiation of radiation clastogenesis by caffeine: nature of possible initial events. *Mutation Research* 14, 61–68.

Kesavan, P.C., Trasi, S. and Ahmad, A. (1973) Modification of barley seed radiosensitivity by post-treatment with caffeine. I. Effect of post irradiation heat shock and nature of hydration. *International Journal of Radiation Biology* 24, 581–587.

Keshava, C., Keshava, N. and Nath, T.O.J. (1998) Protective effect of vanillin on radiation-induced micronuclei and chromosomal aberrations in V79 cells. *Mutation Research* 397, 149–159.

Kihlman, B.A., Sturelid, S., Hartley-Asp, B. and Nilsson, K. (1974) The enhancement by caffeine of the frequencies of chromosomal aberrations induced in plant and animal cells by chemical and physical agents. *Mutation Research* 26, 105–122.

Kim, S.H., Son, C.H., Nah, S.Y., Jo, S.K., Jang, J.S. and Shin, D.H. (2001) Modification of radiation response in mice by *Panax ginseng* and diethyldithiocarbamate. *In vivo* 15(50), 407–411.

Kline, K., Yu, W. and Zhao, B. (1995) Vitamin E succinate: Mechanisms of action as tumour cell growth inhibitor. In: Prasad, K., Santamaria, L. and Williams, R. *Nutrients in Cancer Prevention and Treatment*. Humana Press, New Jersey, USA, pp. 39–55.

Kumar, B., Jha, M.N., Cole, W.C., Bedford, J.S. and Prasad, K.N. (2002) D-Alpha tocopheryl succinate (Vitamin E) enhances radiation–induced chromosomal damage levels in human cancer cells, but reduces it in normal cells. *Journal of American College of Nutrition* 21, 339–343.

Kumar, K.B. and Kuttan, R. (2004) Protective effect of an extract of *Phyllanthus amarus* against radiation-induced damage in mice. *Journal of Radiation Research* 45(1), 133–9.

Kumar, R., Singh, P.K., Arora, R., Chawla, R. and Sharma, R.K. (2007) Radioprotective activities of *Podophyllum hexandrum*: Current knowledge of the molecular mechanisms. *Trees for Life* (in press).

Kumar, R., Singh, P.K., Arora, R., Sharma, A., Prasad, J., Sagar, R.K., Singh, S. and Sharma, R.K. (2005a) Radioprotection by *Podophyllum hexandrum* in the liver of mice: A mechanistic approach. *Environmental Toxicology and Pharmacology* 20, 326–334.

Kumar, R., Singh, P.K., Sharma, A., Prasad, J., Sagar, R., Singh, S., Arora, R. and Sharma, R.K. (2005b) *Podophyllum hexandrum* extract provides radioprotection by modulating the expression of proteins associated with apoptosis: role in radioprotection. *Biotechnology and Applied Biochemistry* 42(1), 81–92.

Kumar, S.S., Chaubey, R.C., Devasagayam, T.P.A., Priyadarsini, K.I. and Chauhan, P.S. (1999) Inhibition of radiation-induced DNA damage in plasmid pBR322 by chlorophyllin and the possible mechanisms. *Mutation Research* 425, 71–79.

Kumar, S.S., Devasagayam, T.P.A., Jayashree, B. and Kesavan, P.C. (2001) Mechanism of protection against radiation–induced DNA damage in plasmid pBR322 by caffeine. *International Journal of Radiation Biology* 77, 617–623.

Kumar, S.S., Ghosh, A., Devasagayam, T.P.A. and Chauhan, P.S. (2000) Effect of vanillin on methylene blue plus light-induced single-strand breaks in plasmid pBR322 DNA. *Mutation Research* 469, 207–214.

Lata, M., Prasad, J., Singh, S., Kumar, R., Singh, L., Choudhary, P., Arora, R., Chawla, R., Tyagi, S., Soni, N.L., Sagar, R.K., Devi, M., Sharma, R.K. and Puri, S.C. (2007) Whole-body protection against lethal ionizing radiation in mice by REC–2001: A semi-purified fraction of *Podophyllum hexandrum*. *Phytomedicine* (Epub advance access).

Luthra, P.M., Kumar, R., Singh, J., Singh, S. and Arora, R. (2008) *Curcuma longa*: A Traditional Herb as an Alternative Medicine. Editors: Watson, Ronald Ross and Preedy, Victor R. *The Encyclopedia of Herbal Medicine in Clinical Practice*, CABI Publishing, Wallingford, Oxon, United Kingdom. p. 21 (in press).

Maisin, J.R. (1998) Chemical radioprotection: past, present and future prospects. *International Journal of Radiation Biology* 73, 443–450.

Maurya, D.K. and Nair, C.K.K. (2006b) Preferential radioprotection to DNA of normal tissues by ferulic acid under *ex vivo* and *in vivo* conditions in tumour bearing mice. *Molecular and Cellular Biochemistry* 285, 181–190.

Maurya, D.K., Balakrishnan, S., Salvi, V.P., Nair, C.K.K. (2005b) Protection of cellular DNA from gamma radiation induced damages and enhancement in DNA repair by troxerutin. *Molecular and Cellular Biochemistry* 280, 57–68.

Maurya, D.K., Devasagayam, T.P.A. and Nair, C.K.K. (2006a) Some novel approaches for radioprotection and the beneficial effect of natural products. *Indian Journal of Experimental Biology* 44, 93–114.

Maurya, D.K., Salvi, V.P. and Nair, C.K.K. (2004) Radioprotection of normal tissues in tumour–bearing mice by troxerutin. *Journal of Radiation Research* 45, 221–228.

Maurya, D.K., Salvi, V.P. and Nair, C.K.K. (2005a) Radiation protection of DNA by ferulic acid under *in vitro* and *in vivo* conditions. *Molecular and Cellular Biochemistry* 280, 209–217.

Mittal, A., Pathania, V., Agrawala, P.K., Prasad, J., Singh, S. and Goel, H.C. (2001) Influence of *Podophyllum hexandrum* on endogenous antioxidant defence system in mice: possible role in radioprotection. *Journal of Ethnopharmacology* 76, 253–262.

Miyanomae, T. and Friendel, E. (1988) Radioprotection of hemopoiesis conferred by *Acanthopanax senticous* (Shigoka) administered before and after irradiation. *Experimental Hematology* 16(9), 801–806.

Morales–Ramirez, P. and Garcia–Rodriquez, M. C. (1994) *In vivo* effect of chlorophyllin on γ-ray-induced sister chromatid exchange in murine bone marrow cells. *Mutation Research* 320, 329–334.

Morales-Ramirez, P.T., Vallarino-Kelly, T. and Rodriquez-Reyes, R. (1984) *In vivo* persistence of sister chromatid exchange (SCE) induced by gamma-rays in mouse bone marrow cells. *Environmental Mutagenesis* 6, 529–537.

Naik, G.H., Priyadarsini, K.I., Naik, D.B., Gangabhagirathi, R. and Mohan, H. (2004) Studies on the aqueous extract of *Terminalia chebula* as a potent antioxidant and a probable radioprotector. *Phytomedicine* 11(6), 530–538.

Nair, C.K.K., Parida, D.K. and Nomura, T. (2001) Radioprotectors in radiotherapy. *Journal of Radiation Research* 42, 21–37.

Nair, C.K.K., Salvi, V., Kagiya, T.V. and Rajagopalan, R. (2004) Relevance of Radioprotectors in Radiotherapy: Studies with Tocopherol Monoglucoside. *Journal of Environmental Pathology, Toxicology, and Oncology* 23, 153–160.

Ong, T.M., Whong, W. Z., Stewart, J. and Brockman, H. E. (1986) Chlorophyllin: a potent antimutagen against environmental and dietary complex mixtures. *Mutation Research* 173, 111–115.

Pahadiya, S. and Sharma, J. (2003) Alteration of lethal effects of gamma rays in Swiss albino mice by *Tinospora cordifolia. Phytotherapy Research* 17(5), 552–554.

Pimentel, E., Cruces, M.P. and Zimmering, S. (1999) On the persistence of the radioprotective effect of chlorophyllin (CHLN) in somatic cells of *Drosophila. Mutation Research* 446, 189–192.

Prasad, K.N. and Edwards–Prasad, J. (1982) Effect of tocopherol (vitamin E) acid succinate on morphological alterations and growth inhibition in melanoma cells in culture. *Cancer Research* 42, 550–555.

Prasad, N.R., Srinivasan, M., Pugalendi, K.V. and Menon, V.P. (2006) Protective effect of ferulic acid on γ-radiation-induced micronuclei, dicentric aberration and lipid peroxidation in human lymphocytes. *Mutation Research* 603, 129–134

Puri, S.C., Nazir, A., Chawla, R., Arora, R., Riyaz-ul-Hassan, S., Amna, T., Ahmed, B., Verma, V., Singh, S., Sagar, R., Sharma, A., Kumar, R., Sharma, R.K. and Qazi, G.N. (2005) The endophytic fungus *Trametes hirsuta* as a novel alternative source of podophyllotoxin and related aryl tetralin lignans. *Journal of Biotechnology* 122(4), 494–510.

Radner, B.S. and Kennedy, A.R. (1986) Suppression of X-ray induced transformation by vitamin E in mouse C3H/10T1/2 cells. *Cancer Letters* 32, 25–32.

Rajagopalan, R., Wani, K., Huilgol, N.G., Kagiya, V.T. and Nair, C.K.K. (2002) Inhibition of gamma radiation induced DNA damage in plasmid pBR322 DNA by TMG, a derivative of vitamin E. *Journal of Radiation Research* 42, 153–159.

Rauth, A. M. (1967) Evidence for dark-reactivation of ultraviolet light damage in mouse L cells. *Radiation Research* 31, 121–138.

Sagar, R.K., Chawla, R., Arora, R., Singh, S., Krishna, B., Sharma, R.K., Puri, S.C., Singh, P., Kumar, R., Sharma, A.K., Singh, S., Prasad, J., Gupta, B., Ahmed, B., Dhar, K.L., Khan, H.A., Gupta, M.L. and Qazi, G.N. (2006) Protection of hematopoietic system by *Podophyllum hexandrum. Planta Medica* 72, 114–120.

Sajikumar, S. and Goel, H.C. (2003) *Podophyllum hexandrum* prevents radiation-induced neuronal damage in postnatal rats exposed *in utero. Phytotherapy Research* 17, 761–766.

Salvi, V.P., Satav, J.G., Kagiya, V.T. and Nair, C.K.K. (2001) Radioprotection of membrane damage by TMG (tocopherol monoglucoside). In: *Proceedings of the International Conference on Natural*

Antioxidants and Free Radicals in Human Health and Radiation Biology, Mumbai, India, July 21–24, p. 188.

Samarth, R.M. and Kumar, A. (2003a) *Mentha piperita* (Linn.) leaf extract provides protection against radiation induced chromosomal damage in bone marrow of mice. *Indian Journal of Experimental Biology* 41, 229–237.

Samarth, R.M. and Kumar, A. (2003b) Radioprotection of Swiss albino mice by plant extract *Mentha piperita* (Linn.). *Journal of Radiation Research* 44, 101–109.

Samarth, R.M., Goyal, P.K. and Kumar, A. (2004) Protection of Swiss albino mice against whole-body gamma irradiation by *Mentha piperita* (Linn.). *Phytotherapy Research* 18, 546–550.

Sandhya, T., Latika, K.M., Pandey, B.N., Bhilwade, H.N., Chaubey, R.C., Priyadarshini, K.I. and Mishra, K.P. (2006) Protection against radiation oxidative damage in mice by Triphala. *Mutation Research* 609, 17–25.

Sarria, A. and Prasad, K.N. (1984) DL-alpha tocopheryl succinate enhances the effect of gamma irradiation on neuroblastoma cells in culture. *Proceedings of the Society for Experimental Biology and Medicine* 175, 88–92.

Sasaki, Y.F., Ohta, T., Imanishi, H., Watanabe, M., Matsumoto, K., Kato, T. and Shirasu, Y. (1990) Suppressing effects of vanillin, cinnamaldehyde, and anisaldehyde on chromosome aberrations induced by X–rays in mice. *Mutation Research* 243, 299–302.

Satyamitra, M., Uma Devi, P., Murase, H. and Kagiya, V.T. (2001) Water soluble vitamin E derivative protects against radiation induced chromosomal changes in mouse bone marrow. In: *Proceedings of the ISRH 2001 Satellite Symposia, Radioprotectors for Human Health*, Kyoto, Japan, July 16, p. 13

Shah, V., Kumar, S., Maurya, D.K. and Nair, C.K.K. (2004) Protection of plasmid pBR322 DNA and membranes of rat liver microsome and mitochondria by *Rubia cordifolia* extract. *Journal of Radiation Research* 1, 46.

Sharma, A., Haksar, A., Chawla, R., Arora, R., Kumar, R., Singh, S., Prasad, J., Gupta, M.L., Lahiri, S.S., Tripathi, R.P., Arora, M.P., Islam, F. and Sharma, R.K. (2007) Behavioural protection by *Mentha spicata* Linn. against radiation-induced conditioned taste aversion: A comparative study with a radioprotective calcium channel-blocking drug. *Indian Journal of Radiation Research* 4(1), 47–56.

Sharma, A., Haksar, A., Chawla, R., Kumar, R., Arora, R., Singh, S., Prasad, J., Islam, F., Arora, M.P. and Sharma, R.K. (2005) *Zingiber officinale* Rosc. modulates gamma radiation–induced conditioned taste aversion. *Pharmacology Biochemistry and Behaviour* 81, 864–870.

Sharma, A.K., Haksar, A., Kumar, R., Arora, R. and Sharma, R.K. (2006) Ginger: Evaluation of its Anti-emetic and Behavioral Radioprotective Properties. In: Govil, J.N., Singh, V.K. and Bhardwaj, R. *Recent Progress in Medicinal Plants*. Studium Press, LLC, USA, Vol.16, 237–264.

Sharma, R.K. and Arora, R. (2006) *Herbal Drugs: A Twenty First Century Perspective*. Jaypee Brothers Medical Publishers Private Ltd., Delhi, India, p. 688.

Singh, A., Singh, H. and Henderson, J.S. (1990) Radioprotection by ascorbic acid, deferral, and mercaptoethylamine. *Methods in Enzymology* 186, 686–696.

Singh, I. and Goyal, P.K. (2004) Response of serum phosphatases activity to whole-body irradiation in mice and its modification by *Emblica officinalis* treatment. *Indian Journal of Radiation Research* 1, 31.

Singh, L., Tyagi, S., Sinha, A.K., Kumar, R., Sharma, R.K. and Rizvi, M.A. (2004) Effects of *Tinospora cordifolia* on macrophage function: A possible role in radioprotection. *Indian Journal of Radiation Research* 1, 45.

Singh, S.P., Abraham, S.K. and Kesavan, P.C. (1995) *In vivo* radioprotection with Garlic extract. *Mutation Research* 345, 147–153.

Singh, S.P., Abraham, S.K. and Kesavan, P.C. (1996) Radioprotection of mice following Garlic pretreatment. *British Journal of Cancer* Supp 27, S102–104.

Tannehill, S.P. and Mehta, M.P. (1996) Amifostine and radiation therapy: past, present and future. *Seminars in Oncology* 23, 69–77.

Tripathi, Y.B., Mehta, R.G., Singh, A.V. and Arjmandi, B.H. (2004) Radioprotective and chemotherapeutic potential of Y Rad A, a patented formulation. *Indian Journal of Radiation Research* 1, 13.

Turner, N.D., Braby, L.A., Ford, J. and Lupton, J.R. (2002) Opportunities for nutritional amelioration of radiation–induced cellular damage. *Nutrition* 18, 904–912.

Uma Devi, P. (2001) Radioprotective, anticarcinogenic and antioxidant properties of the Indian holy basil, *Ocimum sanctum. Indian Journal of Experimental Biology* 3,185–190.

Uma Devi, P. and Ganasoundari, A. (1995) Radioprotective effect of leaf extract of Indian medicinal plant *Ocimum sanctum. Indian Journal of Experimental Biology* 33, 205–209.

Uma Devi. P., Ganasoundari, A., Rao, B.S.S. and Srinivasan, K.K. (1999) *In vivo* radioprotection by *Ocimum sanctum* flavonoids: survival of mice. *Radiation Research* 151, 74–78.

Uma Devi. P., Ganasoundari, A., Vrinda, B. and Srinivasan, K.K. (2000) Radiation protection by *Ocimum sanctum* flavonoids orientin and vicenin–mechanism of action. *Radiation Research* 154, 455–460.

Umegaki, K., Itoh, T. and Ichikawa, T. (1994) Effect of vitamin E on chromosomal damage in bone marrow cells of mice having received low dose of X-ray irradiation. *International Journal of Vitamins and Nutritional Research* 64, 249–252.

Weiss, J.F. and Landauer, M.R. (2003) Protection against ionizing radiation by antioxidant nutrients and phytochemicals. *Toxicology* 189, 1–20.

Zimmering, S., Olivera, O., Hernandez, M. E., Cruces, M. P., Areco, C. and Pimentel, E. (1990) Evidence for radioprotective effect of chlorophyllin in *Drosophila. Mutation Research* 245, 47–49.

Chapter 3

Irradiation, Radioprotection and *Nigella sativa*

M. Cemek, M.E. Büyükokuroğlu, F. Bayıroğlu and M. Koç

Introduction

Radiation may be defined as energy travelling through space. Non-ionizing radiation is essential to life, but excessive exposures cause tissue damage. All forms of ionizing radiation have sufficient energy to ionize atoms that may destabilize molecules within cells and lead to tissue damage. Ionizing radiation consists of energetic particles and electromagnetic radiation, which can penetrate living tissue or cells and result in the transfer of radiation energy to the biological material (Jagetia and Reddy, 2005).

Effect of Radiation on Biological Systems

Ionizing radiation exerts its beneficial and harmful effects on biomolecules directly and/or indirectly (Fig. 3.1). The biologic effects of radiation arise principally from damage to DNA, proteins and lipids which are the critical targets. The atoms of the target itself may be ionized or excited, thus initiating a chain of events that lead to biological changes. This is called direct action of radiation and is the dominant process if radiation with high linear energy transfer (LET), such as neutrons or alpha particles, is considered.

Alternatively, radiation may interact with other atoms or molecules in the cell (particularly water) to produce free radicals that are able to diffuse far enough to reach and damage the critical targets. This is called indirect action of radiation (Vijayalaxmi *et al.*, 2004; Tan *et al.*, 2000; Jhun *et al.*, 1991).

A free radical is a free (not combined) atom or molecule carrying an unpaired orbital electron in the outer shell. For simplicity, it may be considered that radiation interacts with a water molecule, because 80% of a cell is composed of water. The amount of water in the human body, averaging 70%, varies considerably and even from one part of the body to another area (Fig. 3.2) (http://www.bragg.com/books/wtst_excerpt.html).

The effects of low-LET radiations are caused mainly by generation of reactive oxygen species (ROS). Most of the radiation-induced damage to biomolecules in aqueous media, such as those prevailing in living system is caused by the formation of free radicals resulting from the radiolysis of water. Exposure of biological systems to radiation results in radiolytic cleavage of water, giving rise to $e^{-}_{aq}\cdot$, OH· and H· (Prasad *et al.*, 2005; Kamat *et al.*, 2000; Von Sonntag, 1987) (Fig. 3.3). Radiation causes breakage of one of the oxygen-hydrogen covalent bonds in water, leaving a single electron on the hydrogen atom and one on the oxygen atom and creates radicals, especially hydroxyl radicals (Freeman and Crapo, 1982).

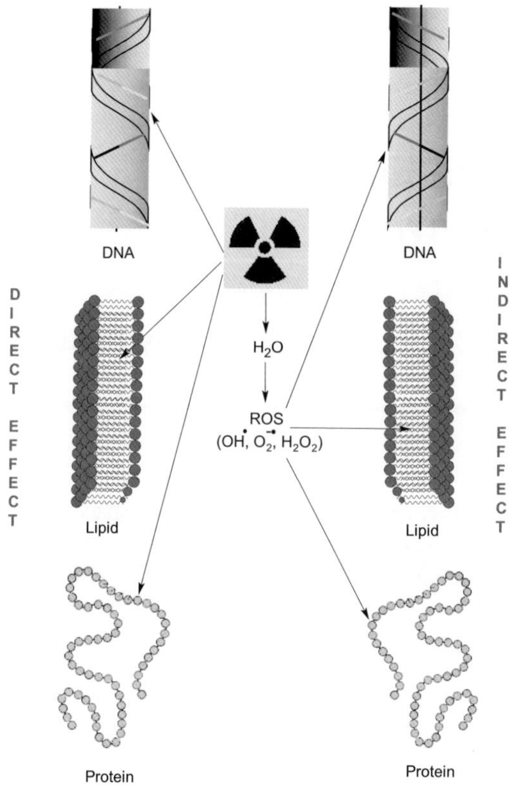

Fig. 3.1. Direct and indirect action of radiation. The structures of biomolecules are shown schematically. In direct action a secondary electron resulting from absorption of an X-ray photon interacts with the biomolecules to produce effect.

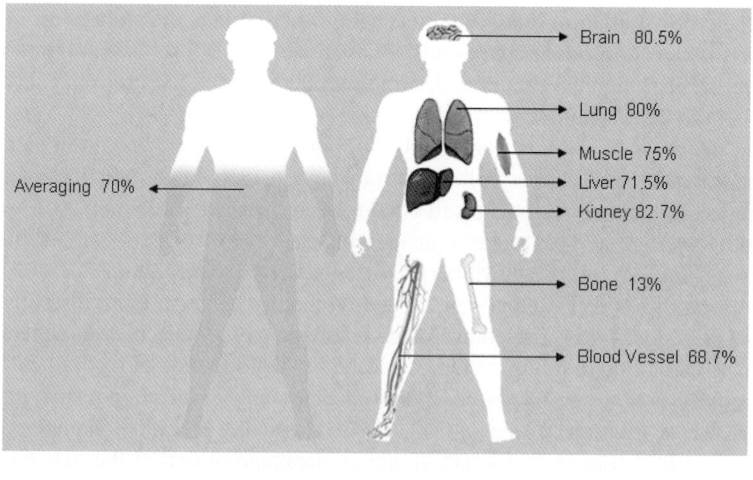

Fig. 3.2. The amount of water (%) in different tissues of the human body.

Molecular oxygen, due to its biradical nature, is the most important electron acceptor in the biosphere. It plays an important role in accepting unpaired electrons, giving rise to a series of partially reduced species like superoxide, hydroxyl, peroxyl and hydroperoxyl radicals. The generation of singlet oxygen, a major reactive species formed by the action of hydroperoxyl radical with superoxide anion, may also have potentially damaging effects on the cell (Fridovich, 1981; Packer and Ong, 1998; Sies, 1996; Von Sonntag, 1987). These radicals then undergo secondary reactions with dissolved O_2, ROS or with cellular solutes (Freeman and Crapo, 1982). However, ionizing radiation can break chemical bonds and cause ionization of biologically important macromolecules such as nucleic acids, membrane lipids and proteins (Ames and Gold, 1991; Daniniak and Tann, 1995; Lett, 1992; Cramp *et al.,* 1994; Ward, 1988; Vijayalaxmi *et al.,* 2004; Dizdaroglu, 1991; Floyd, 1990; Minlotti and Aust, 1987; Davies and Slater, 1987; Stadtman and Berlett, 1997; Dean *et al.,* 1997; Pohl, 1993; Jhun *et al.,* 1991).

Radiation-induced DNA damage

DNA is the most important molecule of the organism that is affected by radiation. This damage to DNA can be either due to direct or indirect effect. Direct and indirect actions are illustrated in Fig. 3.1. For its direct action, ionizing radiation must "hit" DNA. However, because the volume of the DNA is very small, compared with the total volume of the cell, the probability of this occurring is relatively low. This component of radiation damage can be modified by chemical means– either by protectors or sensitizers as opposed to the direct action, which cannot be modified. The indirect actions of radiation occur when it interacts with water molecules in the cell, resulting in the production of highly reactive free radicals (Fig. 3.3).

The half-life of these free radicals is extremely short, of the order of 10^{-6}–10^{-10} seconds. However, they immediately react with any biomolecules in the vicinity and produce highly site-specific oxidative damage. Hydroxyl radicals, the most dangerous ones of the radicals, are generated mostly upon radiolysis of water. An estimated 60–70% of tissue damage induced by ionizing radiation is believed to be caused by OH· radicals (Ward, 1988; Vijayalaxmi *et al.,* 2004; Yokus *et al.,* 2005). Hydroxyl radicals generate multiple products from guanine, adenine, thymine and uracil bases, e.g., 8-hydroxy-guanosine, 8-hydroxyadenine, thymidine glycol, thymine glycol, 5-hydroxymethyluracil (Dizdaroglu, 1991; Floyd, 1990) (Fig. 3.4). The altered DNA can be specifically repaired by DNA glycosylase (Chung *et al.,* 1991). However, if the degree of oxidative stress is too great, DNA repair by glycosylases is circumvented to induce mutations, carcinogenesis and cell death (Gate *et al.,* 1999).

Radiation-induced lipid damage

Apart from DNA, cellular membranes are among the most important targets of radiation damage (Giusti *et al.,* 1998). It is known that the lipid-rich environment of cell membrane, phospholipid acyl chains, phospholipid backbones and cholesterol is constantly subjected to free radical damage. The polyunsaturated hydrocarbon moieties of the phospholipids are particularly sensitive to damage by ROS, generated during radiation (Parasassi *et al.,* 1992; Parasassi *et al.,* 1994; Rice-Evans and Burdon, 1993). Perhaps the most characteristic biological damage caused by radicals is their ability to initiate the free-radical chain reaction known as lipid peroxidation (LPO). This occurs when the hydroxyl radicals are generated close to or within membranes and then attack the fatty-acid side chains of the membrane phospholipids (Halliwell, 1987). Exposure to ionizing radiation leads to generation of free radicals, which increases LPO and also enhances the degradation of haemoglobin, ultimately leading to increase in free cytosolic pool of iron, which acts as a secondary initiator. As secondary initiators, iron ions catalyze hydroxyl radical formation, thereby accelerating LPO (Graf *et al.,* 1984; Biaglow *et al.,* 1997). The extent of initial damage caused by free radicals is further amplified by Fenton reaction generated hydroxyl radicals in the presence of superoxide and hydrogen peroxide (Halliwell and Gutteridge, 1985).

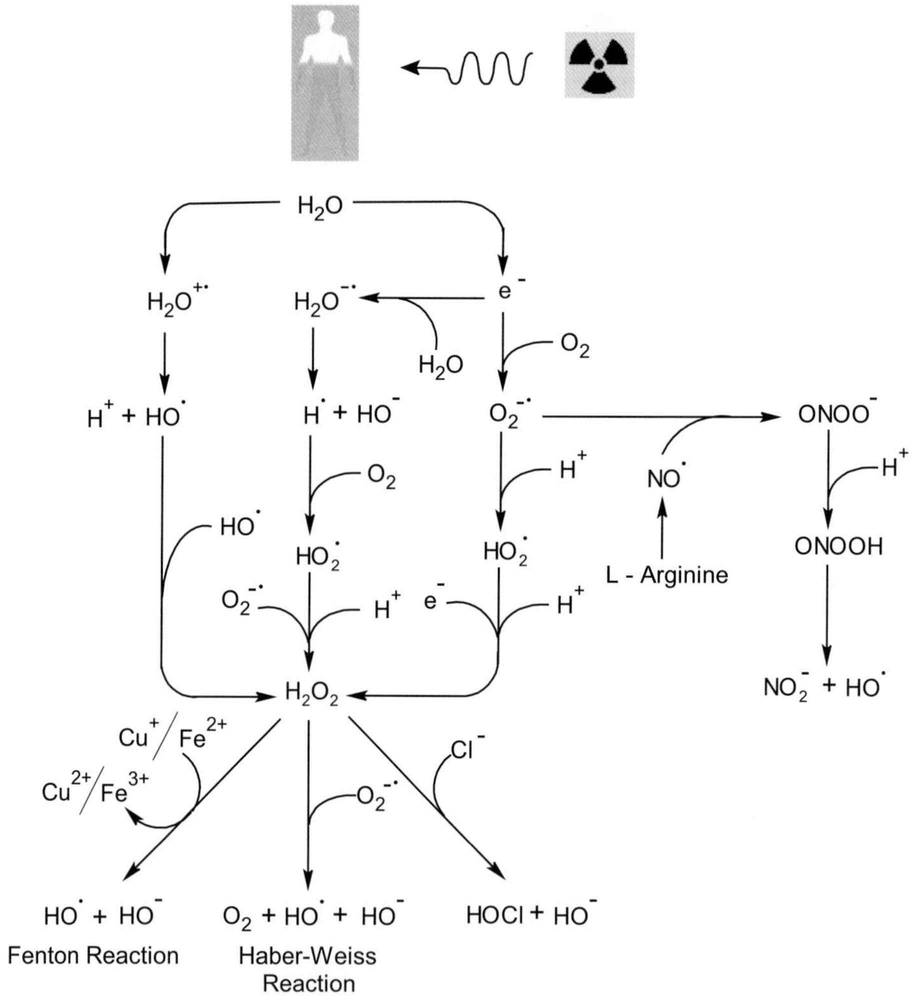

(where OH·: hydroxyl radical, H·: hydrogen radical, $O_2^{-\cdot}$: superoxide radical, $HO_2^\cdot$: hydroperoxyl radical, NO·: nitric oxide radical, ONOO·: peroxynitrite, HOCl: hypochlorous acid)

Fig. 3.3. Radiolysis of water and formation of reactive oxygen species.

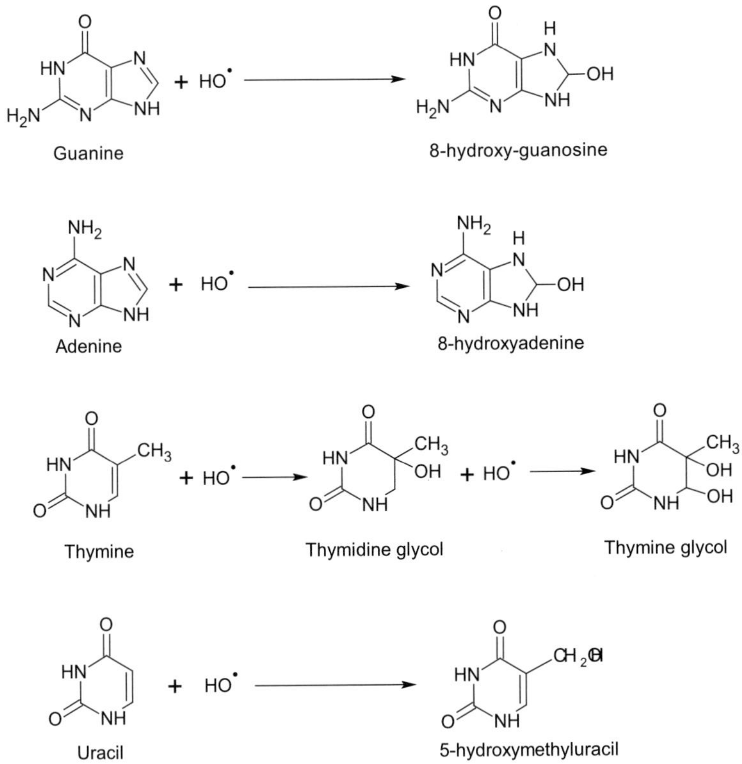

Fig. 3.4. Hydroxyl free radical (OH·) damage and products in DNA bases.

Thus, the redox state and concentration of iron ions in the cellular milieu plays a crucial role in amplification of damage (Minlotti and Aust, 1987), as they interact with membranes to generate alkoxyl and peroxyl radicals, thereby inflicting further damage to the cellular system (Davies and Slater, 1987). Lipid radicals are believed to be formed by the reaction of hydroxyl radicals generated by ionizing radiation with polyunsaturated fatty acids, which subsequently react with oxygen to form lipid peroxyl radical after undergoing molecular rearrangement of conjugation in double bonds and eventually a chain reaction is initiated on irradiation in oxygenated condition (Pandey and Mishra, 2000) (Fig. 3.5). Lipid peroxidation causes changes in membrane permeability as well as degradation of lipids. The most widely used index of lipid peroxidation is malondialdehyde (MDA) formation, often assayed with the thiobarbituric acid (TBA) assay (Parasassi *et al.*, 1994; Giusti *et al.*, 1998).

Radiation-induced protein damage

Proteins are also targets for free radicals. Free radicals generated by exposure to ionizing radiation also induce detectable changes in the structure and function of intrinsic proteins (Pohl, 1993; Jhun *et al.*, 1991). Oxidative damage to proteins, as assessed by formation of carbonyl groups, is also a highly damaging event and may occur in the absence of lipid peroxidation (Stadtman and Berlett, 1997; Dean *et al.*, 1997). Protein oxidation products and carbonyl derivatives of proteins may result from oxidative modifications of amino acid side chains, reactive oxygen-mediated peptide cleavage and from reactions with lipid and carbohydrate oxidation products (de Zwart *et al.,* 1999). Oxidative molecules such as hypochlorous acid (as a result of the interaction of hydrogen peroxide with Cl^- ions in the presence of myeloperoxidase as a catalyst, Fig. 3.3) can induce the production of 3-chlorotyrosine from tyrosine (Domigan *et al.,* 1995), and histidine can be oxidized to 2-oxohistidine in metal-catalyzed oxidative reactions which can occur in the metal binding site of proteins (Lewisch and Levine, 1995). However, oxidation of phenylalanine results in the formation of *ortho*-tyrosine (*o*-Tyr) (Huggins *et al.,* 1993). One-electron oxidation of L-tyrosine generates long-lived tyrosyl radicals, which can react with each other and form dityrosine (DT) (Heinecke *et al.,* 1993) (Fig. 3.6). The modification of proteins by ROS is implicated in the etiology of a number of physiological disorders and diseases (Stadtman and Berlett, 1997; Dean *et al.,* 1997).

Several antioxidant systems have a fundamental role in defending organisms against radiation-induced oxidative stress (Fig. 3.7). Antioxidants are molecules that can prevent or reduce the extent of oxidative destruction of biomolecules when present in small concentrations (Halliwell, 1990; Greenberger and Epperly, 2007; Greenberger *et al.,* 2003). ROS are tightly controlled by antioxidant defence systems, including non-enzymatic radical scavengers and enzymes that can either directly detoxify ROS or indirectly regulate their levels.

As the most important members of the enzymatic defence systems against oxygen radicals superoxide dismutase (SOD), catalase (CAT) and glutathione peroxidase (GPx) have been distinguished (Fig. 3.8).

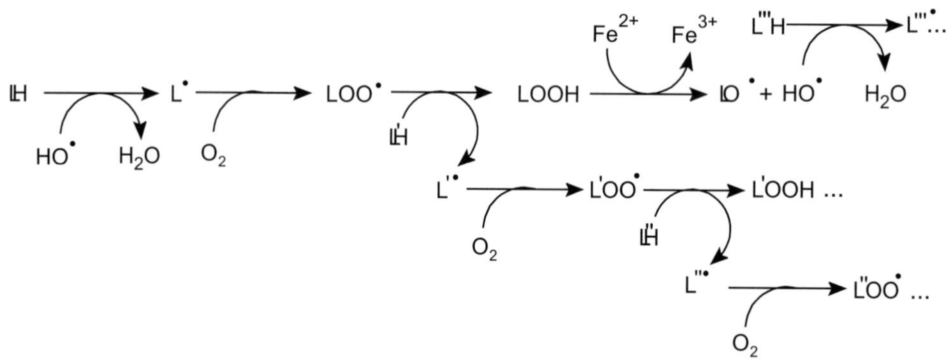

LH: polyunsaturated fatty acids, OH·: hydroxyl radical, L·: lipid radical, LOO·: lipid peroxyl radical, LOOH: lipid hydroperoxide, LO·: lipid alkoxyl radical

Fig. 3.5. Mechanism of lipid peroxidation.

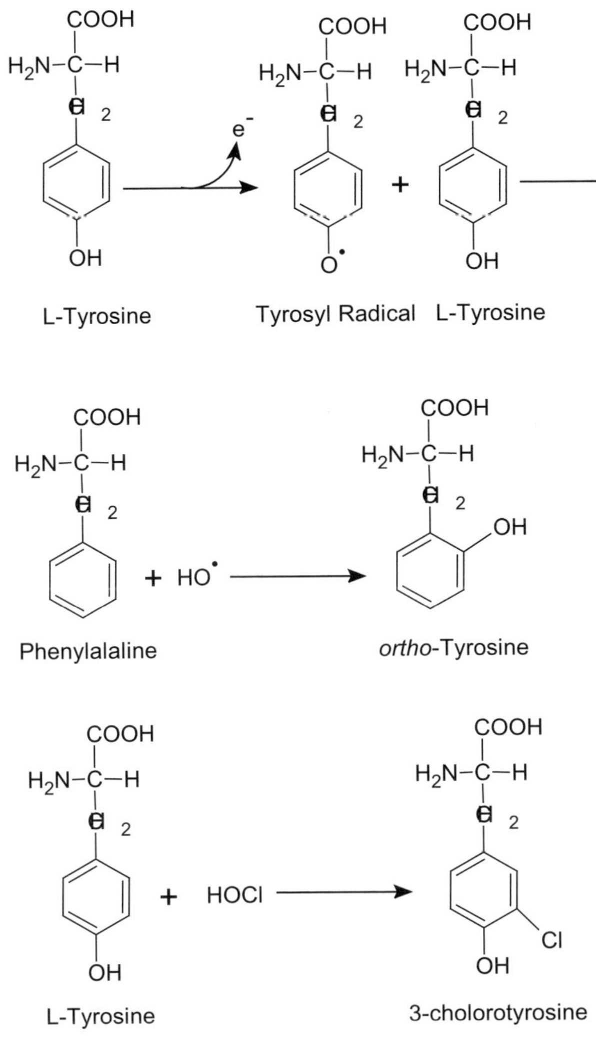

Fig. 3.6. Free radical damage to amino acid.

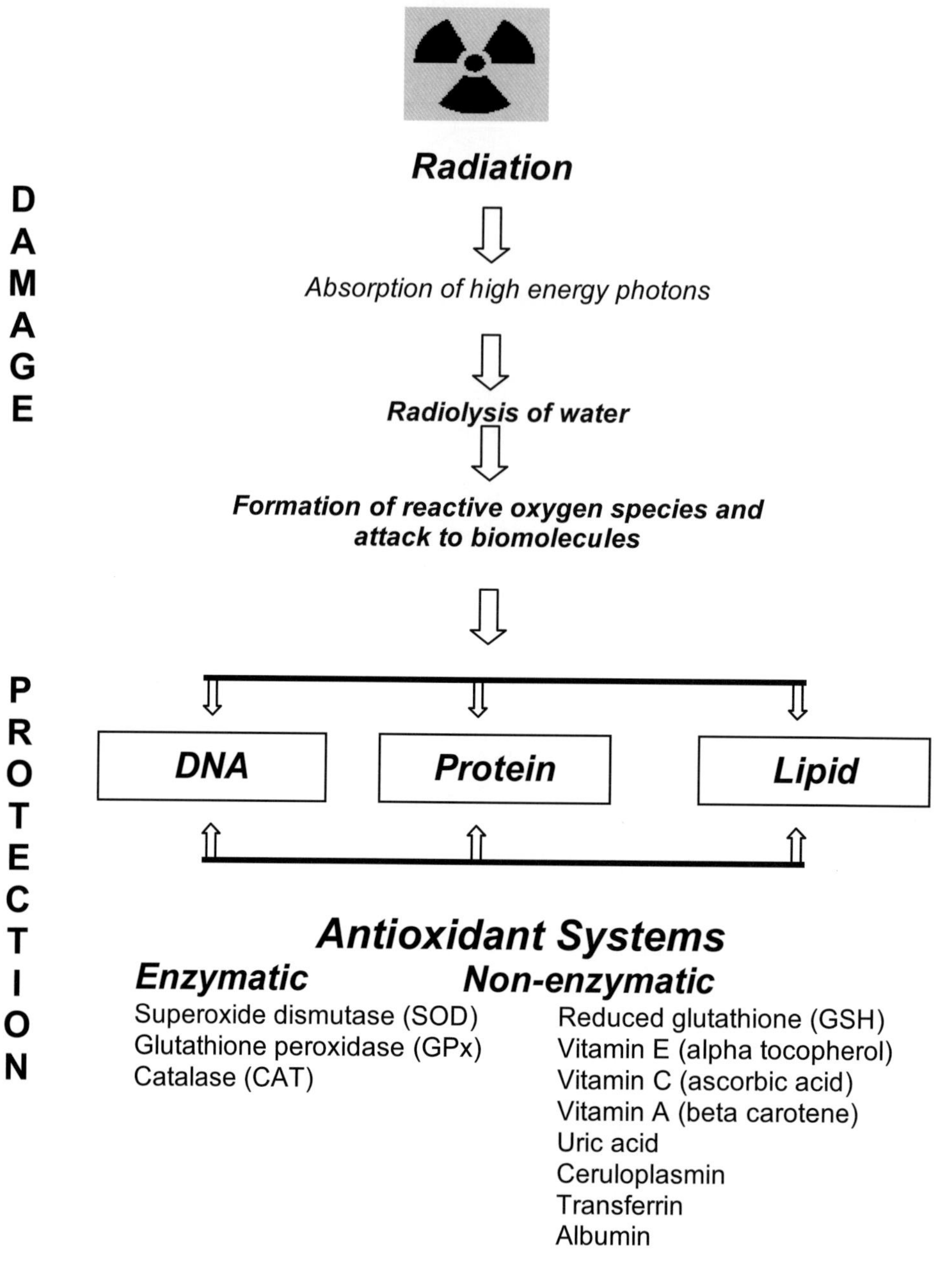

Fig. 3.7. Mechanism of indirect radiation damage and protection by antioxidants.

$$\text{SOD}: \quad 2O_2^{\bullet} + 2H^+ \longrightarrow H_2O_2 + O_2$$

$$\text{CAT}: \quad 2H_2O_2 \longrightarrow 2H_2O + O_2$$

$$\text{GPx}: \quad 2GSH + H_2O_2 \longrightarrow GSSG + 2H_2O$$

(SOD: superoxide dismutase, CAT: catalase, GPx: glutathione peroxidase)

Fig. 3.8. Enzymatic antioxidant defence systems against ROS.

Obviously, assaying these enzymes can offer an indication of the antioxidant status of an individual. Reduced glutathione (GSH), vitamin E (α-tocopherol), vitamin C (ascorbic acid), β-carotene (a precursor of vitamin A), uric acid, ceruloplasmin, transferrin and albumin are some important compounds that function as non-enzymatic antioxidants (Heffner and Repine, 1989; Halliwell, 1996; Fig. 3.9) and have been shown to play a role in radioprotection. Vitamin E, pentoxyfylline and sulfhydryl compounds like amifostine have been shown to render effective radioprotection (Wasserman and Chapman, 2004; Blanco and Chao, 2006). Koç and co-workers reported that melatonin administration prior to exposure to irradiation could protect against the damage inflicted by radiation. In this case the radioprotective effect of melatonin results from the up-regulation of antioxidant enzymes when melatonin is given pre-irradiation, and its ability to act as a scavenger of free radicals generated by ionizing radiation (Koç *et al.*, 2003).

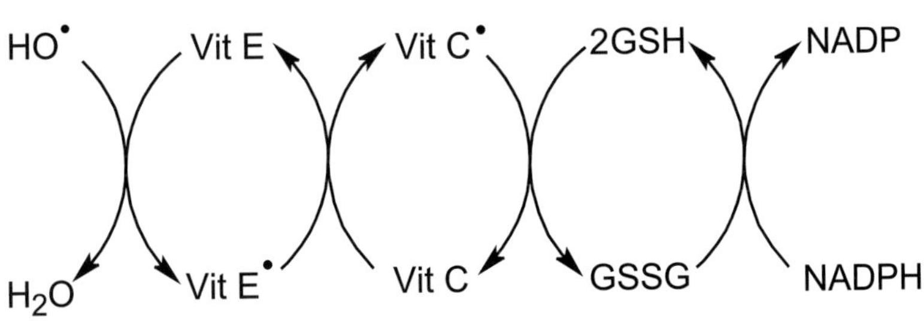

Fig. 3.9. Non-enzymatic antioxidant defence systems against ROS.

Nigella sativa L.: A Plant with Curative Properties

Nigella sativa (NS), commonly known as black seed, is a seed of capsulated plant, and belongs to the Ranunculaceae family. *Nigella sativa* has been employed for thousands of years as a spice and food preservative, as well as a protective and curative remedy for numerous disorders (Sayed, 1980). The historical tradition of black seed in medicine is substantial. Prophet Mohammed referred to black seed as having healing powers. Black seed also finds mention as curative black cumin in the Holy Bible, and is described as the

Melanthion of Hippocrates and Dioscorides and as the Gith of Pliny. *Nigella sativa* is found wild in southern Europe, northern Africa and in the Middle and Far East (Atta-ur-Rahman *et al.,* 1985).

Nigella sativa is a bushy, self-branching plant with white or pale to dark blue flowers. *Nigella sativa* reproduces and forms a fruit capsule, which consists of many white trigonal seeds. Once the fruit capsule has matured, it opens up and the seeds contained within are exposed to the air, becoming black in colour (Schleicher *et al.,* 1998).

Phytochemical profiling of *Nigella sativa*

Phytochemical analyses of *Nigella sativa* demonstrated the presence of volatile oil (important constituents are anethole, p-cymene, limonene, carvone and thymoquinone) and fatty acid (Nickavar *et al.,* 2003). These compounds appear to be responsible for the pharmacological actions of *Nigella sativa.* Eight fatty acids were identified in the extract, which represented about 99.5% of the total fatty acid composition. The volatile oil amount is approximately 0.5%. The extract consisted of four saturated fatty acids (17.0%) and four unsaturated fatty acids (82.5%). Linoleic acid (55.6%), oleic acid (23.4%) and palmitic acid (12.5%) were the major components. Thirty-two compounds, constituting 86.7% of the volatile oil, were identified. The major compounds of the volatile oil were trans-anethole (38.3%), p-cymene (14.8%), limonene (4.3%), carvone (4.0%) and thymoquinone (0.6%). The oil consisted of six phenyl propanoid compounds (46.1%), nine monoterpenoid hydrocarbons (26.9%), four monoterpenoid ketones (6.0%), eight non-terpenoid hydrocarbons (4.0%), three monoterpenoid alcohols (2.7%) and two sesquiterpenoid hydrocarbons (1.0%). *Nigella sativa* seeds contain other compounds, including nutritional components such as carbohydrates, fats, vitamins, mineral elements, proteins and eight of the nine essential amino acids. The seeds contain carotene, which is converted by the liver to vitamin A. The *Nigella sativa seeds* are also a source of calcium, iron and potassium (Nickavar *et al.,* 2003; Omar *et al.,* 1999; Al-Jassir, 1992; Chun *et al.,* 2002; Correa *et al.,* 1986; al-Gaby, 1998; Ali and Blunden, 2003; Burits and Bucar, 2000; Ramadan *et al.,* 2003).

Pharmacological properties of *Nigella sativa*

Nigella sativa has been used for thousands of years as a spice and food preservative. Its seeds have been used both internally and externally in traditional medicine in the Middle and Far East for a wide range of disorders. Ancient Egyptian and Greek physicians also prescribed *Nigella* for the treatment of several diseases. Several pharmacological properties of *Nigella sativa,* including antitumour, antinociceptive, antiinflammatory, antibacterial, antidiabetic, antiviral, antihelminthic, antimicrobial, antihistaminic, immunomodulative and hepatoprotective have been revealed. Furthermore, *in vivo* and *in vitro* antioxidative activities have also been reported (El-Dakhakhny *et al.,* 2002; Peterhans, 1997; Akhter and Riffat, 1991; Morsi, 2000; Worthen *et al.,* 1998; Swamy and Tan, 2000; Kanter *et al.,* 2003).

Thymoquinone is a very important bioactive constituent of the volatile oil of *Nigella sativa* and has been shown to exert several pharmacological activities, including antioxidant, antihistaminic, chemotherapeutic and antiinflammatory activities (Khalife and Lupidi, 2007; Kanter *et al.,* 2006; Norwood *et al.,* 2006; Tekeoglu *et al.,* 2006). Anethole, another constituent of the volatile oil of *Nigella sativa,* is a naturally occurring compound, present in many plants (e.g., *Foeniculum vulgare*, *Artemisia dracunculus*). It also shows a number of biological activities, such as fungicidal, antioxidant, antiinflammatory, ulcer healing and anticonvulsant activities in experimental models (Fujita and Kubo, 2004; Freire *et al.*, 2005; Sayyah *et al.*, 2004).

Radioprotective properties of *Nigella sativa*

Synthetic drugs do have toxicity and, therefore, in recent years a number of plant products have been screened for (Arora *et al.*, 2005; Bhattacharya *et al.*, 2005; Song *et al.*, 2006; Shukla *et al.*, 2006; Chawla *et al.*, 2007; Cemek *et al.*, 2006).

We have earlier shown the radioprotective characteristics of *Nigella sativa* and determined how *Nigella sativa* extract affects the oxidant and antioxidant levels in radiation-exposed rats. We studied MDA (as an indicator of lipid preoxidation) and nitric oxide (nitrate and nitrite) levels, which are the oxidative stress markers. Ascorbic acid, retinol, β-carotene, GSH and ceruloplasmin levels were determined as antioxidants (Cemek *et al.*, 2006).

MDA levels were studied in the groups that were treated with *Nigella sativa* after they were exposed to radiation (6Gy), and also untreated groups. MDA levels were found to be statistically lower in the *Nigella sativa* group than in the untreated group ($p < 0.001$). Moreover in the *Nigella sativa* treated groups nitrite and nitrate levels were found to be low ($p < 0.05$). These results pointed out that *Nigella sativa* reduces oxidative stress induced by radiation exposure.

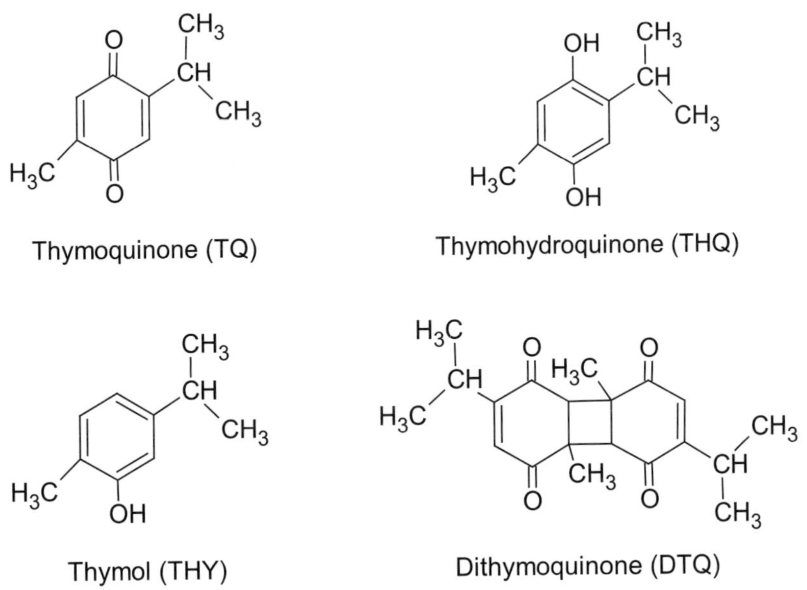

Fig. 3.10. Chemical structures of thymoquinones.

It was observed that *Nigella sativa* treatment had a positive effect on potential antioxidants. Levels of retinol ($p < 0.01$), β-carotene ($p < 0.05$), ceruloplasmin ($p < 0.05$) and reduced glutathione ($p < 0.05$), also known as non-enzymatic antioxidants, were increased as compared to the group that was not treated with *Nigella sativa*.

Antioxidant activity of Thymoquinone (TQ)

Nigella sativa volatile oil is an effective free radical scavenger exhibiting antioxidant activity and protecting against the damage caused by free radicals. Thymoquinone ($C_{10}H_{12}O_2$), the main active constituent of the volatile oil, is extracted from *Nigella sativa* seeds (El Gazzar *et al.*, 2006). *Nigella sativa* volatile oil, thymoquinone (TQ), dithymoquinone (DTQ), which is believed to be nigellone, thymohydroquinone (THQ) and thymol (THY) are considered the main active ingredients (Omar *et al.*, 1999) (Fig. 3.10).

TQ (2-isopropyl-5-methyl-1,4-benzoquinone) has been established as the major component of the oil extracted from *Nigella sativa* plant seeds. Significant differentiations have been detected in terms of TQ amount in volatile oil (range: 0.6–24 %) (El-Mahmoudy *et al.*, 2002; Nickavar *et al.*, 2003).

TQ, by receiving one pair of electrons (2 e^-) and 2 protons (H^+), is induced and converted to THQ. Since this reaction is reversible, THQ again could be oxidized to TQ. These kinds of reversible reactions are abundant in nature (Fig. 3.11).

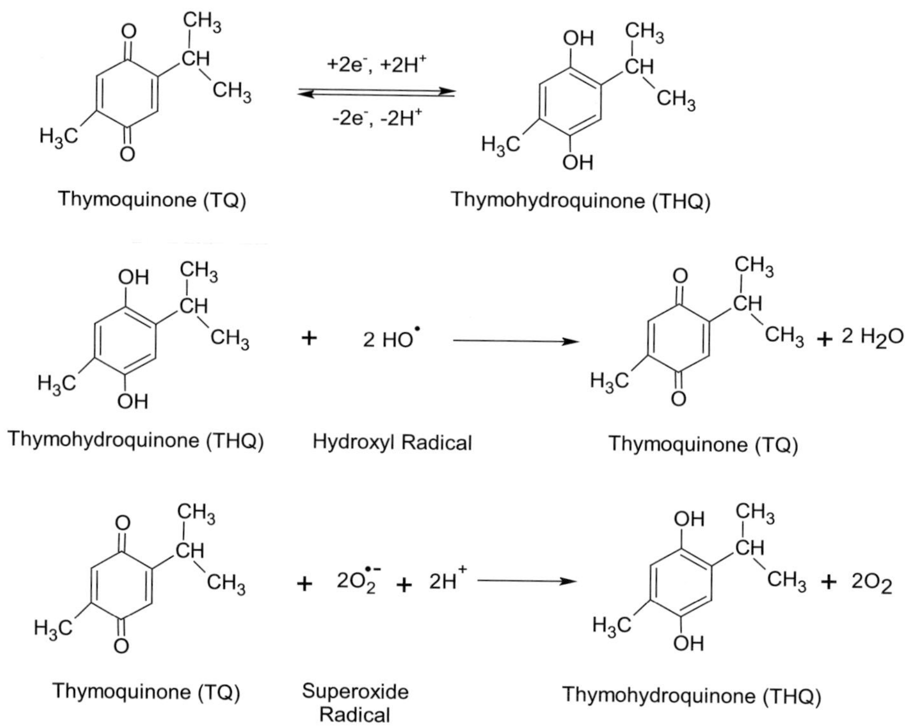

Fig. 3.11. The mechanism of action of TQ and THQ, which build up active part of *Nigella sativa.*

Because of being exposed to radiation, and following radiolytic cleavage of water, plenty of ROS generation takes place. The generated ROS could be neutralized by reduction and oxidation reactions.

Detoxification of ROS by Thymoquinone

The inhibition of lipid peroxidation by thymoquinone as well as its metabolite thymohydroquinone (Nagi *et al.*, 1999) could account for the marked antioxidative stress effect of thymoquinone. Moreover, thymoquinone acts as a scavenger of superoxide anion radical ($O2^{\cdot}$-), hydroxyl radical ($OH^{\cdot}$) and singlet molecular oxygen ($^1O_2$) (Kruk *et al.*, 2000; Badary *et al.*, 2003). The inhibitory effect of thymoquinone against NO production has also been reported (El-Mahmoudy *et al.*, 2002). TQ has been reported to have potent superoxide anion ($O2^{\cdot}$–) scavenging abilities and inhibits iron-dependent microsomal lipid peroxidation (Badary *et al.*, 2003). The generation of superoxide anion by the xanthine/xanthine oxidase system was inhibited by TQ in a dose-dependent manner (IC_{50}: 3.4M). This is promising considering the fact that $O2^{\cdot}$- reacts with protein and non-protein sulfhydryls and polyunsaturated fats and initiates aromatic hydroxylation reactions, thus damaging cells and causing inflammation (Gali-Muhtasib *et al.*, 2006).

The superoxide anion radical induced by radiation exposure could be neutralized by TQ that has antioxidant properties. TQ might convert superoxide radical to molecular oxygen (Fig. 3.11). After receiving 2 electrons from 2 superoxide anion radicals and 2 protons (H^+), TQ is converted to THQ plus 2 molecules of molecular oxygen. In this way, TQ gets converted to THQ by receiving electrons from superoxide radicals. As a consequence of radiolysis of water, hydroxyl radicals are generated which cause more than 60–70% of tissue damage by attacking and damaging biomolecules such as DNA and lipids (Ward, 1988; Vijayalaxmi *et al.*, 2004).

Hydroxyl radicals can be neutralized by THQ. THQ can convert 2 hydroxyl radicals into 2 molecules of water. By giving 2 electrons and 2 H^+ ions to 2 hydroxyl radicals, THQ could be converted to TQ and 2 molecules of water (Fig. 3.11).

Superoxide and hydroxyl radicals are potent in producing destruction of the cell membrane by inducing lipid peroxidation (Bromont *et al.*, 1989). Nitric oxide (NO), which is a diatomic free-radical molecule, has various biological effects. It is produced from its precursor L-arginine by nitric oxide synthase (Fig. 3.3). Excess NO reacts with superoxide to form peroxynitrite, a powerful radical that produces neuronal death after cerebral ischemia. The brain is particularly vulnerable to oxidative stress injury because of its high rate of oxidative metabolic activity, intense production of reactive oxygen species metabolites, and high content of polyunsaturated fatty acids, relatively low antioxidant capacity, low repair mechanism activity and non-replicating nature of its neuronal cells (Evans, 1993).

When TQ has neutralized superoxide radical, nitric oxide would not be able to get converted into peroxynitrite by interacting with superoxide radical. Consequently, TQ would neutralize nitric oxide indirectly (Fig. 3.3).

Conclusions

Total-body irradiation or wide field radical radiotherapy can lead to bone marrow failure (Koç *et al.,* 2002). Irradiation-related myelosuppression causes a decrease in peripheral blood cells, especially leukocytes (neutropenia). *In vitro* cytotoxic screening of extracts of *Nigella sativa* showed that it possesses a potent cytotoxic effect, as well as a potentiating effect, on the cellular immune response (Swamy and Tan, 2000). Administration of black seed significantly decreased total leukocyte and lymphocyte counts, increased heterophil: lymphocyte ratio and lysosomal enzyme activity, and also decreased reticuloendothelial system function induced by oxytetracycline in pigeons (Al-Ankari, 2005). Thus, concomitant *Nigella sativa* administration during radiotherapy can shorten the period of irradiation-related severe neutropenia, improve neutrophil account and reduce morbidity secondary to bacterial and fungal infections. Ionizing radiation-induced inflammation response is an acute phase side effect, which generally appears after a few days or weeks and is followed by a chronic phase characterized by fibrosis that can arise months or years after irradiation. Some studies have shown that *Nigella sativa* extract or thymoquinone exerted antiinflammatory activity, which was characterized by inhibition of cyclooxygenase and 5-lipoxygenase pathways of arachidonate metabolism or inflammatory cytokines (Tekeoglu *et al.,* 2006; El Mezayen *et al.,* 2006; Mansour and Tornhamre, 2004). Thus, supplementing cancer patients, who are undergoing radiotherapy, with *Nigella sativa* may provide an alleviation of the symptoms arising due to radiation-induced tissue inflammation.

It is known that thymoquinone exhibits antiproliferative effect, induces apoptosis, disrupts mitochondrial membrane potential and triggers the activation of caspases 8, 9 and 3 in myeloblastic leukaemia HL-60 cells (El-Mahdy *et al.,* 2005). It is reported that thymoquinone exhibits dose- and time-dependent cytotoxic and apoptotic effects on human cancer cell lines [A549 (lung carcinoma), Hep-2 (larynx epidermoid carcinoma), HT-29 (colon adenocarcinoma) and MIA PaCa-2 (pancreas carcinoma)] (Rooney and Ryan, 2005). The results of one study showed a similar significant decrease in cell number in the group treated with thymoquinone, compared to 5-fluorouracil. Increases in cellular damage were evident after 24, 48 and 72 hours and in all treated groups, compared to control (Norwood *et al.,* 2006). Based on the above-mentioned facts, it may be suggested that *Nigella sativa* can exert adjuvant chemotherapeutic activity after radiotherapy or surgery in some patients with cancer, thereby reducing the risk of cancer recurrence or micrometastasis.

Nigella sativa or some of its bioactive constituents can be used as a radioprotectant. It can be concluded that *Nigella sativa* administration decreases lipid peroxidation, stimulates the antioxidant enzyme activities and reduces inflammatory reactions in radiation-treated cancer patients, and thus alleviates radiation toxicity to the organs and tissues. Due to its non-toxic nature, *Nigella sativa* as a radioprotectant can be safely administered to cancer patients being treated with high dose rates of radiotherapy.

Acknowledgements

We thank Ahmet Büyükben for assistance in drawing the figures.

References

Akhter, M.S. and Riffat, S. (1991) Field trail of *Saussurea lappa* roots against nematodes and *Nigella sativa* seed against cystodes in children. *Journal of Pakistan Medical Association* 41, 185–187.

Al-Ankari, A.S. (2005) Immunomodulating effects of black seed and oxytetracycline in pigeons. *Immunopharmacology Immunotoxicology* 27(3), 515–520.

Al-Gaby, A.M. (1998) Amino acid composition and biological effects of supplementing broad bean and corn proteins with *Nigella sativa* (black cumin) cake protein. *Nahrung* 42, 290–294.

Ali, B.H. and Blunden, G. (2003) Pharmacological and toxicological properties of *Nigella sativa*. *Phytotherapy Research* 17(4), 299–305.

Al-Jassir, M.S. (1992) Chemical composition and microflora of black cumin (*Nigella sativa* L.) seeds growing in Saudi Arabia. *Food Chemistry* 45, 239–242.

Ames, B.N. and Gold, L.S. (1991) Endogenous mutagens and the causes of aging and cancer. *Mutation Research* 250, 3–16.

Arora, R., Gupta, D., Chawla, R., Sagar, R., Sharma, A., Kumar, R., Prasad, J., Singh, S., Samanta, N., Sharma, R.K. (2005) Radioprotection by plant products: Present status and future prospects. *Phytotherapy Research* 19, 1–22.

Atta–ur–Rahman, Malk, S., Cunheng, H. and Clardy, J. (1985) Isolation and structure determination of nigellicine, a novel alkaloid from the seeds of *Nigella sativa*. *Tetrahedron Letters* 26, 2759–2762.

Badary, O.A., Taha, R.A., Gamal el-Din, A.M. and Abdel-Wahab, M.H. (2003) Thymoquinone is a potent superoxide anion scavenger. *Drug and Chemical Toxicology* 26, 87–98.

Bhatia, I.S. and Bajaj, K.L. Tannins in black-plum (*Syzygium cumini* L.) seeds. *Biochemical Journal* 128, 56.

Bhattacharya, S., Subramanian, M., Roychowdhury, S., Bauri, A.K., Kamat, J.P., Chattopadhyay, S. and Bandyopadhyay, S.K. (2005) Radioprotective property of the ethanolic extract of *Piper betel* leaf. *Journal of Radiation Research* (Tokyo) 46(2), 165–171.

Biaglow, J.E., Manevich, Y., Uckun, F. and Held, K.D. (1997) Quantitation of hydroxyl radicals produced by radiation and copper linked oxidation of ascorbate by 2-deoxy-D-ribose method. *Free Radical Biology and Chemistry* 22, 1129–1138.

Blanco, A.I. and Chao, C. (2006) Management of radiation–induced head and neck injury. In: Small, W. and Woloschak, G.E. (eds) *Radiation Toxicity: A Practical Guide*. Springer Science Media Business, New York, pp. 23–42.

Bromont, C., Marie, C. and Bralet, J. (1989) Increased lipid peroxidation in vulnerable brain regions after transient forebrain ischemia in rats. *Stroke* 20, 918–924.

Burits, M. and Bucar, F. (2000) Antioxidant activity of *Nigella sativa* essential oil. *Phytotherapy Research* 14(5), 323–328.

Cemek, M., Enginar, H., Karaca, T. and Unak, P. (2006) *In vivo* radioprotective effects of *Nigella sativa* L. oil and reduced glutathione against irradiation-induced oxidative injury and number of peripheral blood lymphocytes in rats. *Photochemistry and Photobiology* 82(6), 1691–1696.

Chawla, R., Arora, R., Singh, S., Sagar, R.K., Sharma, R.K., Kumar, R., Sharma, A., Gupta, M.L., Singh, S., Prasad, J., Khan, H.A., Swaroop, A., Sinha, A.K., Gupta, A.K., Tripathi, R.P. and Ahuja, P.S. (2007) Radioprotective and antioxidant activity of fractionated extracts of berries of *Hippophae rhamnoides*. *Journal of Medicinal Food* 10(1), 101–109.

Chun, H., Shin, D.H., Hong, B.S., Cho, W.D., Cho, H.Y. and Yang, H.C. (2002) Biochemical properties of polysaccharides from black pepper. *Biological and Pharmaceutical Bulletin* 25, 1203–1208.

Chung, M.H., Kasal, H., Jones, D.S., Inoue, H., Ishikawa, H. and Olatsuka, E. (1991) An endonuclease activity of *E. coli* that specifically removes 8-hydroxyguanine residues from DNA. *Mutation Research* 254, 1–12.

Correa, A.D., Jokl, L. and Carlsson, R. (1986) Amino acid composition of some *Amaranthus* sp. grain proteins and of its fractions. *Archivos Latinoamericanos de Nutricion* 36, 466– 476.

Cramp, W.A., Yatvin, M.B. and Harms-Ringdahl, M. (1994) Recent developments in the radiobiology of cellular membranes. *Acta Oncology* 33, 945–952.

Daniniak, N. and Tann, B.J. (1995) Utility of biological membranes as indicators for radiation exposure: Alterations in membrane structure and function over time. *Stem Cells* 13, 142–152.

Davies, M.J. and Slater, T.F. (1987) Studies on metal ion and lipoxygenase catalyzed breakdown of hydroperoxides using electron-spin-resonance spectroscopy. *Biochemical Journal* 245, 167.

de Zwart, L.L., Meerman, J.H., Commandeur, J.N. and Vermeulen, N.P. (1999) Biomarkers of free radical damage applications in experimental animals and in humans. *Free Radical Biology and Medicine* 26, 202–26.

Dean, R.T., Fu, S., Stocker, R. and Davies, M.J. (1997) Biochemistry and pathology of radical mediated protein oxidation. *Biochemical Journal* 324, 1–18.

Dizdaroglu, M. (1991) Chemical determination of free radical-induced damage to DNA. *Free Radical Biology and Medicine* 10, 225–242.

Domigan, N.M., Charlton, T.S., Duncan, M.W., Winterbourn, C.C. and Kettle, A.J. (1995) Chlorination of tyrosyl residues in peptides by myeloperoxidase and human neutrophils. *Journal of Biological Chemistry* 14, 16542–16548.

El Gazzar, M., El Mezayen, R., Marecki, J.C., Nicolls, M.R., Canastar, A. and Dreskin, S.C. (2006) Antiinflammatory effect of thymoquinone in a mouse model of allergic lung inflammation. *International Immunopharmacology* 6(7), 1135–1142.

El Mezayen, R., El Gazzar, M., Nicolls, M.R., Marecki, J.C., Dreskin, S.C. and Nomiyama, H. (2006) Effect of thymoquinone on cyclooxygenase expression and prostaglandin production in a mouse model of allergic airway inflammation. *Immunology Letters* 106(1), 72–81.

El–Dakhakhny, M., Madi, N.J., Lambert, N. and Ammon, H.P. (2002) *Nigella sativa* oil, nigellone and derived thymoquinone inhibit synthesis of 5-lipoxygenase products in polymorphonuclear leukocytes from rats. *Journal of Ethnopharmacology* 81, 161–164.

El–Mahdy, M.A., Zhu, Q., Wang, Q.E., Wani, G. and Wani, A.A. (2005) Thymoquinone induces apoptosis through activation of caspase-8 and mitochondrial events in p53-null myeloblastic leukemia HL–60 cells. *International Journal of Cancer* 117(3), 409–417.

El–Mahmoudy, A., Matsuyama, H., Brogan, M.A., Shimizu, Y., El-Sayed, M.G., Minamoto, N. and Takewaki, T. (2002) Thymoquinone suppresses expression of inducible nitric oxide synthase in rat macrophages. *International Immunopharmacology* 11, 1603–1611.

Evans, P.H. (1993) Free radicals in brain metabolism and pathology. *British Medical Bulletin* 49, 577–587.

Floyd, R.A. (1990) The role of 8-hydroxyguanine in carcinogenesis. *Carcinogenesis* 11, 1447–1450.

Freeman, B.A. and Crapo, J.D. (1982) Biology of disease; free radicals and tissue injury. *Laboratory Investigation* 47, 412–426.

Freire, R.S., Morais, S.M., Catunda-Junior, F.E. and Pinheiro, D.C. (2005) Synthesis and antioxidant, antiinflammatory and gastroprotector activities of anethole and related compounds. *Bioorganic Medicinal Chemistry* 13, 4353–4358.

Fridovich, I. (1981) The biology of superoxide and of superoxide dismutases–in brief. *Progress in Clinical and Biological Research* 51, 15372.

Fujita, K. and Kubo, I. (2004) Potentiation of fungicidal activities of trans-anethole against *Saccharomyces cerevisiae* under hypoxic conditions. *Journal of Bioscience and Bioengineering* 98(6), 490–492.

Gali–Muhtasib, H., Roessner, A. and Schneider–Stock, R. (2006) Thymoquinone: a promising anticancer drug from natural sources. *International Journal of Biochemistry and Cell Biology* 38(8), 1249–1253.

Gate, L., Paul, J., Nguyen Ba, G., Tew, K.D. and Tapiero, H. (1999) Oxidative stress induced in pathologies: the role of antioxidant. *Biomedicine and Pharmacotherapy* 53, 169–180.

Giusti, A.M., Raimondi, M., Ravagnan, G., Sapora, O. and Parasassi, T. (1998) Human cell membrane oxidative damage induced by single and fractionated doses of ionizing radiation: a fluorescence spectroscopic study. *International Journal of Radiation Biology* 74, 595–605.

Graf, E., Mahoney, J.R., Bryant, R.G. and Eaton, J.W. (1984) Iron catalyzed hydroxyl radical formation: Stringent requirements for free iron coordination site. *Journal of Biological Chemistry* 259, 3620–3624.

Greenberger, J.S. and Epperly, M.W. (2007) Antioxidant gene therapeutic approaches to normal tissue radioprotection and tumour radiosensitization. *In vivo* 21(2), 141–146.

Greenberger, J.S., Epperly, M.W., Gretton, J., Jefferson, M, Nie, S., Bernarding, M., Kagan, V. and Guo, H.L. (2003) Radioprotective gene therapy. *Current Gene Therapy* 3(3), 183–195.

Halliwell, B. (1987) Oxidants and human disease: some new concepts. *FASEB Journal* 1, 358–364.

Halliwell, B. (1990) How to characterize a biological antioxidant. *Free Radical Research Communication* 9, 1–32.

Halliwell, B. (1996) Antioxidants in human health and disease. *Annual Review of Nutrition* 16, 33–50.

Halliwell, B. and Gutteridge, J.M.C. (1985) The importance of free radicals in catalytic metal ions in human diseases. *Molecular Aspects in Medicine* 8, 189–193.

Heffner, J.A. and Repine, J.E. (1989) State of the art: pulmonary strategies of antioxidant defense. *American Review of Respiratory Disease* 140, 531–554.

Heinecke, J., Li, W., Daehnke, H.L. and Goldstein, J.A. (1993) Dityrosine, a specific marker of oxidation, is synthesized by the myeloperoxidase-hydrogen peroxide system of human neutrophils and macrophages. *Journal of Biological Chemistry* 25, 4069–4077.

http://www.bragg.com/books/wtst_excerpt.html

Huggins, T.G., Wells–Knecht, M.C., Detorie, N.A., Baynes, J.W., Thorpe, S.R. (1993) Formation of o–tyrosine and dityrosine in proteins during radiolytic and metal–catalyzed oxidation. *Journal of Biological Chemistry* 17, 12341–12347.

Jagetia, G.C. and Reddy, T.K. (2005) Modulation of radiation-induced alteration in the antioxidant status of mice by naringin. *Life Science* 77, 780–794.

Jhun, E., Jhun, B.H. and Jones, L.R. (1991) Direct effects of ionizing radiation on integral membrane proteins. *Journal of Biological Chemistry* 266, 9403–9407.

Kamat, J.P., Boloor, K.K., Devasagayam, T.P.A. and Venkatachalam, S.R. (2000) Antioxidant properties of *Asparagus racemosus* against damage induced by γ-radiation in rat liver mitochondria. *Journal of Ethnopharmacology* 71, 425–435.

Kanter, M., Coskun, O. and Uysal, H. (2006) The antioxidative and antihistaminic effect of *Nigella sativa and* its major constituent, thymoquinone on ethanol–induced gastric mucosal damage. *Archives of Toxicology* 80(4), 217–224.

Kanter, M., Meral, I., Dede, S, Gunduz, H., Cemek, H., Ozbek, H. and Uygan, I. (2003) Effects of *Nigella sativa* L. and *Urtica dioica* L. on lipid peroxidation, antioxidant enzyme systems and some liver enzymes in CCl_4-treated rats. *Journal of Veterinary Medicine. A, Physiology, Pathology, Clinical Medicine* 50, 264– 268.

Khalife, K.H. and Lupidi, G. (2007) Nonenzymatic reduction of thymoquinone in physiological conditions. *Free Radical Research* 41(2), 153–161.

Koç, M., Buyukokuroglu, M.E. and Taysi, S. (2002) The effect of melatonin on peripheral blood cells during total body irradiation in rats. *Biological and Pharmaceutical Bulletin* 25(5), 656–657.

Koç, M., Taysi, S., Buyukokuroglu, M.E. and Bakan, N. (2003) The effect of melatonin against oxidative damage during total-body irradiation in rats. *Radiation Research* 160(2), 251–255.

Kruk, I., Michalska, T., Lichszteld, K., Kladna, A. and Aboul-Enein, H.Y. (2000) The effect of thymol and its derivatives on reactions generating reactive oxygen species. *Chemosphere* 4, 1059–1064.

Lett, J.T. (1992) Damage to cellular DNA from particulate radiation, the efficacy of its processing and the radiosensitivity of mammalian cells. Emphasis on DNA strand breaks and chromatin break. *Radiation and Environmental Biophysics* 31, 257–277.

Lewisch, S.A. and Levine, R.L. (1995) Determination of 2-oxohistidine by amino acid analysis. *Annals of Biochemistry* 231, 440–446.

Mansour, M., and Tornhamre, S. (2004) Inhibition of 5-lipoxygenase and leukotriene C4 synthase in human blood cells by thymoquinone. *Journal of Enzyme Inhibition and Medicinal Chemistry* 19, 431–436.

Minlotti, G. and Aust, S.D. (1987) The requirement of iron (III) in the initiation of lipid peroxidation by iron (II) and hydrogen peroxide. *Journal of Biological Chemistry* 262, 1098.

Morsi, N.M. (2000) Antimicrobial effect of crude extracts of *Nigella sativa* on multiple antibiotics–resistant bacteria. *Acta Microbiologica Polonica* 49, 63–74.

Nagi, M.N., Alam, K., Badary, O.A., al–Shabanah, O.A., al–Sawaf, H.A. and al-Bekairi, A.M. (1999) Thymoquinone protects against carbon tetrachloride hepatotoxicity in mice *via* an antioxidant mechanism. *Biochemistry and Molecular Biology International* 47, 153–159.

Nickavar, B., Mojab, F., Javidnia, K., Amoli, M.A.R. (2003) Chemical Composition of the fixed and volatile oils of *Nigella sativa* L. from Iran. *Zeitschrift für Naturforschung* (C) 58, 629–631.

Norwood, A.A., Tan, M., May M., Tucci, M. and Benghuzzi, H. (2006) Comparison of potential chemotherapeutic agents, 5-fluoruracil, green tea, and thymoquinone on colon cancer cells. *Biomedical Sciences Instrumentation* 42, 350–356.

Omar, A., Ghosheh, S., Abdulghani, A., Houdi, A. and Crookscor, P.A. (1999) High performance liquid chromatographic analysis of the pharmacologically active quinones and related compounds in the oil of the black seed (*Nigella sativa* L). *Journal of Pharmaceutical and Biomedical Analysis* 19, 757– 762.

Packer, L. and Ong, A.S.H. (1998) *Biological Oxidants and Antioxidants: Molecular Mechanisms and Health Effects*. AOCS Press, Champaign, Illinois, USA.

Pandey, B.N. and Mishra, K.P. (2000) Fluorescence and ESR studies on membrane oxidative damage by γ–radiation. *Applied Magnetic Resonance* 18, 483–492.

Parasassi, T., Giusti, A.M. and Gratton, E. (1994) Evidence for an increase in water concentration in bilayers after oxidative damage of phospholipids induced by ionizing radiation. *International Journal of Radiation Biology* 65, 329–334.

Parasassi, T., Ravagnan, G. and Sapora, O. (1992) Membrane oxidative damage induced by ionizing radiation detected by diphenylhexatriene fluorescence lifetime distributions. *International Journal of Radiation Biology* 61, 791–796.

Peterhans, E. (1997) Oxidants and antioxidants in viral diseases: Disease mechanisms and metabolic regulation. *Journal of Nutrition* 127, 962S–965S.

Pohl, L.R. (1993) An immunochemical approach of identifying and characterizing protein targets of toxic reactive metabolites. *Chemical Research in Toxicology* 6, 786–793.

Prasad, N.R., Menon, V.P., Vasudev, V. and Pugalendia, K.V. (2005) Radioprotective effect of sesamol on γ–radiation induced DNA damage, lipid peroxidation and antioxidants levels in cultured human lymphocytes. *Toxicology* 209, 225–235.

Ramadan, M.F., Kroh, L.W. and Morsel, J.T. (2003) Radical scavenging activity of black cumin (*Nigella sativa* L.), coriander (*Coriandrum sativum* L.), and niger (*Guizotia abyssinica* Cass.) crude seed oils and oil fractions. *Journal of Agricultural and Food Chemistry* 51(24), 6961–6969.

Rice–Evans, C. and Burdon, R. (1993) Free radical-lipid interaction and their pathological consequences. *Progress in Lipid Research* 32, 71–110.

Rooney, S. and Ryan, M.F. (2005) Effects of alpha-hederin and thymoquinone, constituents of *Nigella sativa,* on human cancer cell lines. *Anticancer Research* 25(3B), 2199–2204.

Sayed, M.D. (1980) Traditional medicine in health care. *Journal of Ethnopharmacology* 2(1), 19–22.

Sayyah, M., Nadjafnia, L. and Kamalinejad, M. (2004) Anticonvulsant activity and chemical composition of *Artemisia dracunculus* L. essential oil. *Journal of Ethnopharmacology* 94(2–3), 283–287.

Schleicher, P. and Saleh, M. (1998) *Black Seed Cumin: The Magical Egyptian Herb for Allergies, Asthma, and Immune Disorders*. Vermont Healing Arts Press, Rochester, p. 90.

Shukla, S.K., Chaudhary, P., Kumar, I.P., Samanta, N., Afrin, F., Gupta, M.L., Sharma, U.K., Sinha, A.K., Sharma, Y.K. and Sharma, R.K. (2006) Protection from radiation-induced mitochondrial and genomic DNA damage by an extract of *Hippophae rhamnoides*. *Environmental and Molecular Mutagenesis* 9, 647–656.

Sies, H. (1996) *Antioxidants in Disease, Mechanisms and Therapy*. Academic Press, New York, USA.

Song, L.H., Yan, H.L. and Cai, D.L. (2006) Protective effects of soybean isoflavone against gamma-irradiation induced damages in mice. *Journal of Radiation Research* (Tokyo) 47(2), 157–165.

Stadtman, E.R. and Berlett, B.S. (1997) Reactive oxygen mediated protein oxidation in aging and disease. *Chemical Research in Toxicology* 10, 485.

Swamy, S.M.K. and Tan, B.K.H. (2000) Cytotoxic and immunopotentiating effects of ethanolic extract of *Nigella sativa* L. seeds. *Journal of Ethnopharmacology* 70, 1–7.

Tan, D.X., Manchester, L.C. and Reiter, R.J. (2000) Melatonin directly scavenges hydrogen peroxide: A potentially new metabolic pathway of melatonin biotransformation. *Free Radical Biology and Medicine* 29, 1177–1185.

Tekeoglu, I., Dogan, A. and Demiralp, L. (2006) Effects of thymoquinone (volatile oil of black cumin) on rheumatoid arthritis in rat models. *Phytotherapy Research* 20(10), 869–871.

Vijayalaxmi, Reiter, R.J., Tan, D.X., Herman, T.S. and Thomas, C.R. (2004) Melatonin as a radioprotective agent: a review. *International Journal of Radiation Oncology, Biology and Physics* 59, 639–653.

Von Sonntag, C. (1987) *The Chemical Basis of Radiation Biology*. Francis and Taylor, London, p. 139.

Ward, J.F. (1983) Chemical aspects of DNA radioprotection. In: Nygaard, O.F. and Simic, M.G. (eds) *Radioprotectors and Anticarcinogens*. Academic Press, New York, USA, pp. 73–85.

Ward, J.F. (1988) DNA damage produced by ionizing radiation in mammalian cells: Identities, mechanisms of formation and repairability. *Progress in Nucleic Acid Research and Molecular Biology* 35, 95–125.

Wasserman, T.H. and Chapman, J.D. (2004) Radiation Response Modulation. In: Perez, C.A., Brady, L.W., Halperin, E.C. and Schmidt-Ullrich, R.K. (eds) *Principles and Practice of Radiation Oncology*, Lippincott William and Wilkins, Philadelphia, USA, pp. 663–696.

Worthen, D.R., Ghosheh, O.A. and Crooks, P.A. (1998) The *in vitro* antitumour activity of some crude and purified components of blackseed, *Nigella sativa* L. *Anticancer Research* 18, 1527–1532.

Yokus, B., Cakir, D.U., Akdag, M.Z., Sert, C. and Mete, N. (2005) Oxidative DNA damage in rats exposed to extremely low frequency electro magnetic fields. *Free Radical Research* 39(3), 317–323.

Chapter 4

Modulation of Radiation-induced Damage by Serbian Natural Plant Products: Implications for Radioprotection

Gordana Joksić, Andreja Leskovac and Sandra Petrović

Introduction

Use of radiation for medical examinations is the largest man-made source of radiation exposure and is continuously rising in Europe. The biological effects of low dose radiation on living cells may result in three outcomes:

(i) injured or damaged cells repair themselves without residual damage,
(ii) the cells die or
(iii) cells incorrectly repair themselves, resulting in a biological change.

Chromosomal aberrations constitute measurable end-points that can be related to exposure to ionizing radiations in humans and have been used for the purpose of biological dosimetry. Radiation-induced aberrations can be observed in human lymphocytes within a few hours of exposure. Their frequency is related to the dose and quality of radiation, and can be detected in blood samples taken long after the exposure (IAEA, 2001). DNA damage that is not repaired during cell proliferation represents initiating event that may result in neoplasia. An indirect estimate of chromosomal damage can be obtained by scores of micronuclei (MN). The formation of MN in dividing cells is the result of unrepaired or mis-repaired chromosomal breaks, chromosome malsegregation, inappropriate expression of genes involved in cell cycle checkpoints or lack of essential cofactors such as magnesium, calcium and folate. The causal link between MN induction and cancer development is supported by a number of observations such as elevated baseline level of micronuclei in subjects affected by cancer-prone congenital diseases and in cancer patients (Bonassi *et al.*, 2004, 2007).

An early finding of Patt *et al.* (1949), that the amino acid cysteine possesses radioprotective properties, was a starting point for synthesizing radioprotective sulfhydryl compounds. However, due to the high toxicity of thiols at the optimum protective doses searches for alternative, less-toxic compounds was required. Plant products are widely used throughout the world for possible drug identification and development, and as dietary supplements due to their empirically observed medicinal properties. Numerous studies have shown that natural antioxidants in plants are closely related to their biofunctionalities such as reduction of chronic diseases and inhibition of pathogenic bacteria growth, which are often associated with the termination of free radical propagation in biological systems.

Certain plants and phytochemical compounds like ascorbic acid, beta-carotene, vitamin E, curcumin, caffeine, chlorogenic acid, ellagic acid, vitamin A, bixin, polyphenols and a variety of flavonoids display strong antioxidant abilities in biological systems (Arora *et al.,* 2005a,b,c, 2006a,b,c, 2007a,b; Chawla *et al.,* 2004, 2005a,b, 2006, 2007; Lata *et al.,* 2007;). Some could be used as non-toxic radioprotectors (Thresiamma *et al.*, 1998). An important feature of flavonoids is their scavenging of oxygen-derived free radicals. When present in low concentrations, in relation to oxidizable substrates, they significantly inhibit or delay oxidative processes, while often themselves being oxidized (Nijveldt *et al.,* 2001). Polyphenols have demonstrated significantly greater protection against free radicals and free-radical-induced lipid peroxidation and DNA damage than vitamins C, E and β-carotene (Bagchi *et al.,* 2000). These compounds can scavenge both the hydroxyl radical and lipid radicals, suggesting that they act to reduce toxicity by interfering with lipid peroxidation.

Lipid peroxidation generates a complex variety of products, many of which are reactive electrophiles. Some of these react with protein and DNA and as a result are toxic and mutagenic. One of the lipid peroxidation product– malondialdehyde (MDA)– reacts with nucleic acid bases to form multiple adducts. Besides, MDA can be formed independent of lipid peroxidation by direct oxidation of DNA by agents that abstract the 4´hydrogen atom at the sugar backbone (Benamira *et al.,* 1995). Metal ions, such as Fe (II), directly attack deoxy-D-ribose, leading to the formation of malondialdehide (MDA), and amplification of radiation-induced oxidative stress (Chawla *et al.*, 2006). Increasing concentration of MDA leads to more MDA-dG adducts and consequently increased mutations.

Natural antioxidants, particularly polyphenols, may function as chain breakers or chelating agents, reducing the redox potential, thereby stabilizing the oxidized form of the metal ions, levels of nitric oxide and hydroxyl radicals. Nitric oxide is considered an important molecule in cellular signalling, which in higher concentrations induces the nitric oxide synthase, triggering inflammatory responses associated with radiation-induced oxidative stress (Gupta *et al.,* 2003). Radiation-induced hydroxyl radical also attacks DNA, causing oxidative damage to sugar and base residues, which later can be converted to strand breaks and chromosomal aberrations. Polyphenols may function as peroxidation protectors weakening oxidative stress, which to some extent can modulate radioresponse (Sagar *et al.,* 2006). The beneficial effects of polyphenols are mainly attributed to their strong antioxidant properties primarily depending on their chemical structures. Their action as peroxidation protector is related to their binding capacity.

Nowadays it is known that formation of chromosomal aberrations is a very complex process involving numerous enzymes, which respond to oxidative stress induced by ionizing radiation. More than 100 genes have been identified for DNA repair alone (Wood *et al.,* 2002). The discovery of DNA repair gene polymorphisms (for example in XRCC1, Apel, XRCC3, Ku70, Ku80, DNAPKcs, Ligase IV, XRCC4, BRACA1, BRACA2) raises the issue of their importance in radiation susceptibility.

A considerable number of medicinal plants have shown protective effects against ionizing radiation (Arora and Goel, 2000; Arora, 2006; Arora *et al.,* 2003; Arora *et al.,* 2005a, 2006a,b, 2007a,b). Radioprotective and radiorecovery properties of *Podophyllum hexandrum, Ocimum sanctum, Ginkgo biloba, Mentha piperita, Rhodiola imbricata, Hippophae rhamnoides* etc. have been attributed to the presence of antioxidant molecules, mainly polyphenols (Arora *et al.,* 2005a, 2007b, 2008; Kumar *et al.,* 2005a,b; Swaroop *et al.,* 2005). Plant extracts, such as abana, *Syzygium cumini*, *Spirulina platensis* and *Moringa oleifera*, have been reported to inhibit the radiation-induced micronuclei formation *in vivo* and *in vitro* (Jagetia and Aruna, 1997; Jagetia and Baliga, 2002; Qishen *et al.,* 1989; Rao *et al.,* 2001).

Serbian Plants for Radioprotection: An Appraisal

Serbia has a rich repository of medicinal plants that find use in traditional medicine. In traditional Serbian medicine, leaves of *Cornus mas* L. (Cornaceae), as well as fruits of *Crataegus monogyna* Jacq. (Rosaceae), are used as cardioprotective substances, which improve circulation and inhibit platelet aggregation. Aerial parts of *Gentianella austriaca* (A. Kern. & Jos. Kern.) Holub are not in use, but some other species from the Gentianaceae family such as *Gentianella achalensis*, *Gentianella nitida* and *Swertia japonica* find use in folk medicine in Peru, Pakistan, Japan and other Asian countries. *Gentiana lutea* is used in Serbian traditional medicine to reduce stomach pain and improve digestion of food. *Equisetum arvense* L. (Equisetaceae), *Ononis spinosa* L. (Farbaceae) and *Arctostaphylos uva-ursi* L. (Ericaceae) display antibacterial properties and are used for the treatment of urinary tract infections. The diverse medicinal properties attributed to *C. monogyna*, *C. mas*, *G. austriaca*, *E. arvense*, *G. dinarica*, *O. spinosa* and *A. uva-ursi* inspired us to investigate their radioprotective properties.

In vitro Studies

One approach to identify non-toxic and effective radioprotective compounds that can reduce adverse effects of radiation are *in vitro* experiments, using human lymphocytes as a model system and the most explored radiation dose in humans is 2Gy gamma or X-rays (Joksic *et al.,* 2003; Leskovac *et al.,* 2004, 2007). Protective effect of extracts of the following medicinal plants *viz.,* *C. monogyna*, *C. mas*, *G. austriaca*, *E. arvense*, *G. dinarica*, *O. spinosa* and *A. uvae ursi* was examined on cultured human peripheral blood lymphocytes *in vitro*, after irradiation with 2Gy of ^{60}Co γ-rays. Incidence of micronuclei (Fenech, 1993), origin of micronuclei (Becker *et al.,* 1990), apoptosis and lipid peroxidation products were examined. One of the nutraceutics, Gonebazol– a mixture of carefully selected natural compounds, was also examined exploiting the same experimental approach.

Plant Material

Plants were collected from Mt. Maljen in Serbia. The plant material was air-dried, powdered and the total phenolic content was analyzed. *C. monogyna* fruit was rich in procyanidins and flavonoids. In *C. mas* the dominant compounds present in leaves were ellagic and gallic acid, whereas the main constituents of *G. austriaca* aerial parts were γ-pyrones (flavonoids and xanthones) and secoiridoids. Polyphenols found in *G. dinarica* root include xanthones, C-glucoflavones and secoiridoids. In *E. arvense,* the saponoside equisetonin was found, besides flavonoids (isoquercitrin galuteoline and equisetrine) and small amounts of vitamin C, malic acid, oxalic acid, linoleic and oleic acids. *O. spinosa* root was rich in ononin (7-oxy-4-metoxyisoflavonglucoside), whereas triterpene saponosides were found in traces.

In *A. uva-ursi* leaves, the dominant compound present was quercetin. Aqueous-ethanolic extracts of all the plants were used for treatment of irradiated human lymphocytes and amifostine was employed as a positive control.

Results

The effects of various concentrations of water-soluble extracts of these plants on unirradiated human lymphocytes are presented as percents of control (Fig. 4.1). No significant changes in baseline incidence of micronuclei in samples treated with extracts of *C. mas*, *C. monogyna*, *G. austriaca* and *A. uva-ursi* were found, whereas extracts of *G. dinarica*, *E. arvense* and particularly *O. spinosa* significantly enhanced the incidence of micronuclei compared to control.

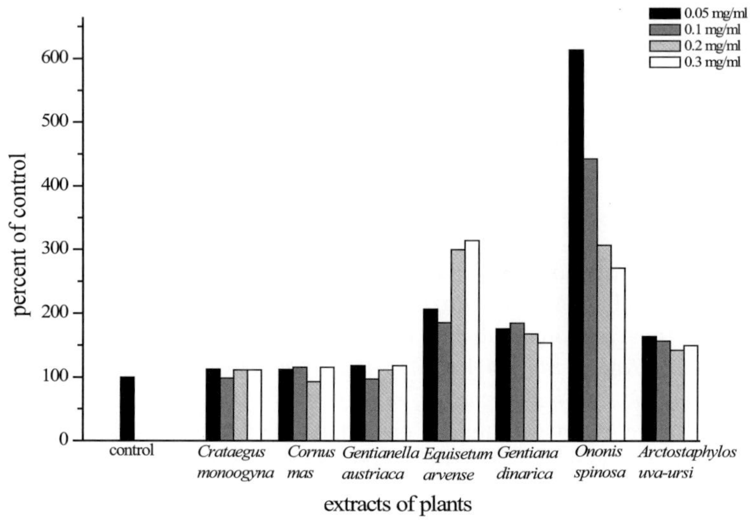

Fig. 4.1. Incidence of baseline micronuclei in unirradiated human lymphocytes treated with different concentrations of plant extracts (expressed as percent of control).

The effects of these plants on irradiated human lymphocytes are presented as percents of control (Fig. 4.2). These include the incidence of radiation-induced micronuclei (MN), cell proliferation index (CBPI), level of malondialdehyde (MDA) and the percent of cells undergoing apoptosis (AP).

C. monogyna and *C. mas* reduced the yield of radiation-induced micronuclei, mainly by stimulating apoptosis (three and two fold, respectively), which is rather a radiorecovery effect. This effect is certainly important, particularly for "long lived" chromosomal aberrations, since it can lead to over expression of oncogenes and cancer development. Biologically active compounds found in these plants enhanced apoptosis of irradiated cells. Therefore, in irradiated tissue, the damaged cells could be removed emphasizing a

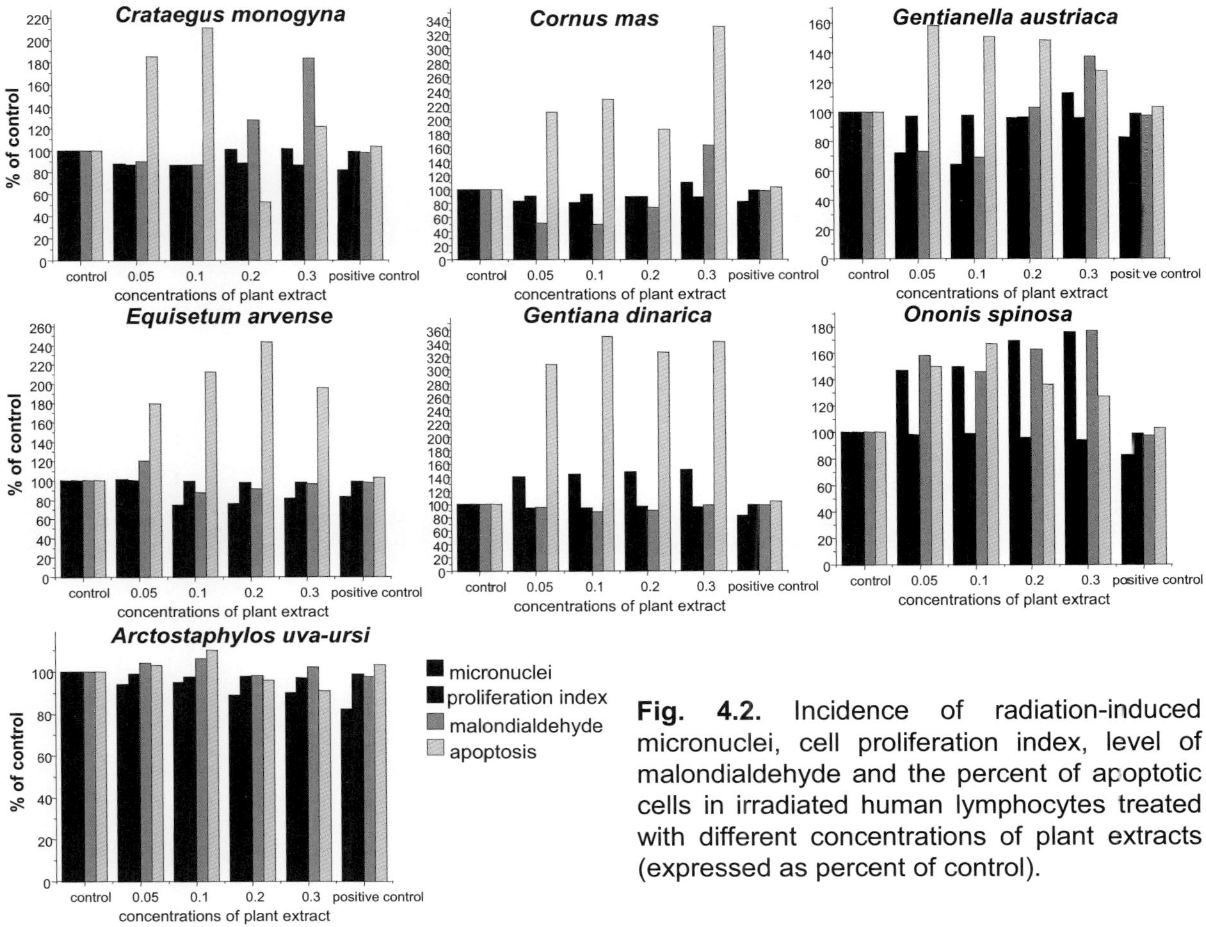

Fig. 4.2. Incidence of radiation-induced micronuclei, cell proliferation index, level of malondialdehyde and the percent of apoptotic cells in irradiated human lymphocytes treated with different concentrations of plant extracts (expressed as percent of control).

physiological mechanism of cell death with no inflammation. Apoptosis is a programmed, active, highly selective mechanism of cell death that facilitates removal of cells that are excessively damaged. Apoptosis starts with a lethally insulted, but otherwise normal, cell, and ends in organized lysis of the cell. Irradiation enhances the level of free radicals, possibly by mediating the balance of the intracellular oxidant/antioxidant status, the level of ATP in the cell, the extent of induced membrane damage, which leads cells to necrosis rather than apoptosis (Petrovic *et al.*, 2005). *C. monogyna* and *C. mas* flavonoids certainly preserve appropriate intracellular oxidant/antioxidant status balance enabling apoptosis.

These two medicinal plants also exhibited a mild suppressive effect, reducing cell proliferation of irradiated cells. Slowing down the proliferation allows more time for the action of DNA repair enzymes. Simultaneously, prolonged time for DNA repair and elimination of heavily damaged cells *via* apoptosis enables faster homeostasis among cells in irradiated tissue.

G. austriaca possesses remarkable protective properties. It acts as a strong antioxidant, significantly reducing the level of radiation-induced micronuclei, slightly enhancing apoptosis, with no perturbation of the cell cycle. Polyphenols of *Gentianella* significantly reduced MDA, which in turn decreased the level of radiation-induced DNA damage. The polyphenols present in *Gentianella* break chain reactions by donating an electron to the peroxyl radical of the fatty acid thereby stopping the propagation steps of lipid peroxidation. In addition, they chelate transition metal ions and, therefore, inhibit free radical formation. Investigation of its biological activity could be important for many diseases where lipid peroxidation products have been reported (Seghrouchni *et al.*, 2002; Das, 2002). Statistically significant positive correlations between the incidence of micronuclei and the level of MDA were found for these three plant extracts. The most demanding was for *G. austriaca.* In harvested human lymphocytes, β-carotene has been reported to prevent X-ray-induced micronuclei (Umegaki *et al.*, 1994), as well as orientin and vicenin (flavonoids of *Ocimum sanctum*) (Vrinda and Uma Devi, 2001). Using the same experimental model system, Jagetia and co-workers reported significant radioprotective effects of *Aegle marmelos* (Jagetia *et al.*, 2003).

E. arvense extract increased the incidence of micronuclei in unirradiated samples, indicating its weak toxicity.

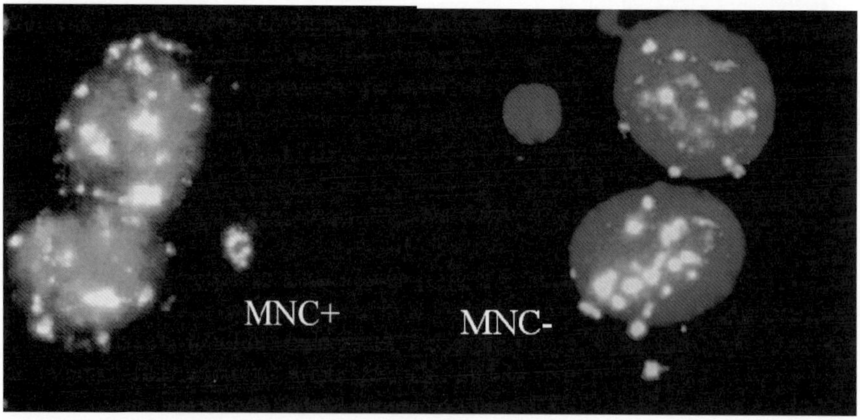

Fig. 4.3. Examples of centromeric positive and centromeric negative micronulei.

In all the treated samples centromeric negative micronulei (MNC-) prevailed as centromeric positive micronulei (MNC+) signifying clastogenic mode of micronuclei origin (Fig. 4.3).

Treatment of irradiated cells with the extract of *E. arvense* lowered the incidence of radiation-induced micronuclei at all the employed concentrations, which was not associated with mitotic inhibition. *E. arvense* extract slightly reduced levels of MDA and significantly enhanced the percent of cells undergoing apoptosis. There is a need to examine the biologically active compound(s) that are responsible for lowering micronuclei after irradiation.

Extract of *G. dinarica* exhibited clastogenic properties i.e., it significantly increased the incidence of radiation-induced MN, which was accompanied with a decrease in the proliferative capacity of the cells and enhancement of apoptosis. Evaluation of the origin of micronuclei showed that MNC- were dominant, indicating that the micronuclei originated by clastogenic mode of action.

O. spinosa behaves as a strong radiosensitizer. Strong clastogenic properties were evidenced as a 5–6-fold increase of the micronuclei incidence in unirradiated sample, as compared to control. Majority of the counted micronuclei were MNC- indicating that they originate by clastogenic mode of action. Treatment of irradiated samples with *O. spinosa* potentiated the micronucleus frequency up to 1.7 fold. Majority of the counted micronuclei were MNC- (77.5%), signifying that they originate by clastogenic mode of action. The mechanism underlying such potentiation of the micronuclei is not fully understood. It is possible that treatment with this extract converts the excision-reparable lesions to micronuclei. The extract of *O. spinosa* contains ononin (7-oxy-4-metoxyisoflavonglucoside) and triterpene saponosides. These components or ononin alone bring about such effects. The other possibility is that ononin enhances the intensity of oxidative base damage in DNA, which gets converted to micronuclei. In samples treated with the highest concentration of *O. spinosa* a portion of binucleated cells with apoptotic figures was observed suggesting that chromosomal rearrangements reached very high incidence that triggers apoptosis. The extract behaves as a strong radiosensitizer and possibly can be used for potentiation of radiotherapy.

A. uva-ursi is seen as a biologically inactive extract. It did not affect the incidence of spontaneously occurring or radiation-induced micronuclei, which could be due to quick degradation of the active compounds. *In vitro* effects of the nutraceutic (Gonebazol) is seen as a very efficient product for protection against ionizing radiation. It reduced the radiation-induced micronuclei by 69%, and malondialdehyde by 50%, significantly enhancing apoptosis. Its protective properties were almost three fold better when compared with amifostine (Fig. 4.4).

The nutraceutic Gonebazol ("Biofarm", Belgrade) consists of plant extracts (*Laminaria digitata* (Huds.) J.V. Lamour. (Family: Phaeophyceae): 1mg; *Echinacea purpurea* (L.) Moench (Family: Asteraceae): 75mg; Germanium-132: 0.50mg; β-carotene: 0.10mg; micronized clinoptiolite: 75mg; trimethylglycine: 2g; piridoxin: 0.50mg; Tocopherol: 30mg; vitamin C: 12mg; pollen; propolis and Ca^{++}, all preserved in honey.

In vivo Studies

Although *in vitro* investigations are a faster method for obtaining information on biological activity, *in vivo* effects can hardly be extrapolated based on the findings of *in vitro* investigations. Our *in vivo* studies focused on the chronic professional exposure of invasive cardiologists, who represent the extreme far end (with the highest exposure) of the spectrum of medical radiation workers (Joksic *et al.,* 2006).

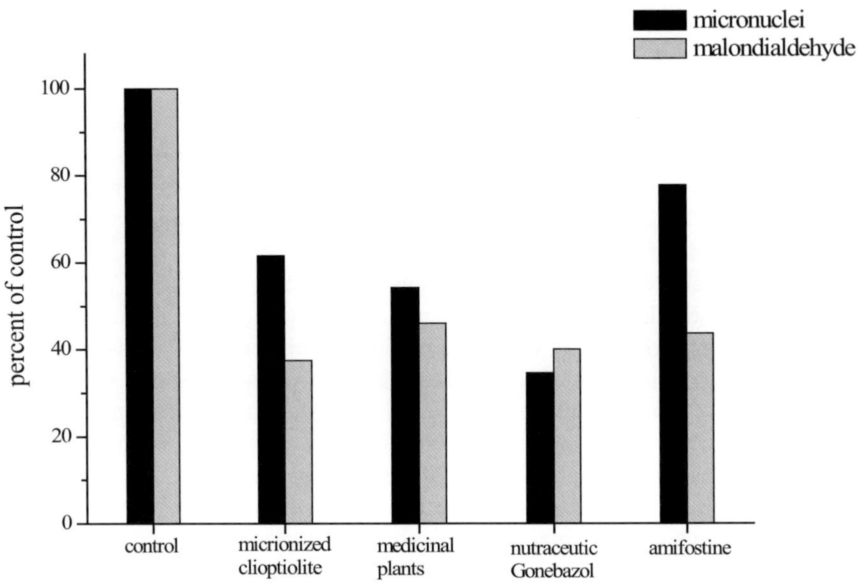

Fig. 4.4. Incidence of micronuclei and level of malondialdehyde in irradiated human lymphocytes treated with micronized clinoptiolite, Serbian medicinal plants and the nutraceutic Gonebazol (expressed as percent of control).

Invasive cardiologists have two– to three times higher exposure to ionizing radiation per head per year as compared to other radiologists (Cordella *et al.,* 2000). The source of exposure is medical X-rays, which are about 4-fold more mutagenic per cGy than A-bomb radiation and gamma rays (ICRP, 2005).

Previous investigations have shown that 25% of the cardiologists performing catheterization carry dicentrics in their lymphocytes (Andreassi *et al.,* 2005). In most of them a significantly decreased ability of leukocytes to undergo apoptosis was observed (Joksic and Petrovic, 2004). Invasive cardiologists are a large and expanding population in Europe. Low and protracted dose radiation exposure gives a small yet excess risk of cancer, besides additional ill-defined non-cancer health risks for atherosclerotic disease. Proof of a link between radiation exposure and adverse health effects is made even more difficult by two major factors, possibly diluting the association between physical dose exposure and clinical damage:

(i) individual variability (due to genetic and environmental factors) and
(ii) the presence of defensive responses by which oxidative stress conditioning from irradiation elicits a response against damage to subsequent stress such as challenging irradiation (Miura, 2004).

Individual variability and poorly understood adaptive mechanisms may further weaken the link between physical dose and observed damage. The currently available biomonitoring system is based mostly on the cytogenetic markers of chromosomal aberrations and micronuclei in circulating lymphocytes. These biomarkers may be useful to identify individuals more vulnerable to radiation damage who might represent the target of preventive measures (by radiation sparing policies or attempts at pharmacological or dietary radioprotection).

For that purpose we evaluated radioprotective properties *in vivo,* of the previously *in vitro* examined nutraceutical Gonebazol, which is composed of natural compounds ["Biofarm", Belgrade]. Members of medical staff performing invasive radiological diagnostics, who were identified as carrying dicentrics in their lymphocytes, were advised to consume Gonebazol for three weeks at a dosage of 20g daily (2 x 2 tea spoon dissolved in a glass of water). Lymphocyte aberrations, micronuclei, apoptosis and haematological parameters were analyzed prior to (day 0) and after 10 and 21 days of the treatment. Apoptosis of leukocytes was analyzed at the beginning and at the end of the treatment. Out of 47 subjects 12 were identified as carrying dicentric chromosomes in their lymphocytes. Lymphocyte aberrations in these subjects were accompanied by haematological disorders (mild leucopenia or leukocytosis, and an imbalance in the ratio between lymphocytes, monocytes and granulocytes). Ten of them, all women, were interested in further investigation and were willing to consume the nutraceutic for 3 weeks at a dosage of 20g daily. This group was 38.7 ± 3.4 years old on the average, with an average duration of occupational exposure of 14.6 ± 3.4 years. The radiation doses measured by TLD dosimeters on the chest were below the annual limits of 20mSV.

Parameters under consideration

Haematological parameters (red blood cell number (RBC), white blood cell number (WBC) and counts of neutrophils (NpG), lymphocytes (Ly) and monocytes (Mo)) were analyzed on the haematology analyzer CELL-DYN 3700. Chromosomal aberrations were analyzed as described by the International Atomic Energy Agency (IAEA, 2001), whereas for micronuclei the method described by Fenech (1993) was followed. Apoptosis of leukocytes was analyzed according to the method of Crompton and Ozsahin (1997) (Fig. 4.5).

Individual data on leukocyte parameters and lymphocyte findings are listed in Table 4.1. These are numbers of leukocytes and percentage and absolute number of lymphocytes, granulocytes and monocytes. The sum of exchange aberrations (dicentrics or rings) and excess acentrics is expressed as the incidence of breakages per cell. Lymphocyte micronuclei are expressed as incidence of micronuclei per binucleated cell. Apoptosis is expressed as the percentage of cells displaying apoptotic granularity.

At the beginning of the treatment, the absolute number of leukocytes correlated positively with the incidence of chromosomal aberrations, and negatively with micronuclei and apoptosis, although the results were statistically insignificant. At the end of the treatment, a statistically significant correlation was found between the absolute number of granulocytes and incidence of lymphocyte micronuclei ($r = 0.68$, $p < 0.05$).

Table 4.1. Leukocyte parameters and lymphocyte findings (chromosome aberrations and micronuclei) in radiation workers at the beginning (I) and after three weeks of treatment (II).

Subject number		No of leukocytes	Lymphocyets %	Granulocytes %	Monocytes %	Chromosome aberrations	Incidence of breakage per cell	Incidence of MN per cell	Percent of apoptotic leukocytes
1.	I	5500	45.00	44.10	10.90	1 dic, 1 AF, 1 chr break	0.025	0.033	55.00
	II	5400	56.80	34.40	8.76	2 ring, 1 chr break	0.025	0.028	62.60
2.	I	7800	41.10	54.70	4.20	2 dic, 1AF, 1 chr break	0.035	0.039	11.70
	II	5500	27.60	61.20	11.20	-	0.000	0.038	18.80
3.	I	5500	45.00	44.00	11.00	1 dic, 1 AF, 1 chr breaks	0.025	0.033	26.10
	II	5400	56.80	34.40	8.76	1 pericen inv, 1 ring, 1 chr break	0.025	0.028	44.70
4.	I	7820	41.20	54.60	4.20	1 dic, 1 AF, 3 chr breaks	0.035	0.039	53.50
	II	5500	27.60	61.20	11.20	-	0.000	0.038	59.30
5.	I	5300	37.40	56.00	6.60	1 AF, 1 chr break	0.015	0.024	52.90
	II	4900	38.70	52.20	9.06	1 AF	0.007	0.010	80.20
6.	I	4800	44.90	50.90	4.20	1 dic, 1 chr break	0.015	0.045	67.20
	II	5600	48.20	44.70	7.11	1 chr break	0.005	0.018	70.10
7.	I	4200	51.10	43.80	5.10	1 dic, 1 chr break	0.015	0.038	28.00
	II	4600	60.20	34.00	5.85	-	0.000	0.011	33.80
8.	I	7900	40.50	55.70	3.80	1 AF	0.010	0.016	81.20
	II	7700	47.40	46.80	5.79	1 chr break	0.005	0.034	80.30
9.	I	2800	47.20	48.40	4.40	1 AF, 1 chr break	0.015	0.025	91.00
	II	3100	47.40	43.90	8.74	-	0.000	0.011	96.50
10.	I	7100	36.40	60.10	3.50	1 sym interch, 1 inv, 1 chr break	0.025	0.023	68.70
	II	4600	40.40	51.10	8.49	-	0.000	0.021	67.30

dic-dicentric chromosome; ring-ring chromosome; AF-acentric fragment; chr break-chromosome break; inv-chromosome inversion; pericen inv- pericentric chromosome inversion; sym interch-symmetrical chromosome interchange

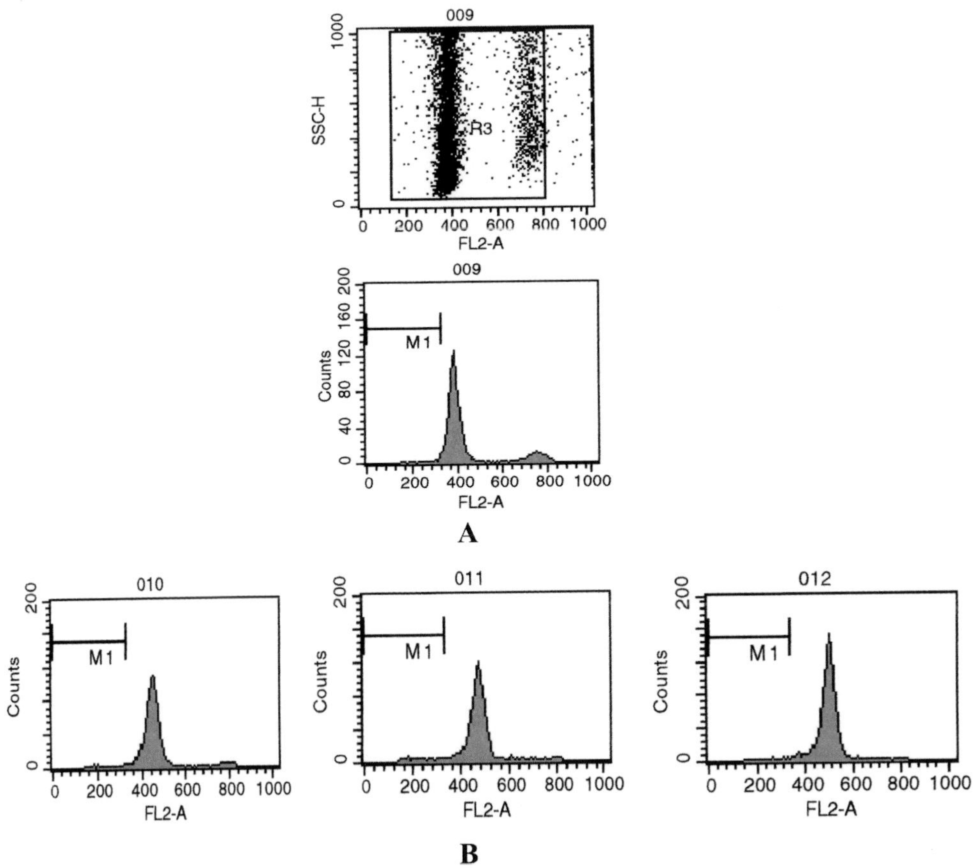

Fig. 4.5. An example of hypodiploid flow cytometric analysis: **A**. control sample, **B**. sample treated with plant extract.

An inverse correlation at the border of significance was found between the number of monocytes and percentage of leukocytes dying via apoptosis (r = –0.59, $p < 0.06$). Statistically significant differences were found between the incidence of micronuclei at the beginning versus the end of treatment, between the percentage of apoptotic leukocytes at the beginning and at the end of treatment. The difference between the number of monocytes was at the border of significance.

Six out of ten persons involved in this study carried dicentrics and rings in their lymphocytes and leukocyte formula disturbances (mild granulocytosis, accompanied by mild lymphopenia). After 10 days of consuming the nutraceutic Gonebazol, the number of leukocytes slightly increased in the treated subjects; particularly an increase in lymphocytes and monocytes was observed, whereas the number of granulocytes remained almost the same. Concurrently, the incidence of chromosomal aberrations was slightly higher. In some subjects, "long lived" chromosomal exchanges appeared in peripheral blood lymphocytes, which was not found at the beginning of the investigation. These chromosomal aberrations could represent a consequence of the cumulative effects of chronic exposure to low doses of ionizing radiation. It is well known that radiation-induced chromosomal aberrations persist in peripheral lymphocytes for as long as each lymphocyte lives. After three weeks, the chromosomal aberrations disappeared from peripheral blood lymphocytes in 9 out of 10 persons, whereas incidence of micronuclei decreased by 50–80 percent when compared with the status at the beginning of treatment. The number of leukocytes decreased, but still stayed in physiological range. The absolute number of granulocytes was lower, whereas an increased number of monocytes, achieved after 10 days of treatment, remained unchanged.

The adverse health effects of ionizing radiation are still a very important field of investigation. The scientific community in the field of radiation protection agrees that the cumulative effects of chronic exposure to ionizing radiation cannot be predicted or totally avoided. DNA and chromosomal changes can give rise to malignancies either by activation of oncogenes or by the loss of suppressor genes.

Radiation can activate oncogenes through a number of mechanisms, including point mutation, chromosomal rearrangement or chromosomal translocations. Low doses of ionizing radiation induce oxidative stress, which can significantly modulate some transcription factors. Once activated, these transcription factors might drive transcription of survival-related proteins. For instance, ionizing radiation modulates more than 100 genes in human endothelial cells, causing up-regulation of genes involved in coagulation and peroxidase activity (Lanza *et al.,* 2005). It also triggers production of signals that mobilize the innate and adaptive immune systems to deal with the intrusion and effect tissue repair with the goal of maintaining the integrity of the tissue and the body. Little is known about the role of signals in tissue responses to this agent. This signal could mediate the pathogenesis of or recovery from radiation damage and alter intrinsic cellular radiosensitivity. This study has shown that the nutraceutics, consisting of natural products rich in flavonoids and methyl-donating compounds significantly reduce the incidence of radiation-induced chromosomal aberrations and micronuclei in persons occupationally overexposed to ionizing radiation. Homeostasis among leukocytes and stability of the genome were obtained in a short period of time (3 weeks). No side effects were observed, so we conclude that physiological mechanisms were augmented in the elimination of accumulated DNA damage. Powerful radioprotective capacities partly are achieved by the scavenging ability of vitamins. Biologically active compounds from *Laminaria digitata* like oligoglucan laminarin and vitamin B12, which display immunomodulatory properties, enhance monocyte-macrophage activity and potentate apoptosis. Another component present in the nutraceutic Gonebazol is micronized clinoptiolite, which represents the source of oxygen needed for oxidative modification of proteins and lipids in cells undergoing apoptosis. Safe and powerful radioprotective nutraceuticals should be applied in clinical practice for patients undergoing interventional procedures with long screening periods and multiple image acquisitions because doctors are unaware of the radiobiological risk and grossly underestimate the doses of radiation even for the most commonly requested radiological examinations (Shiralkar *et al.,* 2003). Research should decrease the uncertainty associated with the potential health effects of low doses of radiation. This is especially important in Europe, where each citizen receives on an average the radiation equivalent of at least 100 chest X-rays per head per year (data refers to the radiological year 1997 in Germany; Regulla *et al.,* 2003).

The present investigation has pointed out once more the importance of developing therapeutic formulations with appropriate proportions of the bioactive constituents in order to achieve desired beneficial effects (radioprotection).

Acknowledgements

The authors are thankful to Director of the Institute for Medicinal Plant Research "Dr. Josif Pancic" and "Biofarm" Belgrade for providing plants and nutraceutical products for research, and the Serbian Ministry of Science and Environmental Protection, Project No. 143046 for financial support.

References

Andreassi, M.G., Cioppa, A., Botto, N., Joksic, G., Manfredi, S., Federici, C., Ostojic, M., Rubino, P. and Picano, E. (2005) Increased micronucleus frequency in interventional cardiologists: A case–control study. *FASEB Journal* 19, 998–999.

Arora, R. (2006) Phytodefence Molecules for Defence Forces: Utility and Prospects. In: Gupta, M., Kumar, I.P. and Arora, R. (eds) *Herbs for Health: Application in Armed Forces*. Publisher: Institute of Nuclear Medicine and Allied Sciences, Delhi, pp. 58–62.

Arora, R. and Goel, H.C. (2000) Herbal radioprotectors. *Proceedings of the International Conference on Radiation Biology (Radiobiology 2000)* Organized under the aegis of Indian Society for Radiation Biology by Regional Cancer Centre (RCC), Thiruvananthapuram, India, from 17–19 Feb, 2000, p. 87.

Arora, R., Chawla, R., Prasad, J., Singh, S., Kumar, R., Sharma, A., Sagar, R.K., Singh, S. and Sharma, R.K. (2005c) Radioprotective activity of *Rhodiola imbricata*- a high altitude medicinal plant. *Molecular and Cellular Biochemistry* 273, 209–223.

Arora, R., Chawla, R., Puri, S.C., Sagar, R.K., Singh, S., Kumar, R., Sharma, A.K., Prasad, J., Singh, S., Kaur, G., Chaudhary, P., Qazi, G.N. and Sharma, R.K. (2005b) Radioprotective activity of low altitude *Podophyllum hexandrum*. *Journal of Environmental Pathology, Toxicology and Oncology* 24(4), 299–314.

Arora, R., Chawla, R., Singh, S., Kumar, R., Sharma, A., Puri, S.C., Sinha, A.K., Tripathi, R.P. and Sharma, R.K. (2006a) Radioprotection by Himalayan high-altitude region plants. In: Sharma, R.K. and Arora, R. (eds) *Herbal Drugs: A Twenty First Century Perspective*. Jaypee Brothers Medical Publishers Private Ltd., New Delhi, India, pp. 301–325.

Arora, R., Chawla, R., Singh, S., Sagar, R.K., Kumar, R., Sharma, A., Prasad, J., Singh, S., Gurudutta, G.U. and Sharma, R.K. (2006b) Bioprospection for radioprotective molecules from indigenous flora. In: Govil, J.N., Singh, V.K. and Bhardwaj, R. (eds), *Recent Progress in Medicinal Plants*. Vol. 16, Studium Press, LLC, USA, pp. 179–219.

Arora, R., Gupta, D., Chawla, R., Sagar, R., Sharma, A.K., Prasad, J., Singh, S., Samanta, N. and Sharma, R.K. (2005a) Radioprotection by plant products: present status and future prospects. *Phytotherapy Research* 9, 1–22.

Arora, R., Gupta, D., Prasad, J., Sharma, A.K., Singh, S., Sagar, R., Samanta, N., Kumar, R. and Goel, H.C. (2003) Bioprospection of Himalayan high-altitude region flora for radioprotective biomolecules. *Proceedings of the National Seminar on New Millennium Strategies for Quality, Safety and GMPs for Herbal Drugs/Products*, National Botanical Research Institute, Lucknow, India, p. 51.

Arora, R., Sagar, R., Singh, S., Kumar, R., Sharma, A.K., Prasad, J., Singh, S., Gupta, M., Sharma, R.K., Puri, S.C., Krishna, B., Siddiqui, M.S., Lahiri, S.S., Tripathi, R.P., Qazi, G.N. (2007a) Cytoprotective effect of *Podophyllum hexandrum* against gamma radiation is mediated via hemopoietic system stimulation and up-regulation of haeme-oxygenase-1 and the pro-survival multidomain protein Bcl-2. *Integrative Cancer Therapies* 6(1), 54–65.

Arora, R., Singh, S., Puri S.C. and Sharma, R.K. (2008) Himalayan Mayapple (*Podophyllum hexandrum* Royle): Traditional Uses, Clinical Indications and Future Prospects. In: Watson, R.R. and Preedy, V.R. *The Encyclopedia of Herbal Medicine in Clinical Practice*, CABI Publishing, Wallingford, Oxon, United Kingdom (in press).

Arora, R., Singh, S., Sagar, R.K., Chawla, R., Kumar, R., Puri, S.C., Singh, S., Prasad, J., Gupta, M.L., Krishna, B., Siddiqui, S.M., Sharma, A.K., Tripathi, R.P., Qazi, G.N. and Sharma, R.K. (2007b) Radiomodulatory and free radical scavenging activity of the fractionated aquo–alcoholic extract of the adaptogenic nutraceutical (*Rhodiola*)– A comparative *in vitro* assessment with ascorbate. *Journal of Herbal Pharmacotherapy* 7(4).

Bagchi, D., Bagchi, M., Stohs, S., Das, D., Ray, S. and Kuszynski, C. (2000) Free radicals and grape proanthocyanidin extract: importance in human health disease and prevention. *Toxicology* 148, 187–197.

Becker, P., Scherthan, H. and Zankl, H. (1990) Use of a centromere–specific DNA probe (p82H) in nonisotopic *in situ* hybridisation for classification of micronuclei. *Genes, Chromosomes and Cancer* 2, 59–62.

Benamira, M., Johnson, K., Chaudhary, A., Bruner, K., Tibbetts, C. and Marnett, L.J. (1995) Induction of mutations by replication of malondialdehyde-modified M13 DNA in *Escherichia coli*: determination of the extent of DNA modification, genetic requirements for mutagenesis, and types of mutations induced. *Carcinogenesis* 16, 93–99.

Bonassi, S., Znaor, A., Ceppi, M., Lando, C., Chang, W.P., Holland, N., Kirsch-Volders, M., Zeiger, E., Ban, S., Barale, R., Bigatti, M.P., Bolognesi, C., Cebuleska–Wasilewska, A., Fabianova, E., Fucic, A., Hagmar, L., Joksic, G., Martelli, A., Migliore, L., Mirkova, E., Scarfi, M.R., Zijno, A., Norppa, H. and Fenech, M. (2007) An increased micronucleus frequency in peripheral blood lymphocytes predicts the risk of cancer in humans. *Carcinogenesis* 28(3), 625–631.

Bonassi, S., Znaor, A., Norppa, H. and Hagmar, L. (2004) Chromosomal aberrations and risk of cancer in humans: an epidemiologic perspective. *Cytogenetic and Genome Research* 104(1–4), 376–382.

Chawla, R., Arora, R., Kumar, R., Sharma, A.K., Prasad, J., Singh, S., Sagar, R.K., Chaudhary, P., Shukla, S., Kaur, G., Sharma, R.K., Puri, S.C., Handa, G., Gupta, V.K. and Qazi, G.N. (2005a) Antioxidant activities of fractionated extracts of high altitude *Podophyllum hexandrum* in radiation protection. *Molecular and Cellular Biochemistry* 273, 193–208.

Chawla, R., Arora, R., Puri, S.C., Kumar, R., Sharma, A., Prasad, J., Singh, S., Sagar, R., Kaur, G. and Sharma, R.K. (2004) A comparative study of the antioxidant potential of various fractions of *Podophyllum hexandrum vis-à-vis* radiation protection. *Indian Journal of Radiation Research* 1(2), 40–41

Chawla, R., Arora, R., Sagar, R.K., Singh, S., Puri, S.C., Kumar, R., Singh, S., Sharma, A.K., Prasad, J., Khan, H.A., Sharma, R.K., Dhar, K.L., Spiteller, M. and Qazi, G.N. (2005b) 3-O-β-D- galactopyranoside of quercetin as an active principle from high–altitude *Podophyllum hexandrum* and evaluation of its radioprotective properties. *Zeitschrift für Naturforschung* 60C (9/10), 728–738.

Chawla, R., Arora, R., Singh, S., Sagar, R.K., Sharma, R.K., Kumar, R., Sharma, A., Tripathi, R.P., Puri, S.C., Khan, H.A., Shawl, A.S., Sultan, P., Krishan, T. and Qazi, G.N. (2006) *Podophyllum hexandrum* offers radioprotection by modulating free radical flux: role of aryl-tetralin lignans. *Evidence-based Complementary and Alternative Medicine* 3(4), 503–511.

Chawla, R., Arora, R., Singh, S., Sagar, R.K., Sharma, R.K., Kumar, R., Sharma, A., Gupta, M.L., Singh, S., Prasad, J., Khan, H.A., Swaroop, A., Sinha, A.K., Gupta, A.K., Tripathi, R.P. and Ahuja, P.S. (2007) Radioprotective and antioxidant activity of fractionated extracts of berries of *Hippophae rhamnoides*. *Journal of Medicinal Food* 10(1), 101–109.

Cordella, J., Faulkner, K., Hopewell, J., Nakamura, H., Rehani, M., Rosenstein, M., Sharp, C., Shope, T., Vano, E., Worgul, B. and Wuchere, M. (2000) Avoidance of radiation injuries from medical interventional procedures. Extracted from *International Commission on Radiaton Protection. ICRP Publication* 85; Ann ICRP 30 (2).

Crompton, N.E. and Ozsahin, M. (1997) A versatile and rapid radiosensitivity assay of peripheral blood leukocytes based on DNA and surface–marker assessment of cytotoxicity. *Radiation Research* 147, 55–60.

Das, U. (2002) A radical approach to cancer. *Medical Science Monitor* 8, 79–92.

Fenech, M. (1993) The cytokinesis-block micronucleus technique: a detailed description of the method and its application to genotoxicity studies in human populations. *Mutation Research* 285, 35–44.

Gupta, D., Arora, R., Garg, A.P. and Goel, H.C. (2003) Radiation protection of HepG2 cells by *Podophyllum hexandrum. Molecular and Cell Biochemistry* 250, 25–40.

International Atomic Energy Agency (2001) Cytogenetic analysis for Radiation Dose Assessment. Technical Report Series No. 405, Vienna, Austria.

International Commission on Radiological Protection (2005) *ICRP Recommendations.* Chapter 2, Part 7. Recommendation Section 3.4.1.

Jagetia, G.C. and Aruna, R. (1997) The herbal preparation abana protects against the radiation-induced micronuclei in mouse bone-marrow. *Mutation Research* 391, 157–163.

Jagetia, G.C. and Baliga, M.S. (2002) *Syzygium cumini* (jamun) reduces the radiation–induced DNA damage in the cultured human peripheral blood lymphocytes: a preliminary study. *Toxicology Letters* 132, 19–25.

Jagetia, G.C., Venkatesh, P. and Baliga, M.S. (2003) Evaluation of the radioprotective effect of *Aegle marmelos* (L.) Correa in cultured human peripheral blood lymphocytes exposed to different doses of γ-radiation: a micronucleus study. *Mutagenesis* 18, 387–393.

Joksic, G. and Petrovic, S. (2004) Lack of adaptive response of human lymphocytes exposed *in vivo* to low doses of ionizing radiation. *Journal of Environmental Pathology, Toxicology and Oncology* 3 (23), 195–206.

Joksic, G., Ilic, N. and Spasojevic-Tisma, V. (2006) Radioprotective properties of nutraceutical Gonebazol: *In vivo* study. *Archive of Oncology* 14(1–2), 15–18.

Joksic, G., Stankovic, M. and Novak, A. (2003) Antibacterial medicinal plants *Equiseti herba* and *Ononidis radix* modulate micronucleus formation in human lymphocytes *in vitro. Journal of Environmental Pathology, Toxicology and Oncology* 22(1), 41–48.

Kumar, R., Singh, P.K., Arora, R., Sharma, A., Prasad, J., Sagar, R.K., Singh, S. and Sharma, R.K. (2005a) Radioprotection by *Podophyllum hexandrum* in the liver of mice: A mechanistic approach. *Environmental Toxicology and Pharmacology* 20, 326–334.

Kumar, R., Singh, P.K., Sharma, A., Prasad, J., Sagar, R., Singh, S., Arora, R. and Sharma, R.K. (2005b) *Podophyllum hexandrum* extract provides radioprotection by modulating the expression of proteins associated with apoptosis role in radioprotection. *Biotechnology and Applied Biochemistry* 42(1), 81–92.

Lanza, V., Pretazzoli, V., Olivieri, G., Pasccarella, G., Panconesi, A. and Negri, R. (2005) Transcriptional response of human umbilical vein endothelial cells to low doses of ionizing radiation. *Journal of Radiation Research* 46, 265–276.

Lata, M., Prasad, J., Singh, S., Kumar, R., Singh, L., Choudhary, P., Arora, R., Chawla, R., Tyagi, S., Soni, N.L., Sagar, R.K., Devi, M., Sharma, R.K. and Puri, S.C. (2007) Whole-body protection against lethal ionizing radiation in mice by REC-2001: A semi–purified fraction of *Podophyllum hexandrum.* Phytomedicine (epub advance access).

Leskovac, A., Joksic, G. and Petrovic, S. (2007) Radiprotective effects of medicinal plants' polyphenols on human lymphocytes *in vitro. Proceedings of the European Journal of Human Genetics* 15(1), 119.

Leskovac, A., Krinulovic, K. and Joksic, G. (2004) Polyphenols isolated from *Gentianella austriaca* and *Gentianella dinarica* significantly modulate micronuclei formation in human lymphocytes *in vitro. Proceedings of the 6th Balkan Meeting on Human Genetics (BMHG),* Thessaloniki, Greece, p. 107.

Miura, Y. (2004) Oxidative stress, radiation–adaptive responses, and aging. *Journal of Radiation Research* 45, 357–372.

Nijveldt, R.J., van Nood, E., van Hoorn, D.E.C., Boelens, P.G., van Norren, K. and van Leeuwen, P.A.M. (2001) Flavonoids: a review of probable mechanisms of action and potential applications. *American Journal of Clinical Nutrition* 74, 418–425.

Patt, H.M., Tyree, E.B., Straube, R.L. and Smith, D.E. (1949) Cysteine protection against X-irradiation. *Science* 110, 213–214.

Petrovic, S., Leskovac, A. and Joksic, G. (2005) Positive correlation between micronuclei and necrosis of lymphocytes in medical personnel occupationally exposed to ionizing radiation. *Archive of Oncology* 13(2), 65–68.

Qishen, P., Guo, B.J. and Kolman, A. (1989) Radioprotective effects of extract from *Spirulina plantenis* in mouse bone–marrow cells studied by using the micronucleus test. *Toxicology Letters* 48, 165–169.

Rao, A.V., Devi, P.U. and Kamath, R. (2001) *In vivo* radioprotective effect of *Moringa oleifera* leaves. *Indian Journal of Experimental Biology* 39, 858–863.

Regulla, D., Griebel, J., Nosske, D., Bauer, B. and Brix, G. (2003) Acquisition and assessment of patient exposure in diagnostic radiology and nuclear medicine. *Zeitschrift für Medizinische Physik* 13, 127–135.

Sagar, R.K., Chawla, R., Arora, R., Singh, S., Krishna, B., Sharma, R.K., Puri, S.C., Singh, P., Kumar, R., Sharma, A.K., Singh, S., Prasad, J., Gupta, V., Ahmed, B., Dhar, K.L., Khan, H.A., Gupta, M.L. and Qazi, G.N. (2006) Protection of the hemopoietic system by *Podophyllum hexandrum* against gamma radiation-induced damage. *Planta Medica* 72, 114–120.

Seghrouchni, I., Drai, J., Bannier, E., Riviere, J., Calmard, P., Garcia, I., Orgiazzi, J. and Revol, A. (2002) Oxidative stress parameters in type I, type II and insulin–treated type 2 diabetes mellitus; insulin treatment efficiency. *Clinica Chimica Acta* 321, 89–96.

Shiralkar, S., Rennie, A., Snow, M., Galland, R.B., Lewis, M.H. and Gower–Thomas, K. (2003) Doctor's knowledge of radiation exposure: questionnaire study. *British Medical Journal* 327, 371–372.

Swaroop, A., Sinha, A.K., Chawla, R., Arora, R., Sharma, R.K. and Kumar, J.K. (2005) A new triglyceride with potent antioxidant activity isolated from fruits of *Hippophae rhamnoides*. *Chemical and Pharmaceutical Bulletin* 53(8), 1021–1024.

Thresiamma, K.C., George, J. and Kuttan, R. (1998) Protective effect of curcumin, ellagic acid and bixin on radiation induced genotoxicity. *Journal of Experimental and Clinical Cancer Research* 17, 431–434.

Umegaki, K., Ikegami, S., Inoue, K., Ichikawa, T., Kobayashi, S., Soeno, N. and Tomabechi, K. (1994) Beta-carotene prevents X-ray induction of micronuclei in human lymphocytes. *American Journal of Clinical Nutrition* 59(2), 409-412.

Vrinda, B. and Uma Devi, P. (2001) Radiation protection of human lymphocyte chromosomes *in vitro* by orientin and vicenin. *Mutation Research* 498, 39–46.

Wood, R.D., Mitchell, M., Sgourous, J. and Lindahl, T. (2002). Human DNA repair genes. *Science* 291, 1284–1289.

Chapter 5

Phytoceuticals for Radioprotection with Special Reference to Egyptian Flora

N.M. Abdel-Hamid

Introduction

Many plants have been reported to have a radioprotective benefit against radiation, when used as adjuvant therapy to radiation modalities. This chapter focuses on select herbals, mainly from the Egyptian flora, that hold promise as radiomodulatory drugs. These radioprotective herbs include:

Scutellaria baicalensis Georgi (Family: Lamiaceae)

Scutellaria baicalensis (Baical skull cap) has been shown to suppress the formation of phosphatidylcholine hydroperoxide in the liver, lung and kidney (Lim *et al.,* 1999) and possesses strong anticancer activity. It strongly inhibits cell growth in all cancer cell lines tested, including prostate and breast cancer cells, which are more sensitive than other types of cancer cells. It also inhibits prostaglandin E (PGE2) production, indicating that suppression of tumour cell growth may be due to its ability to inhibit cyclooxygenase (COX-2) activity. This study supports the notion of using *Scutellaria baicalensis* as a novel anticancer agent to treat various cancers, and would most likely be beneficial as a radioprotector (Ye *et al.,* 2002). Dry roots of *Scutellaria baicalensis* are used, especially as alcohol extracts. Flavonoids isolated from *Scutellaria* also have beneficial effects in hepatitis. Flavonoids derived from *Scutellaria baicalensis* possess antioxidative, antineoplastic and cardiomyocyte-protective activity (Kowalczyk *et al.*, 2006).

Baicalein (5,6,7-trihydroxy-2-phenyl-4H-1-benzopyyran-4-one) is a naturally occurring flavone that is present in various *Scutellaria* species *viz.*, *Scutellaria rivularis*, *S. radix* and *S. baicalensis*. Baicalein possesses several therapeutic properties, including antioxidant, antiallergic, antiproliferative, antitumour and antigenotoxic properties (Gao *et al.,* 1996; Shieh *et al.,* 2000). The radioprotective effects of baicalein (protection of mitochondrial membrane and various mitochondrial enzymes such as succinate dehydrogenase, superoxide dismutase, glutathione peroxidase and glutathione reductase) against radiation damage have been reported by Tilak and Devasagayam (2004). Baicalein (5M) was shown to protect against gamma radiation-induced single strand breaks in DNA.

Coptis chinensis (Coptis) (Family: Ranunculaceae)

It is commonly used in Chinese medicine. Its whole extract potentially inhibits the growth of gastric, breast and colon cancers by directly suppressing cyclin B1 protein expression and inhibits cyclin-dependant kinase 1 activity. *Coptis chinensis* also showed radioprotective activity (Li *et al.,* 2000). Interestingly, the two major compounds of *C. chinensis*, berberine and coptisine, showed a strong inhibition on the proliferation of both hepatoma and leukemia cell lines (Ching *et al.,* 2004).

Gardenia jasminoides Ellis (Family: Rubiaceae)

The fruit extract of *Gardenia* showed detoxification capacity, antioxidant and anticarcinogenic properties. Geniposide (an alkaloid isolated from Gardenia seeds) increased the activity of glutathione S-transferase in hepatocytes of rat liver cell line (Kuo *et al.*, 2004). A new iridoid (gardaloside) and a new safranal-type monoterpene (jasminoside G) together with ten known compounds, including nine iridoids and a second safranal-type monoterpene were isolated from the fruits of *Gardenia jasminoides*. The structures of these compounds have been established. Of these compounds, geniposide, 6 alpha-hydroxygeniposide, ixoroside and shanzhiside showed significant inhibition of IL-2 secretion and anti-CD28 monoclonal antibody induced by inflammation inducers in human peripheral blood T cells in *in vitro* studies (Chang *et al.,* 2005).

Ambrosia maritima L (Family: Compositae)

Ambrosia maritima (known locally in Egypt as Damsisa) is a wild Egyptian herb. The alcoholic extract of this plant was used in the prevention of gamma irradiation-induced neurotransmitter disturbances, which control the mood. This experimental study implicated that alcoholic extract of this herb can be used as a protective against radiation-induced mood disturbances (Abdel-Hamid and Tarabanko, 2004). In addition, the alcoholic extract of the plant leaves showed antimycobacterial and cytotoxic activity (Abou El Seoud *et al.,* 2003). The biochemical alterations resulting from hepatotoxic drug administration were inhibited by pre-treatment with *A. maritima* L. extract. The data suggested that this plant might act as a hepatoprotective and antioxidant agent (Ahmed and Khater, 2001). Additional metabolic-modulating activities of the herb have also been stressed upon by Egyptian researchers as the hypoglycemic effects may be exerted through the inhibition of glucose absorption, increased sensitivity of receptors to insulin, insulinase inhibiting effect, stimulation of B cells of pancreas to secret insulin or stimulation of peripheral tissues, uptake of glucose (Ammar *et al.,* 1993). Metabolic-modulating activities may be of relevance to radiosensitization, since modulators of glucose uptake, transport and glycolysis like 2-deoxy-D-glucose have been shown to affect radioresponse (Dwarakanath and Jain, 1989).

Myristica fragrans Houtt (Family: Myristicaceae)

M. fragrans, a spice used in several traditional systems of medicine, possesses antifungal, hepatoprotective and antioxidant properties. Its radioprotective effect against 6, 8 and 10Gy gamma radiation was proved by a 30-day survival assay. It significantly enhanced hepatic glutathione (GSH) and decreased testicular lipid peroxidation (LPO) level, whereas acid phosphatase (ACP) and alkaline phosphatase (ALP) activities were not significantly altered. *M. fragrans* extract (MF) pre-treatment effectively protected against radiation-induced biochemical alteration, as reflected by a decrease in LPO level and ACP activity, and an increase in GSH and ALP activity. The study showed implications for the potential use of *M. fragrans* as a radioprotector (Sharma and Kumar, 2007).

Rosmarinus officinalis Linn (Rosemary) (Family: Lamiaceae)

Leaf extract of *Rosmarinus officinalis* has been investigated in Swiss albino mice against gamma radiation (Sancheti and Goel, 2007). In irradiated group, glutathione level was registered low in the blood, whereas a significant elevation was estimated in rosemary pre-treated animals. An increase in lipid peroxidation level above normal was evident in serum of irradiated mice, while a significant decrease in such values was noted in rosemary pre-treated group. This suggests a possible radioprotective ability of rosemary extract (Sancheti and Goyal, 2007). Rosemary extract proved to be an effective radioprotector in animals exposed to higher gamma irradiation doses, through modulation of oxidative stress parameters as glutathione and lipid peroxidation (Jindal *et al.*, 2006).

Conclusion

Egyptian flora has hundreds of herbs with proven antioxidant activity that could be considered as radioprotective or radiosensitizing drugs. However, attention to assessment of radiomodulatory action of these plants remains at a lower level. Mostly, this can be attributed to the little interest at clinical level for herbal therapy. Nonetheless, many trials have been made at experimental level calling for the use of herbal medicine and encouraging the legislation. It is now established that herbals with antioxidant activity exhibit beneficial radioprotective potential. In addition, they also can be prescribed as radioprotectors and radiosensitizers. The antioxidative potential of these drugs seems to be responsible for its efficacy. It is worth mentioning that several coloured plants, which are common in Egypt e.g., Hibiscus, many spices grown around the Nile valley, fruits and flowers, Papas, pumpkins and many other Sinai plants of historically accepted folk use, hold immense potential as radiomodulatory drugs and need further investigation at experimental and clinical level.

References

Abdel-Hamid, N.M. and Tarabanko, V.E. (2004) Radioprotective effect of an Egyptian wild herb *Ambrosia maritima* L. (Damsisa): Biochemical study on neurotransmitters. *Journal of Chemistry of Plant Raw Materials* 4, 53–58.

Abou El Seoud, K.A.E.H., Bibby, M.C., Shoeib, N. and Wright C.W. (2003) Evaluation of some Egyptian plant species for in vitro antimycobacterial and cytotoxic activities. *Pharmaceutical Biology* 41(6), 463–465.

Ahmed, B. and Khater, M.R. (2001) Evaluation of the protective potential of *Ambrosia maritima* extract on acetaminophen–induced liver damage. *Journal of Ethnopharmacology* 75(2–3), 169–174.

Ammar, N.M., Al Okbi, S.Y. and Badawy, I.H (1993) The hypoglycemic effect of different extracts of *Ambrosia maritima* L. Compositae. *Journal of Islamic Academy of Sciences* 6(4), 298–301.

Chang, W.L, Wang, H.Y, Shi, L.S, Lai, J.H. and Lin, H.C (2005) Immunosuppressive iridoids from the fruits of *Gardenia jasminoides*. *Journal of Natural Products* 68(11), 1683–1685.

Ching, L.C., Chai, L.C., Teik, N.L., Fang, H.F. and En, S.D. (2004) Cytotoxic effects of *Coptis chinensis* and *Epimedium sagittatum* extracts and their major constituents (berberine, coptisine and icariin) on hepatoma and leukaemia cell growth. *Clinical and Experimental Pharmacology and Physiology* 31(1/2), 65–69.

Dwarakanath, B.S. and Jain, V. (1989) Energy linked modifications of the radiation response in a human cerebral glioma cell line. *International Journal of Radiation Oncology, Biology and Physics* 17, 1033–1040.

Gao, D., Sakurai, K., Katoh, M., Chen, J. and Ogiso, T. (1996) Inhibition of microsomal lipid peroxidation by baicalein: a possible formation of an iron–baicalein complex. *Biochemistry and Molecular Biology International* 39, 583–587.

Jindal, A., Soyal, D., Sancheti, G. and Goyal, P.K. (2006) Radioprotective potential of *Rosmarinus officinalis* against lethal effects of gamma radiation: a preliminary study. *Journal of Environmental Pathology, Toxicology and Oncology* 25(4), 633–642.

Kowalczyk, E., KrzesiÅ, S.P., Kura, M., Niedworok, J., Kowalski, J.B.Å. and Aszczyk, J. (2006) Pharmacological effects of flavonoids from *Scutellaria baicalensis*. *Przeglad Lekarski* 63(2), 95–96.

Kuo, W.H., Wang, C.J., Young, S.C., Sun, Y.C., Chen, Y.J. and Chou, F.P. (2004) Differential induction of the expression of GST subunits by geniposide in rat hepatocytes. *Pharmacology* 70(1), 15–22.

Li, X.K., Motwani, M., Tong, W., Bornmann,W. and Schwartz, G.K. (2000) Huanglian, a Chinese herbal extract, inhibits cell growth by suppressing the expression of cyclin B1 and inhibiting CDC2 kinase activity in human cancer cells. *Molecular Pharmacology* 58, 1287–1293.

Lim, B.O., Yu, B.P., Kim, S.C. and Park, D.K. (1999) The antioxidative effect of ganhuangenin against lipid peroxidation. *Phytotherapy Research* 13(6), 479–483.

Sancheti, G. and Goyal, P.K. (2007) Prevention of radiation-induced haematological alterations by medicinal plant *Rosmarinus officinalis* in mice. *African Journal of Traditional, Complementary and Alternative Medicine* 4(2), 165–172.

Sharma, M. and Kumar, M. (2007) Radioprotection of Swiss albino mice by *Myristica fragrans Houtt.* *Journal of Radiation Research* 48(2), 135–141.

Shieh, D.E., Liu, L.T. and Lin, C.C. (2000) Antioxidant and free radical scavenging effects of baicalein, and wogonin. *Anticancer Research* 20, 2861–2866.

Tilak, J.C. and Devasagayam, T.P.A. (2004) Radioprotective properties of baicalein. *BARC Newsletter* 249, 98–104.

Ye, F., Xui, L., Yi, J., Zhang, W. and Zhang, D.Y. (2002) Anticancer activity of *Scutellaria baicalensis* and its potential mechanism. *Journal of Alternative and Complementary Medicine* 8, 567–572.

Chapter 6

Melatonin Mitigates the Damaging Effects of Ionizing Radiation

Russel J. Reiter, Dun-Xian Tan, Hiroshi Tamura, Lucien C. Manchester and M. Pilar Terron

Introduction

The biological consequences that are the result of exposure to ionizing radiation are attributable to damage to essential molecules. Electrons in molecules absorb energy from high-energy photons or electrons that are most commonly used therapeutically; this results in the ionization or excitation of molecules. Ionizations generate highly reactive species, molecules or portions thereof that possess an unpaired electron in their valence orbital; these reactants are unstable and are known as free radicals.

Free radicals are often derivatives of ground state oxygen (O_2). A single reduction of O_2 produces the superoxide anion radical ($O_2^{\bullet-}$), a product that is not highly reactive. It is dismutated by the antioxidative enzyme superoxide dismutase to hydrogen peroxide (H_2O_2), which, in the presence of a transition metal, is converted to the very reactive hydroxyl radical (•OH). This latter product has a very brief half-life and pummels any molecule that is in the vicinity of where it is produced. H_2O_2, since it does not possess an unpaired electron, is not a free radical but is classified as a reactive oxygen species (ROS) along with $O_2^{\bullet-}$ and the •OH (Halliwell, 1996).

In addition to the ROS, there are several reactive nitrogen species (RNS) that are potentially damaging to essential molecules when they are produced. Nitric oxide (NO•) is synthesized by a variety of isoforms of nitric oxide synthase in numerous tissues. While NO is not particularly reactive in terms of oxidizing bystander molecules, it readily combines with the $O_2^{\bullet-}$ to produce the peroxynitrite anion ($ONOO^-$), a non-radical species that is, nevertheless, a potent oxidizing agent (Packer *et al.*, 2007). Additionally, its fate within cells may include metabolism to the •OH, which, as noted above, is also highly reactive. The combined actions of ROS/RNS can be devastating not only to the structure of molecules but to their functions as well.

Biomolecules are readily damaged by the most reactive radicals and, once mangled, molecules may function abnormally or not at all. Injury that results from the ionization of a target molecule is referred to as the direct effects of ionizing radiation, while molecular mutilation to a target molecule that is a consequence of the reaction with other radiolytic products is referred to as an indirect effect.

Ionizing Radiation-induced Damage

One commonly used therapeutic dose of radiation, i.e., 2Gy, is estimated to produce 2μM free radicals (Von Sonntag, 1987). This corresponds to about one radical for every 10 million molecules (with a molecular weight of 100). This number may seem minuscule considering that even normal aerobic metabolism generates free radicals in unirradiated cells (Chance *et al.,* 1979). Within cells, estimated steady state concentrations of H_2O_2 are on the order of 10^{-8}M while the $O_2^{\bullet-}$ is at a concentration of 10^{-11}M. In this regard, it is interesting that estimates suggest there are roughly 10,000 oxidative attacks on the DNA in each cell every day from normal metabolism (Ames *et al.,* 1993). This number is the same order of magnitude as the number of DNA hits that result from 2Gy of ionizing radiation (Ward, 1990), yet the biological effects of these latter hits are more damaging. Thus, it must be assumed that there is something qualitatively different about free radicals generated as a consequence of ionizing radiation versus those produced by normal metabolic activity.

A likely difference between free radical damage resulting from ionizing radiation and that from aerobic metabolic activity may be due, at least in part, to the fact that radiation produces clusters of radicals (Fig. 6.1). As energy is deposited during exposure to ionizing radiation, each absorption produces several radicals within a very small volume (Brenner and Ward, 1992).

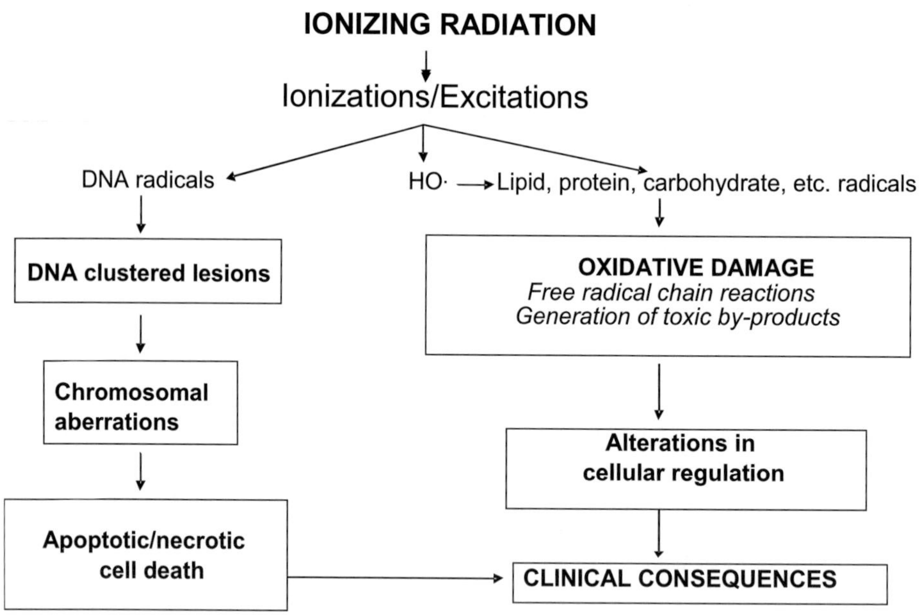

Fig. 6.1. Mechanisms of DNA damage by ionizing radiation illustrating the role of oxidative stress. A common clinical endpoint of chromosomal damage is cancer.

When the areas of these radicals overlap a segment of a DNA double helix, both strands can be damaged. While the primary cytotoxic effects of ionizing radiation are accounted for by these double breaks, other molecules, e.g., lipids, proteins, etc., are likewise destroyed when cells are exposed to high-energy radiation. In terms of the resulting cell death that occurs as a consequence of ionizing radiation exposure, the non-DNA damage that occurs is complete within 10msec. Interestingly, however, the addition of antioxidants (e.g., vitamin E) more than 10msec after the imposed radiation does provide some protection against the biological toxicity (Ramakrishnan *et al.*, 1989) indicating that not only the DNA mutilation but other molecular damage as well plays some role in the cytotoxic and apoptosis-inducing effects of ionizing radiation.

While damaging mechanisms by which ionizing radiation kills cells can often be traced to free radicals, some of the toxic actions of radiation are related only secondarily to these oxidizing species. As an example, clastogenic factor, which is capable of promoting DNA damage in non-irradiated cells, is often elevated in the blood for prolonged periods in individuals who have been exposed to ionizing radiation (Emerit *et al.*, 1995; Morgan, 2003). The presumed source of clastogenic factor in these situations is from the inflammatory response of cells resulting from the radiation. Thus, the production and release of this potential toxic factor is unrelated to the initial radiochemical events. However, the damage that clastogenic factor eventually induces is, in fact, also mediated by free radicals.

Radioprotection

The formation of free radicals during the exposure of tissues to ionizing radiation is a major event that causes molecular damage and cell death. In particular, the •OH is a critical contributor in the toxicity of high energy radiation. The reaction time of the •OH, once it is formed, is in the nanosecond range. Thus, any free radical scavenger that prevents the molecular abuse by a •OH must be in the immediate vicinity of where the radical is produced. For example, an antioxidant located in the cytosol of a cell is inept in protecting nuclear DNA from damage by a •OH.

A plethora of molecules have been tested as radioprotective agents. For example, non-protein sulfhydryls, aminothiols, stable free radicals, DNA-binding bisbenzimidazoles, antioxidant vitamins, superoxide dismutase, etc., have been examined for their ability to attenuate molecular damage resulting from exposure to ionizing radiation. While many of these agents function at least partially as primary for secondary free radical scavengers, other mechanisms of protection including oxygen depletion, immunomodulation, hydrogen atom donation, etc., have also been described.

One of the most widely used radioprotective agents is the aminothiol, WR-2721 (S-2-(3-aminopropylamino) ethyl-phosphorothionic acid) (also called amifostine) (Kouvaris *et al.*, 2007). Interest in this agent stems from its selectivity; thus, it protects normal cells but not cancer cells from the damage inflicted by ionizing radiation (Yuhas *et al.*, 1980). There are several shortcomings to the use of WR-2721 as a radioprotective agent. First, toxic side effects including nausea, vomiting, hypotension and hypocalcemia (Kligerman *et al.*, 1994) occur with the use of high doses. Second, it has to be given intravenously. Both these drawbacks make its use impractical under conditions where large numbers of people are exposed to toxic levels of ionizing radiation, e.g., a nuclear explosion on the battlefield or a nuclear accident such as occurred at Chernobyl.

A significant factor that determines the efficacy of radioprotective agents in alleviating the lesions by ionizing radiation is its concentration at the site of free radical generation. The overall concentration of an antioxidant within an organism is less important in terms of its protection of a specific molecule. For example, any free radical scavenger that protects a molecule from oxidation by a •OH or the ONOO- must, in essence, be straddling that molecule. The concentration of the antioxidant at that specific site is more important than its concentration throughout the cell. Thus, a free radical scavenger that is differentially distributed intracellularly and that is positioned to afford maximal protection would have a protective advantage over a molecule that may be in a higher concentration but without the positional advantage.

Melatonin as a Free Radical Scavenger and Anti-inflammatory Agent

Given that many radioprotectors do so via their ability to detoxify ROS/RNS, melatonin's ability to function in this capacity may also relate to its efficacy as a free radical scavenger. Melatonin is, in fact, a powerful •OH scavenger. This was documented in the first report in which melatonin was examined for its antioxidative function. Using a combination of a spin trapping agent and electron spin resonance spectroscopy (ESR), Tan and co-workers (1993a) showed that melatonin competed very effectively with the spin trap, in this case 5,5-dimethyl-pyrroline oxide (DMPO), in neutralizing the •OH. Furthermore, these researchers showed that each molecule of melatonin actually scavenges two •OH with the resultant formation of cyclic 3-hydroxymelatonin (cyclic 3-OH-Mel). The use of a spin trapping agent combined with ESR is the most reliable method to document the radical scavenging activity of an antioxidant (Towell and Kalyanaraman, 1991). Using the same methodologies with a different spin trapping agent (5-diethoxyphosphoryl-5-methyl-1-pyrroline-N-oxide, DEPMPO), Ebelt and colleagues (2000) confirmed melatonin's ability to scavenge the •OH. While the studies of Tan *et al.* (1993a) and Ebelt and co-workers (2000) were performed in pure chemical systems, Li *et al.* (1997) documented that melatonin functioned similarly *in vivo*. In the study in question, using microdialysis combined with the salicylate trapping method, they showed that melatonin reduced dihydroxybenzoic acid in dialysates collected from the rat brain subjected to ischemia/reperfusion injury. During the last decade many other methodologies have been used to document melatonin's high efficacy in detoxifying the highly damaging •OH (Matuszek *et al.,* 1997; Stascia *et al.,* 1998; Reiter *et al.,* 2001; Poeggeler *et al.,* 2002; Allegra *et al.,* 2003).

Once a •OH is formed, its fate is determined by at least two processes, i.e., it is either neutralized by an antioxidant or it damages an adjacent molecule. Hence, if its precursor, i.e., H_2O_2 (Fig. 6.2), could be removed it would limit •OH formation and the resulting damage. Melatonin also has this capability since it directly interacts with this non-radical species (Tan *et al.,* 2000) in a dose-dependent manner; the product of that interaction is N^1-acetyl-N^2-formyl-5-methoxykynuramine (AFMK). While H_2O_2 is a non-radical and only sluggishly reacts with the majority of bioorganic molecules, its removal by melatonin ensures that it does not give rise to the highly reactive •OH. Certainly, the scavenging of H_2O_2 by melatonin indirectly minimizes molecular damage normally produced by the •OH.

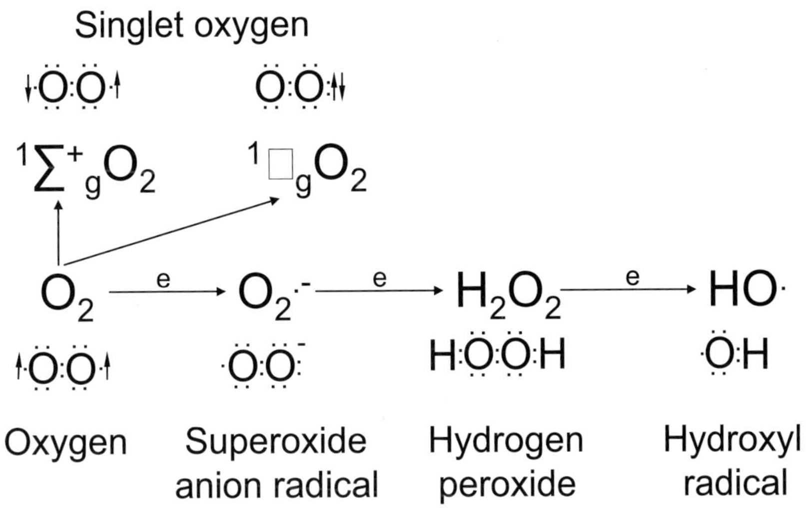

Fig. 6.2. The three electron reduction of melatonin yields the highly reactive hydroxyl radical. Ionizing radiation generates numerous hydroxyl radicals when molecules are split in vivo resulting in massive oxidative damage. The dots represent the electron configuration in each of the species. Both the superoxide anion and the hydroxyl radical possess an unpaired electron (single dot). Ground-state oxygen is a biradical with both unpaired electrons having the same direction of spin (arrows). Singlet oxygen, a reactive species with some toxicity, is inconsequential in ionizing radiation damage.

Melatonin has been reported to be an effective scavenger of the $O_2^{\bullet -}$ (Marshall *et al.*, 1996; Zang *et al.*, 1998); both these studies were conducted *in vitro*. While $O_2^{\bullet -}$ is generally not considered to inflict much cellular damage, its major toxicity stems from its eventual conversion to the •OH (Fig. 6.2) and its rapid coupling with NO• to form $ONOO^-$ which is essentially equivalent to the •OH in terms of destructive potential. The significance of melatonin's detoxification of the $O_2^{\bullet -}$ in protecting against oxidative damage *in vivo* remains undefined.

More important than the reported effects of melatonin in directly neutralizing the $O_2^{\bullet -}$ may be prevention of its formation at the level of the mitochondria, in a process referred to as radical avoidance (Hardeland, 2005; Hardeland *et al.*, 2005; Leon *et al.*, 2005). During the transfer of electrons through the complexes of the mitochondrial electron transport chain (ETC), some are fumbled and reduce nearby O_2 molecules. Thus, the mitochondria are a major source of free radical formation. The complexes of the ETC could well be damaged by ionizing radiation resulting in the increased release of electrons and elevated $O_2^{\bullet -}$ formation; this could contribute to secondary molecular damage associated with high-energy radiation.

The coupling of $O_2^{\bullet -}$ with NO•, two relatively unreactive free radicals, is extremely rapid and the reaction produces a much more reactive species, $ONOO^-$. Both Zhang *et al.* (1999) and Blanchard *et al.* (2000) found that melatonin is a substrate for $ONOO^-$. Additionally, melatonin may directly remove NO• to reduce $ONOO^-$ formation (Turjanski *et al.*, 2000). The elimination of these reactants by melatonin may be important in reducing some of the destruction induced by ionizing radiation.

Singlet oxygen (Fig. 6.2) is a well-known reactive species capable of damaging macromolecules. This reactant is a high-energy form of O_2, which is generated within cells via several means. As first noted by Poeggeler *et al.* (1994), and subsequently confirmed by others (Zang *et al.*, 1998; Roberts *et al.*, 2000), melatonin scavenges singlet oxygen. Whether this has relevance to the protection by melatonin against ionizing radiation damage has not been examined. The multiple scavenging actions of melatonin have been summarized in several recent reviews (Hardeland *et al.*, 2007; Reiter *et al.*, 2007; Tan *et al.*, 2007).

A potentially important aspect of melatonin's ability to arrest oxidative damage in molecules during exposure to ionizing radiation may relate to the fact that not only melatonin but also several of its metabolites (which are formed when melatonin detoxifies ROS/RNS) are likewise very efficient scavengers. These products include cyclic 3-OH-Mel, AFMK and N^1-acetyl-5-methoxykynuramine (AMK) (Hardeland, 2005; Tan *et al.*, 2001, 2007; Rosen *et al.*, 2006). The sequence of reactions whereby melatonin and its metabolites neutralize progressively more radicals is referred to as melatonin antioxidant cascade (Tan *et al.*, 2002). These sequential reactions make melatonin highly proficient in reducing oxidative/nitrosative damage anytime these toxic reactants are generated in excess, such as during ionizing radiation.

In addition to their direct scavenging actions, melatonin and its metabolites promote the activities of antioxidative enzymes and inhibit prooxidative enzymes thereby depressing the oxidizing potential within cells. In particular, melatonin stimulates the enzymatic dismutation of O_2•- to H_2O_2 and enhances the removal of the latter molecule by promoting the activity of glutathione peroxidase (Rodriguez *et al.*, 2004; Tomas-Zapico and Coto-Montes, 2005). Melatonin also elevates concentrations of another important intracellular antioxidant, glutathione (GSH), by stimulating its rate-limiting enzyme gamma-glutamylcysteine synthase (Urata *et al.*, 1999; Winiarska *et al.*, 2006). The prooxidative enzymes that are restrained by melatonin and/or its metabolites are the lipooxygenases and nitric oxide synthases (Hardeland *et al.*, 2005). It was recently shown that the melatonin derivative, AMK, is responsible for the observed inhibition of nitric oxide synthase by melatonin (Leon *et al.*, 2006). The reported changes in enzyme activities due to melatonin and its derivatives could come into play in situations where exposure to ionizing radiation was anticipated and melatonin was taken in advance of the exposure.

Cellular damage induced by ionizing radiation commonly initiates a post-irradiation inflammatory response, which leads to the secondary destruction of essential molecules, which then causes functional deterioration (Patchen, 1998). Although melatonin has never been tested for its ability to quench post-irradiation inflammation, given its marked anti-inflammatory actions (Guerrero and Reiter, 2002; Carrillo-Vico *et al.*, 2005), it would be expected to do so.

Melatonin as a Radioprotector

Radiobiologists have long realized that the genome is readily damaged by ionizing radiation and that the widespread destruction of DNA is a consequence of both the direct and indirect effects of the radiation (Laayoun *et al.*, 1998; Karbownik and Reiter, 2000; Zhou *et al.*, 2006). It is also estimated that up to 70% of the damage that occurs as a result of exposure to high-energy radiation is mediated by the •OH formed from the radiolysis of water. The battering of DNA by ionizing radiation causes the formation of oxidized bases, abasic sites, DNA-DNA intrastrand adducts, DNA single and double strand breaks and DNA-protein crosslinks. These are most often the result of an interaction of the •OH with DNA bases and to the lesser extent with DNA sugars. Given the high potency of melatonin in scavenging the •OH, it was anticipated that this indoleamine may have significant potential as a radioprotective agent.

This was essentially verified in the first study that identified melatonin as a •OH scavenger. Thus, Tan and co-workers (1993a) showed that the metabolite formed, i.e., cyclic 3-OH-Mel, when melatonin scavenges two •OH was elevated in the urine of mice subjected to whole-body ionizing radiation. While DNA damaged products were not specifically measured in this investigation, the protective actions could be inferred from this study and others (Tan *et al.,* 1993b, 1994) published at essentially the same time, which documented melatonin's ability to reduce DNA damage that is a consequence of toxic ROS/RNS. There is evidence that melatonin is normally in close proximity to nuclear DNA (Menendez-Pelaez *et al.,* 1993).

While many compounds have been tested as radioprotective agents, their efficacies have not been optimal and there is still an urgent need to identify novel, non-toxic and effective products that will reduce damage and that can be conveniently administered to large groups of individuals who may be subjected to ionizing radiation, either accidentally or in the process of doing their job. One such molecule may be melatonin since it is highly efficacious as a radioprotector, it has virtually no acute or chronic toxicity and it can be taken orally. Melatonin has been reported to be present in several edible plants (Reiter and Tan, 2002). In the mid-1990s studies were initiated to determine if melatonin protected against the damaging effects of ionizing radiation. In a series of three *in vitro* studies, Vijayalaxmi *et al.,* (1995a, 1995b, 1996a) documented the cytoprotective effectiveness of melatonin in human blood lymphocytes exposed for 20 minutes to 150cGy gamma radiation. In these studies, melatonin significantly depressed the incidence of abnormal cells expressing genetic damage, exchange type of aberrations, acentric fragments and micronuclei as compared to those abnormalities seen in similarly-irradiated cells not treated with melatonin. Of further importance is that melatonin itself has minimal cytotoxic and genotoxic effects in normal cells. These observations were extended to an *in vivo*/*in vitro* study in which adult humans were given melatonin orally and thereafter blood lymphocytes were collected and exposed to gamma irradiation; the lymphocytes were harvested one and two hours after melatonin administration. The blood cells were exposed to 150cGy gamma radiation, which increased the frequency of chromosomal aberrations and micronuclei; these abnormal changes were reduced by 60–65% in the lymphocytes collected from individuals who had ingested melatonin (Vijayalaxmi *et al.,* 1996b).

The first *in vivo* investigation was carried out by Blickenstaff and co-workers (1994) who lethally irradiated mice (950cGy whole-body irradiation) and examined the ability of melatonin to reduce the death rate. Within 12 days of the radiation exposure, all the control mice were dead while at 30 days, 43% of the irradiated mice that had received melatonin continued to survive. This group also tested melatonin homologues, which were also found to have radioprotective effects. In a similar study, we (Vijayalaxmi *et al.*, 1999a) observed that melatonin greatly improved the 30-day survival of mice subjected to 815cGy ionizing radiation. Also, the injection of 5 or 10 mg/kg melatonin prior to whole-body radiation attenuated damage to bone marrow cells (Vijayalaxmi *et al.,* 1999b). These observations are of particular importance given the high sensitivity of bone marrow cells to ionizing radiation (Malaker, 1998).

A commonly used index of free radical-damaged DNA is 8-hydroxy-2-deoxyguanosine. Likewise, cell membrane rigidity results as a consequence of the peroxidation of lipids in this structure. When rats were exposed to whole-body radiation, levels of hepatic 8-hydroxy-2-deoxyguanosine and elevated membrane rigidity were noted; these effects were almost totally counteracted when the animals were treated with melatonin prior to their irradiation (Fig. 6.3) (Karbownik *et al.,* 2000).

There are several reviews which have summarized the protective actions of melatonin against the cytotoxicity and lethality of ionizing radiation that were observed in earlier studies (Vijayalaxmi, 1998, 2004; Karbownik and Reiter, 2000; Shirazi *et al.,* 2007). The results of several recent studies that have been published since those reviews were written are evaluated in the following paragraphs.

In addition to bone marrow, intestinal epithelial cells are readily damaged by ionizing radiation with sloughing of the gut lining cells often contributing to the morbidity. Monobe and co-workers (2005) tested whether melatonin would preserve intestinal epithelial integrity in mice exposed to whole-body gamma irradiation (7–21Gy). This purely morphological study examined the epithelial cells in the intestinal crypts as well as the number of crypts. The authors concluded that melatonin in a dose-response manner reduced damage to the gut when it was given in advance of the radiation exposure. These findings require verification using additional endpoints of epithelial integrity and gastrointestinal pathophysiology. In another morphological study it was shown that exogenously administered melatonin protects rat thyrocytes from gamma irradiation (8Gy) (Kundurovic and Sofic, 2006).

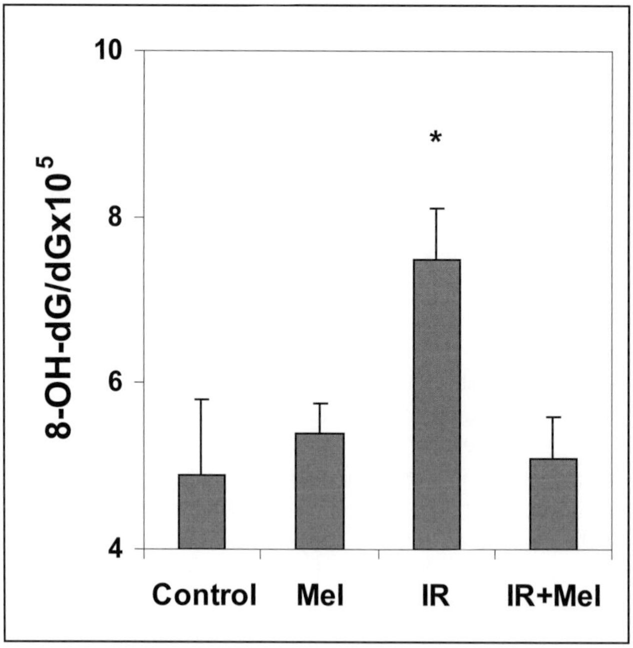

Fig. 6.3. 8-Hydroxy-2-deoxyguanosine (8-OH-dG) levels in hepatic DNA from rats subjected to 800cGy whole-body ionizing radiation (IR). 8-OH-dG is a damaged DNA product and a sensitive indicator of exposure to IR. Melatonin (Mel), given in advance of IR, almost totally prevented the DNA-damaging effects of IR (from Karbownik *et al.*, 2000).

Yilmaz and Yilmaz (2006) exposed rats to 720cGy whole-body irradiation and compared the ability of either melatonin or vitamin E (both given at doses of 100 mg/kg intraperitoneally before radiation exposure) to protect skeletal muscle from lipid peroxidation (as malondialdehyde). Unlike vitamin E, melatonin prevented the rise in lipid peroxidation products and the antioxidative enzyme changes that occurred as a result of radiation exposure. The implication of these findings is that, *in vivo*, melatonin is a more effective antioxidant and radioprotective agent than vitamin E. The ability of melatonin to more effectively reduce *in vivo* free radical-mediated molecular damage is a common finding (Qi *et al.*, 2000; Baydas *et al.*, 2002; Aktoz *et al.*, 2007).

Melatonin has also been compared with amifostine (WR-2721) in terms of their abilities to reduce the genotoxicity of gamma radiation in human lymphocytes *in vitro* (Kapjar *et al.*, 2006). By evaluating the incidence of micronuclei and sister chromatid exchanges, the authors concluded that melatonin and amifostine are equally effective in reducing irradiation-mediated genomic damage. An improvement of the radioprotective action was seen when the cells were treated with a combination of melatonin and amifostine. They surmised that if the two molecules would be used in combination, it would be highly effective and melatonin may reduce the side effects of amifostine. However, given that melatonin was equally as beneficial as amifostine, giving it in combination with the latter molecule may not afford any advantage.

Zhou and colleagues (2006) examined the protective actions of melatonin against damage to Chinese hamster cells irradiated with high-linear energy transfer (LET) radiation; typically chemical protection against LET irradiation is more difficult to achieve. Cellular inactivation was measured with a conventional colony-forming assay.

Both carbon beam and X-ray induced cell inactivation, hprt gene mutation and cell cycle G2 blockade in a dose-dependent manner; carbon beam exposure caused more pronounced effects. Melatonin, also in a dose-response manner, exhibited protective effects against these radiations. In terms of the number of cells that died, melatonin improved survival in cultures exposed to 8Gy or larger of X-rays and 6Gy or larger of carbon beam.

Melatonin also depressed hprt induction and prevented G2 blockade induced by the carbon beam but not those resulting from exposure to X-irradiation. These results suggest that melatonin reduces that direct interaction of particles with cells; the mechanisms as to how this may be achieved, however, were not investigated. Since LET radiation is a major component of space irradiation, these findings have implications for long-duration, deep space missions.

That some of the radioprotective effects of melatonin may be due to a melatonin metabolite was supported by a recent study (Manda *et al.*, 2007). Recall, as mentioned above, the metabolites of melatonin, like the parent molecule itself, are potent •OH scavengers (Hardeland, 2005; Tan *et al.*, 2007). To examine whether AFMK would reduce oxidative damage to DNA, proteins and lipids, mice were exposed to whole-body X-irradiation (0.55Gy/minute for 10.9 minutes). With the aid of the comet assay (alkaline microgel electrophoresis), Manda *et al.* (2007) reported a dramatic reduction in DNA damage (Fig. 6.4) by AFMK. Likewise, the amine was similarly effective in limiting protein and lipid damage resulting from X-ray exposure.

Using a spin trap (DMPO) and ESR they also documented that AFMK is a powerful •OH scavenger. This may well have been one of the mechanisms that permitted AFMK to be so protective against ionizing radiation. These findings leave open the possibility that the metabolites of melatonin, particularly AFMK, in addition to or rather than melatonin itself may be the agent(s) that alleviates molecular damage due to ionizing radiation.

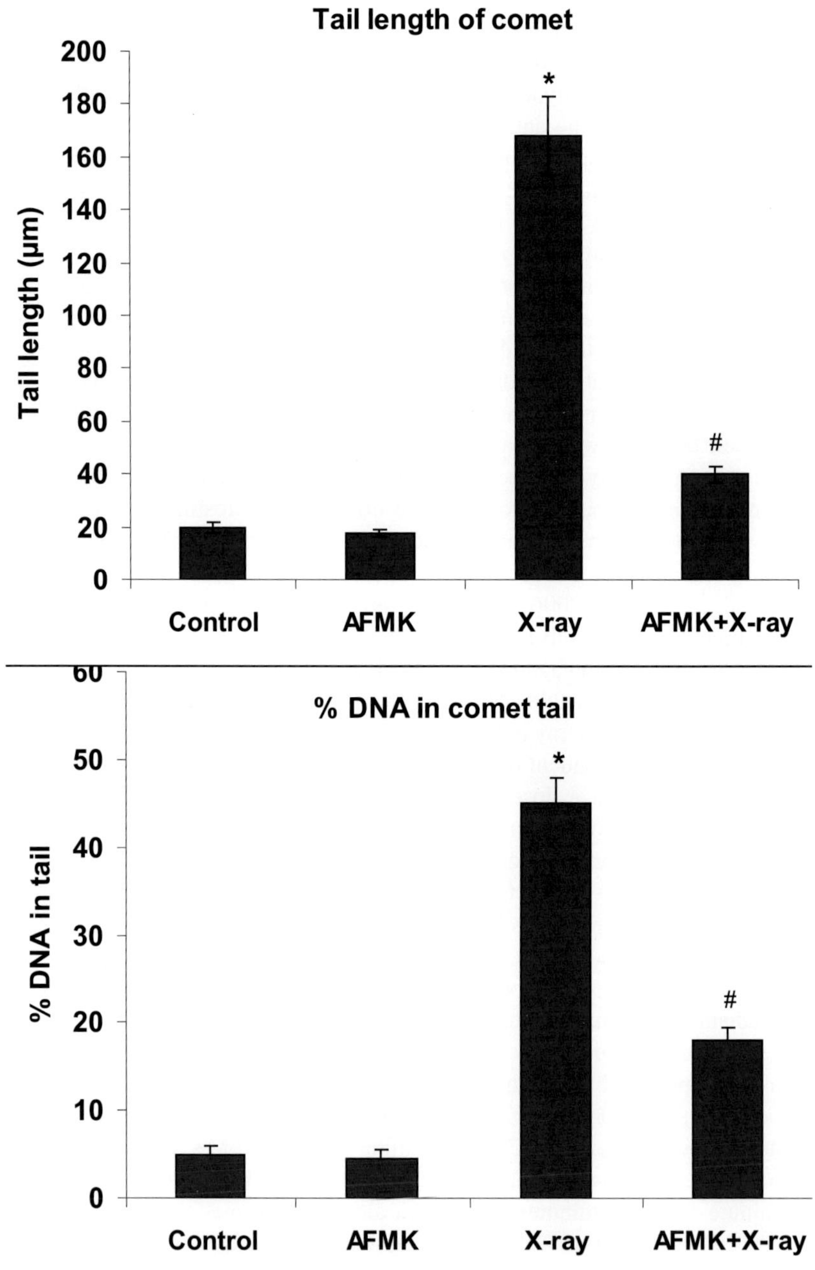

Fig. 6.4. DNA damage, as assessed using the comet assay (alkaline microgel electrophoresis), in the cerebral cortex of mice subjected to whole-body X-irradiation. Both the length of the comet tail and the %DNA damage were significantly reduced when AFMK was injected prior to irradiation exposure. AFMK is a melatonin metabolite (from Manda *et al.*, 2007).

Concluding Remarks

There is now a substantial number of reports providing unequivocal documentation of the beneficial effects of melatonin in protecting against ionizing radiation. The protection has been proven in both the reduction in cytotoxicity and animal lethality. Besides being highly effective as a radioprotector, melatonin has advantages over other agents with a similar function because of its virtual absence of side effects and its very wide safety margin in terms of dose. Also, unlike a commonly used radioprotective agent amifostine, it can be taken via any route including orally. This means, of course, that it could be carried by individuals who are in jeopardy of being accidentally or intentionally exposed to toxic doses of ionizing radiation and taken as needed. This is also supported by its very long shelf life and its lack of a requirement for refrigeration.

References

Aktoz, T., Aydoglu, N., Alagol, B., Yalcin, O., Huseyinova, G. and Atakan, I.H. (2007) The protective effects of melatonin and vitamin E against renal ischemia–reperfusion injury in rats. *Renal Failure* 29, 535–542.

Allegra, M., Reiter, R.J., Tan, D.X., Gentile, C., Tesoriere, L. and Livrea, M.A. (2003) The chemistry of melatonin's interaction with reactive species. *Journal of Pineal Research* 34, 1–10.

Ames, B.N., Shinenaga M.K. and Hagen, T.M. (1993) Oxidants, antioxidants, and the degenerative diseases of aging. *Proceedings of the National Academy of Science of the United States of America* 90, 7915–7922.

Baydas, G., Canatan, A. and Turkoglu, A. (2002) Comparative analysis of the protective effects of melatonin and vitamin E on streptozocin-induced diabetes mellitus. *Journal of Pineal Research* 32, 255–230.

Blanchard, B., Pompon, D. and Ducroq, C. (2000) Nitrosation of melatonin by nitric oxide and peroxynitrite. *Journal of Pineal Research* 29, 184–192.

Blinkenstaff, R.T., Brandstadter, S.M., Reddy, S. and Witt, R. (1994) Potential radioprotective agents: 1. Homologues of melatonin. *Journal of Pharmaceutical Science* 83, 216–218.

Brenner, D.J. and Ward, J.F. (1992) Constraints on energy deposition and target size of multiple damaged sites associated with DNA double strand breaks. *International Journal of Radiation Biology* 61, 737–748.

Carrillo-Vico, A., Guerrero, J.M., Lardone, P.J. and Reiter, R.J. (2005) A review of the multiple actions of melatonin on the immune system. *Endocrine* 27, 189–200.

Chance, B., Sies, H. and Boveris, A. (1979) Hydroperoxide metabolism in mammalian organs. *Physiological Reviews* 59, 527–605.

Ebelt, H., Peschke, D., Brömme, H.J., Mörke, W., Blume, R. and Peschke, E. (2000) Influence of melatonin on free–radical induced changes in rat pancreatic beta–cells *in vitro*. *Journal of Pineal Research* 28, 65–72.

Emerit, I., Arutyunyan, R., Oganisian, M., Levy, A., Cernjavski, L., Sarkisian T, Pogosian, A. and Asrian, K. (1995) Radiation–induced clastogenic factors: anticlastogenic effect of *Ginkgo biloba* extract. *Free Radical Biology and Medicine* 18, 985–991.

Guerrero, J.M. and Reiter, R.J. (2002) Melatonin-immune system relationships. *Current Topics in Medicinal Chemistry* 2, 167–180.

Halliwell, B. (1996) Free radicals, proteins and DNA: oxidative damage versus redox regulation. *Biochemical Society Transactions* 24, 1023–1027.

Hardeland, R. (2005) Antioxidative protection by melatonin: multiplicity of mechanisms from radical detoxification to radical avoidance. *Endocrine* 27, 119–130.

Hardeland, R., Backhaus, C., Fadavi, A. and Hess, M. (2007) N^1-acetyl-5-methoxykynuramine contrasts with other tryptophan metabolites by a peculiar type of NO scavenging: cyclization to a cinnolinone prevents formation of unstable nitrosamines. *Journal of Pineal Research* 43, 104–105.

Hardeland, R., Pandi-Perumal, S.R. and Cardinali D.P. (2005) Melatonin. *International Journal of Biochemistry and Cell Biology* 38, 313–316.

Kapjar, N., Miocic, S., Ramic, S., Milic, M. and Viculin, T. (2006) Assessment of the radioprotective effects of amifostine and melatonin on human lymphocytes irradiated with gamma-rays *in vitro*. *Archiv Higado Roda Toksiologie* 57, 155–163.

Karbownik, M. and Reiter, R.J. (2000) Antioxidative effects of melatonin in protection against cellular damage caused by ionizing radiation. *Proceedings of the Society for Experimental Biology and Medicine* 225, 9–22.

Karbownik, M., Reiter, R.J., Qi, W., Garcia, J.J., Tan, D.X. and Manchester, L.C. (2000) Protective effects of melatonin against oxidation of guanine bases in DNA and decreased microsomal membrane fluidity in rat liver induced by whole-body ionizing radiation. *Molecular and Cellular Biochemistry* 211, 137–144.

Kligerman, N.M., Glover, D.J., Turrisi, A.T., Norfleet, A.L., Youhas, J.M, Coia, L.R., Simone, C., Glick, J.H. and Goodman, R.L. (1994) Toxicity of WR-2721 administered in single and multiple doses. *International Journal of Radiation Oncology and Biological Physics* 10, 1773–1776.

Kouvaris, J.R., Kouloulias, V.E. and Vlahos, L.J. (2007) Amifostine: the first selective-target and broad-spectrum radioprotector. *Oncologist* 12, 738–747.

Kundurovic, Z. and Sofic, E. (2006) The effects of exogenous melatonin on the morphology of thyrocytes in pinealectomized and irradiated rats. *Journal of Neural Transmission* 113, 48–49.

Laayoun, A., Lhomme, J., Berger, M. and Cadet, J. (1998) Protection against radiation damage to DNA bases. In: Bump, E.A. and Malaker, K. (eds) *Radioprotectors*, CRC Press, Boca Raton, Florida, pp. 167–183.

Leon, J., Acuna–Castroviejo, D., Escames, G., Tan, D.X. and Reiter, R.J. (2005) Melatonin mitigates mitochondrial malfunction. *Journal of Pineal Research* 37, 1–9.

Leon, J., Escames, G., Rodriguez, M.I., Lopez, L.C., Tapias, V., Entrena, A., Camacho, E., Carrion, M.D., Gallo, M.A., Espinosa, A., Tan, D.X., Reiter, R.J. and Acuna–Castroviejo, D. (2006) Inhibition of neuronal nitric oxide synthase activity by N^1–acetyl–5–methoxykynuramine, a brain metabolite of melatonin. *Journal of Neurochemistry* 98, 2023–2033.

Li, X.J., Zhang, L.M., Gu, J., Zhang, A.Z. and San, F.Y. (1997) Melatonin decreases the production of hydroxyl radical during cerebral ischemia–reperfusion. *Acta Pharmacologica Sinica* 18, 394–396.

Malaker, K. (1998) Clinical experience with radioprotectors. In: Bump, E.A. and Malaker, K. (eds) *Radioprotectors*, CRC Press, Boca Raton, Florida, pp. 373–409.

Manda, K., Ueno, M. and Anzai, K. (2007) AFMK, a melatonin metabolite, attenuates X–ray–induced oxidative damage to DNA, proteins and lipids. *Journal of Pineal Research* 42, 386–393.

Marshall, K.A., Reiter, R.J., Poeggeler, B., Aruoma, O.I. and Halliwell, B. (1996) Evaluation of the antioxidant activity of melatonin *in vitro*. *Free Radical Biology and Medicine* 21, 307–315.

Matuszak, Z., Reszka, K. and Chignell, C.F. (1997) Reaction of melatonin and related indoles with hydroxyl radicals: EPR and spin trapping investigations. *Free Radical Biology and Medicine* 23, 367–372.

Menendez–Pelaez, A., Poeggeler, B., Reiter, R.J., Barlow–Walden, L., Pablos, M.I. and Tan, D.X. (1993) Nuclear localization of melatonin in different mammalian tissues: immunocytochemical and radioimmunoassay evidence. *Journal of Cellular Biochemistry* 53, 373–382.

Monobe, M., Hino, M., Sumi, M., Uzawa, A., Hirayama, R., Ando, K. and Kojima, S. (2005) Protective effects of melatonin on gamma–ray induced intestinal damage. *International Journal of Radiation Biology* 81, 855–860.

Morgan, W.F. (2003) Is there a common mechanism underlying genomic instability, bystander effects and other nontargeted effects of exposure to ionizing radiation? *Oncogene* 22, 7094–7099.

Packer, P., Beckman, J.S. and Liaudet, J. (2007) Nitric oxide and peroxynitrite in health and disease. *Physiological Reviews* 87, 315–424.

Patchen, M.L. (1998) Immunomodulators and cytokines: their use in the mitigation of radiation-induced hemopoietic injury. In: Bump, E.A., and Malaker, K. (eds) *Radioprotectors*, CRC Press, Boca Raton, Florida, pp. 213–236.

Poeggeler, B., Saarela, S., Reiter, R.J., Tan, D.X., Chen, L.D., Manchester, L.S. and Barlow–Walden, L. (1994) Melatonin – a highly potent endogenous scavenger and electron donor: new aspects of its oxidation chemistry of this indole assessed *in vitro*. *Annals of the New York Academy of Sciences* 29, 94–99.

Poeggeler, B., Thuermann, S., Dose, A., Schoenke, M., Burkhardt, S. and Hardeland, R. (2002) Melatonin's unique scavenging properties–roles of its functional substituents as revealed by a comparison with its structural analogues. *Journal of Pineal Research* 33:20–30.

Qi, W., Reiter, R.J., Tan, D.X., Garcia, J.J., Manchester, L.C., Karbownik, M. and Calvo, J.R. (2000) Chromium (III)-induced 8-hydroxydeoxyguanosine in DNA and its reduction by antioxidants: comparative effects of melatonin, ascorbate and vitamin E. *Environmental Health Perspectives* 108, 399–402.

Ramakrishnan, N., McClain, D.E. and Catravas, G.N. (1989) Membranes as sensitive targets in thymocyte apoptosis. *International Journal of Radiation Biology* 63, 693–701.

Reiter, R.J. and Tan, D.X. (2002) Melatonin: an antioxidant in edible plants. *Annals of the New York Academy of Sciences* 957, 341–344.

Reiter, R.J., Tan, D.X., Manchester, L.C. and Qi, W. (2001) Biochemical reactivity of melatonin with reactive oxygen and nitrogen species. *Cell Biochemistry and Biophysics* 34, 237–256.

Reiter, R.J., Tan, D.X., Terron, M.P., Flores, L.J. and Czarnocki, Z. (2007) Melatonin and its metabolites: new findings regarding their production and their radical scavenging activities. *Acta Biochimica Polonia* 54, 1–9.

Roberts, J.E., Hu, D.N. and Martinez, L. (2000) Photophysical studies on melatonin and its receptor agonists. *Journal of Pineal Research* 29, 94–99.

Rodriguez, C., Mayo, J.C., Sainz, R.M., Antolin, I., Herrera, F., Martin, V. and Reiter, R.J. (2004) Regulation of antioxidative enzymes: a significant role for melatonin. *Journal of Pineal Research* 36, 1–9.

Rosen, J., Than, N.N., Koch, D., Poeggeler, B., Laatsch, H. and Hardeland, R. (2006) Interaction of melatonin and its metabolites with thc ABTS cation radical: extension of the radical scavenger cascade and formation of a novel class of oxidation products, C2-substituted 3-indolinones. *Journal of Pineal Research* 41, 374–381.

Shirazi, A., Ghobadi, G. and Ghazi-Khansari, M. (2007) A radiobiological review of melatonin: a novel radioprotector. *Radiation Research* 48, 263–272.

Stasica, P., Ulanski, P. and Rosiak, J.M. (1998) Melatonin as a hydroxyl radical scavenger. *Journal of Pineal Research* 25, 65–66.

Tan, D.X., Chen, L.D., Poeggeler, B., Manchester, L.C. and Reiter, R.J. (1993a) Melatonin: a potent, endogenous hydroxyl radical scavenger. *Endocrine Journal* 1, 57–60.

Tan, D.X., Manchester, L.C., Burkhardt, S., Sainz, R.M., Mayo, J.C., Kohen, R., Shohami, E., Huo, Y.X., Hardeland, R. and Reiter, R.J. (2001) N^1-acetyl-N^2-formyl-5-methoxykynuramine, a biogenic amine and melatonin metabolite, functions as a potent antioxidant. *FASEB Journal* 15, 2294–2296.

Tan, D.X., Manchester, L.C., Reiter, R.J., Plummer, B.F., Limson, J., Weintraub, S.T. and Qi, W. (2000) Melatonin directly scavenges hydrogen peroxide: a potentially new metabolic pathway of melatonin biotransformation. *Free Radical Biology and Medicine* 29, 1177–1185.

Tan, D.X., Manchester, L.C., Terron, M.P., Flores, L.L. and Reiter, R.J. (2007) One molecule, many derivatives: a never-ending interaction of melatonin with reactive oxygen and reactive nitrogen species? *Journal of Pineal Research* 42, 28–42.

Tan, D.X., Poeggeler, B., Reiter, R.J., Chen, L.D., Chen, S., Manchester, L.C. and Barlow-Walden, L. (1993b) The pineal hormone melatonin inhibits DNA–adduct formation induced by the chemical carcinogen safrole *in vivo*. *Cancer Letters* 70, 65–71.

Tan, D.X., Reiter, R.J., Chen, L.D., Poeggeler, B., Manchester, L.C. and Barlow-Walden, L. (1994) Both physiological and pharmacological levels of melatonin reduce DNA adduct formation induced by the carcinogen safrole. *Carcinogenesis* 15, 215–218.

Tan, D.X., Reiter, R.J., Manchester, L.C., Yan, M.T., El–Sawi, M., Sainz, R.M., Mayo, J.C., Kohen, R., Allegra, M. and Hardeland, R. (2002) Chemical and physical properties and potential mechanisms: melatonin as a broad-spectrum antioxidant and free radical scavenger. *Current Topics in Medicinal Chemistry* 2, 181–197.

Tomas–Zapico, C. and Coto–Montes, A. (2005) A proposed mechanism to explain the stimulatory effect of melatonin on antioxidative enzymes. *Journal of Pineal Research* 39, 99–104.

Towell, J. and Kalyanaraman, B. (1991) Detection of radical adducts of 5,5-dimethyl-pyrroline n-oxide by the combined use of high performance liquid chromatography with electrochemical detection and electron spin resonance. *Journal of Pineal Research* 28, 65–72.

Turjanski, A.G., Leonik, F., Rosenstein, R.E., Pastore, A. and Estrin, D.A. (2000) Scavenging of NO by melatonin. *Journal of the American Chemical Society* 122, 10468–10469.

Urata, Y., Honma, S., Goto, S., Todoroki, S., Iida, T., Cho, S., Honma, K. and Kondo, T. (1991) Melatonin induces gamma–glutamylcysteine synthetase mediated by activator protein–1 in human vascular endothelial cells. *Free Radical Biology and Medicine* 27, 383–847.

Vijayalaxmi, Meltz, M.L, Reiter, R.J., Herman, T.S. and Kumar, S. (1999a) Melatonin and protection from whole–body irradiation: survival studies in mice. *Mutation Research* 425, 21–27.

Vijayalaxmi, Meltz, M.L., Reiter, R.J. and Herman, T.S. (1999b) Melatonin and protection from genetic damage in blood and bone marrow: whole-body irradiation studies in mice. *Journal of Pineal Research* 27, 221–225.

Vijayalaxmi, Reiter, R.J. and Meltz, M.L. (1995a) Melatonin protects human blood lymphocytes from radiation–induced chromosome damage. *Mutation Research* 346, 23–31.

Vijayalaxmi, Reiter, R.J. and Meltz, M.L. (1996b) Melatonin and radioprotection from genetic damage: *in vivo*/*in vitro* studies with human volunteers. *Mutation Research* 371, 221–228.

Vijayalaxmi, Reiter, R.J., Lela, B.Z. and Meltz, M.L. (1996a) Effect of melatonin on mitotic and proliferation indices and sister chromatid exchanges in human blood lymphocytes. *Mutation Research* 351, 187–192.

Vijayalaxmi, Reiter, R.J., Meltz, M.L. and Herman, T.S. (1998) Melatonin: possible mechanisms involved in its "radioprotective" effect. *Mutation Research* 404, 187–189.

Vijayalaxmi, Reiter, R.J., Sewerynek, E., Poeggeler, B., Leal, B.Z. and Meltz, M.L. (1995b) Marked reduction of radiation–induced micronuclei in human blood lymphocytes pretreated with melatonin. *Radiation Research* 143, 102–106.

Vijayalaxmi, Reiter, R.J., Tan, D.X., Herman, P.S. and Thomas, C.R. Jr. (2004) Melatonin as a radioprotective agent: a review. *International Journal of Oncology and Biological Physics* 59, 639–653.

Von Sonntag, C. (1987) *The Chemical Basis of Radiation Biology*, Taylor and Francis, London, UK.

Ward, J.F. (1990) The yield of DNA double-strand breaks produced intracellularly by ionizing radiation: a review. *International Journal of Radiation Biology* 57, 1141–1150.

Winiarska, K., Fraczyk, T., Malinska, D., Prozak, J. and Bryla, J. (2006) Melatonin attenuates diabetes–induced oxidative stress in rabbits. *Journal of Pineal Research* 40, 168–176.

Yilmaz, S. and Yilmaz, E. (2006) Effects of melatonin and vitamin E on oxidative–antioxidative status in rats exposed to irradiation. *Toxicology* 222, 1–7.

Yuhas, J.M., Spellman, J.M. and Culo, F. (1980) The role of WR-2721 in radiotherapy and/or chemotherapy. *Cancer Clinical Trials* 3, 211–216.

Zang, L.Y., Cosma, G., Gardner, H. and Vallyathan, V. (1998) Scavenging of reactive oxygen species by melatonin. *Biochimica Biophysica Acta* 1425, 467–477.

Zhang, H., Squadrito, G.L., Uppu, R. and Pryor, W.A. (1999) Reaction of peroxynitrite with melatonin: a mechanistic study. *Chemical Research in Toxicology* 12, 526–534.

Zhou, G., Kawata, T., Furukawa, Y., Aoki, M., Hirayama, S., Ando, K. and Ito, H. (2006) Protective effects of melatonin against low- and high-LET irradiation. *Journal of Radiation Research* 47, 175–181.

Chapter 7

Radioprotective Effect of Citrus and Hawthorn Extracts against Genotoxicty Induced by Gamma Irradiation

Seyed Jalal Hosseinimehr

Introduction

Ionizing radiation upon interaction with water present in cells produces reactive free radicals, such as hydroxyl radical, hydrogen radical and a toxic substance like hydrogen peroxide that can damage critical biomacromolecules. The free radicals can interact with critical macromolecules, such as DNA, proteins and membranes, to induce cellular damage, biological dysfunction and ultimately death. It is evidenced that damage to DNA may be the most important factor in cell death (Karbownik and Reiter, 2000). Damage to DNA is followed by altered cell division, genotoxic effects and cell death. Although our bodies are equipped with endogenous enzymes (e.g., superoxide dismutase) that are able to detoxify and remove the products of water radiolysis, when these reactive oxygen species increase in the biological system (e.g., during exposure to irradiation), then the endogenous system can not defend the cells from the hazardous effects of free radicals. Radioprotective agents are synthetic or natural origin compounds that are administrated before irradiation to reduce injuries caused by ionizing radiation. Sulfhydryl compounds have been among the first radioprotective agents (Hosseinimehr, 2001, 2002). Amifostine as a thiol compound is a powerful radioprotective agent, but its application is clinically limited due to side effects and toxicity (Turrisi *et al.*, 1988). It is commonly believed that natural compounds provide functional antioxidants, such as vitamins, minerals and enzymes. Reduction of oxidative damage by natural antioxidants provides a degree of protection against ionizing radiation injury. We have earlier reported that herbal medicines can protect mouse bone marrow cells against gamma irradiation (Hosseinimehr, 2003; Hosseinimehr *et al.*, 2007). Citrus and hawthorn are two common herbal medicines that can be used to mitigate genotoxicty induced by gamma irradiation in mouse bone marrow cells. In this respect, the present chapter highlights the results of radioprotective effects of these plants and discusses the approaches for the application of these agents.

Materials and Methods

Chemicals

Hesperdin, chlorogenic acid, (-)- epicatechin, rutin, hyperoside and 1,1-diphenyl-2-picryl hydrazyl radical (DPPH) were purchased from Fluka Chemical Co. (Buchs, Switzerland). All other chemicals and acetonitrile (HPLC grade) were obtained from Merck Company.

Preparation of *Crataegus* and *Citrus* extract

The ripe fruits of *Crataegus microphylla* were collected from Neka in the north of Iran. The peels of hawthorn were dried at room temperature and powdered in a grinder. Aqueous methanolic solution (75%) was added to the powdered peels (20g) and stirred for one hour. The mixture was kept at room temperature for 72h. After filtration, methanol was evaporated under reduced pressure at 40°C. The remaining water extract was freeze-dried. The yield obtained was 6g.

Citrus aurantium var. amara, which is consumed as a ripe fruit in Iran, was collected from Amol region, Iran. Dried peels of the plant (100g) were extracted at room temperature with 75% aqueous ethanol (750ml) for 24h. After filtration, the solvent was evaporated under reduced pressure at 40°C, until the ethanol was removed. The aqueous portion was shaken with chloroform to remove liposoluble substances, and the aqueous layer was evaporated under reduced pressure. In this way, 18g of extract, in the form of a dried powder, was obtained (18% w/w).

HPLC analysis

The phenolics present in the hawthorn fruits were analyzed by HPLC method as described by Zhang and co-workers with brief modification (Zhang *et al.,* 2001).

Treatments

All the solutions were freshly prepared before treatment of the animals. Three test doses (250, 500 and 1000mg/kg) of freshly prepared *Citrus aurantium* were administered to the experimental animals intraperitoneally (i.p.). In case of hawthorn extract, four doses (25, 50, 100 and 200mg/kg) were administrated to the experimental animals 1h before gamma irradiation. Amifostine, used as a control, was dissolved in distilled water and injected at a dose 100mg/kg in mice. The control animals received the same volume of distilled water. Five mice were used for each treatment group.

Irradiation

Whole-body irradiation was performed with a cobalt-60 γ-radiation source (Theratron 780, Canada). Mice were placed in ventilated Plexiglas cages and irradiated in groups of five mice simultaneously.

Micronucleus assay

The mouse bone marrow micronucleus test was carried out according to the method described by Schmid for evaluation of the chromosomal damage in experimental animals (Schmid, 1975). Animals were sacrificed by cervical dislocation 24h after irradiation. The bone marrow from both femurs was flushed in the form of a fine suspension into a centrifuge tube containing foetal calf serum (FCS). The cells were dispersed by gentle pipetting and collected by centrifuge at 400g for 5min at 4°C. The cell pellet was resuspended in a drop of FCS, and bone marrow smears were prepared. The slides were coded to avoid observer bias. After 24h air-drying, smears were stained with May-Grunwald/Giemsa as described by Schmid. With this method, polychromatic erythrocytes (PCEs) stained reddish-blue, and normochromatic erythrocytes (NCEs) stained orange, while nuclear material stained dark purple. For each experimental point, five mice were used, and a total of 5000 PCEs were scored per experimental point to determine the percentage of micronucleated polychromatic erythrocytes (MnPCEs), micronucleated normochromatic erythrocytes (MnNCEs) and the ratio of PCE to (PCE + NCE). The ratio of PCE to (PCE + NCE) was determined for each experimental group to assess radiation effects with or without hawthorn extract on bone marrow proliferation (Mozdarani and Gharabi, 1993).

Measurement of free radical scavenging activity

The free radical-scavenging capacity of extracts was determined as bleaching of the stable 1,1-diphenyl-2-picrylhydrazyl radical (DPPH) (Koleva *et al.*, 2002). Different concentrations of extracts in equal volume were added to methanolic solution of DPPH (100μm). After 15 min at room temperature, the absorbance was recorded at 517nm. The experiment was performed in triplicate. BHT was used as an antioxidant standard.

Statistical analysis

The data are presented as mean ± SD. One-way ANOVA analysis and Tukey's HSD test were used for multiple comparisons of data.

Results

The effects of gamma irradiation with or without *Citrus* and *Crataegus* extracts on the induction of MnPCEs and the PCE/PCE+NCE ratio in bone marrow cells, 24h after γ-irradiation, are shown in Tables 7.1. and 7.2. respectively.

Group	Treatment	MnPCE/PCE (%)*	MnNCE/NCE (%)*	PCE/PCE+NCE (%)*
1	Control	1.35 ± 0.44	1.23 ± 0.37	53.56 ± 3.4
2	Irradiation	4.77 ±1.9 [a]	4.35 ± 1.5 [a]	50.1 ± 2.7
3	250mg/kg Citrus + irradiation	2.28 ± 1.05 [b]	2 ± 0.98 [b]	53.58 ± 1.9
4	500mg/kg Citrus + irradiation	2.60 ± 0.55 [b, c]	1.86 ± 0.6 [b, c]	52.6 ± 2.7
5	1000mg/kg Citrus + irradiation	1.90 ± 0.92 [b, c]	2.7 ± 1.04 [b, c]	53.9 ± 3.5
6	1000mg/kg Citrus	1.38 ± 0.98 [d]	1.53 ± 0.8 [d]	60.9 ± 5.5

Table. 7.1. Effects of citrus extract on the formation of radiation-induced micronuclei PCE and NCE and ratio of PCE/PCE+NCE in Balb/c male mice bone marrow exposed to 1.5Gy γ-radiation.

* Values are means ± SD for each group of five mice.
a: $p < 0.005$ compared to control, b: $p < 0.05$ compared to irradiation alone.
c: no significant difference compared to 250mg/kg citrus + irradiation group.
d: no significant difference compared to control.

Group	Treatment	MnPCE/PCE (%)*	PCE/PCE+NCE (%)*
1	Control	0.25 ± 0.19	52.6 ± 0.97
2	Irradiation	11.50 ± 0.87[a]	43 ± 1.73[a]
3	25mg/kg Hawthorn + irradiation	6.73 ± 0.46[b]	47.25 ± 0.5[e]
4	50mg/kg Hawthorn + irradiation	5.76 ± 0.96[b]	48.20 ± 1.3[e]
5	100mg/kg Hawthorn + irradiation	5.97 ± 1.10[b]	48.67 ± 2.08[e]
6	200mg/kg Hawthorn + irradiation	2.02 ± 0.49[b,c]	48.3 ± 0.84[e]
7	100mg/kg Amifostine + irradiation	2.38 ± 0.22[b]	48.43 ± 1.77[e]
8	100mg/kg Hawthorn	0.43 ± 0.2[d]	52 ± 2.45
9	200mg/kg Hawthorn	0.46 ± 0.22[d]	50 ± 0.73
10	100mg/kg Amifostine	0.27 ± 0.12[d]	52 ± 1

Table 7.2. Effects of hawthorn fruits extract on the formation of radiation-induced micronuclei PCE and the ratio of PCE/PCE+NCE in NMRI male mice bone marrow exposed to 2Gy gamma irradiation.

* Values are means ± SD for each group of five mice. [a] $p<0.0001$ compared to control, [b] $p < 0.001$ compared to irradiation alone, [c] $p < 0.001$ compared to 100mg/kg hawthorn + irradiation group. [d] no significant difference compared to control. [e] $p < 0.01$ compared to irradiation alone.

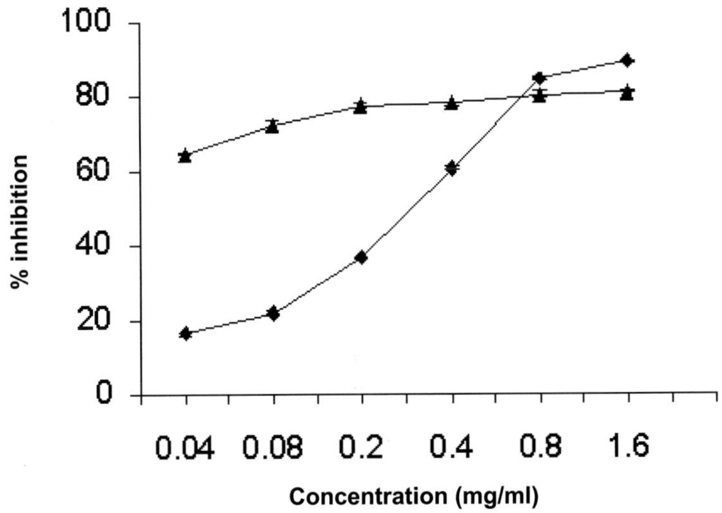

Fig. 7.1. Scavenging effect of different concentrations (0.04 to 1.6mg/ml) of citrus extract (■) and BHT (▲) on the 1,1-diphenyl-2-picrylhydrazyl (DPPH) free radical. The absorbance was recorded at 517nm after 15 min at room temperature. The experiment was performed in triplicate.

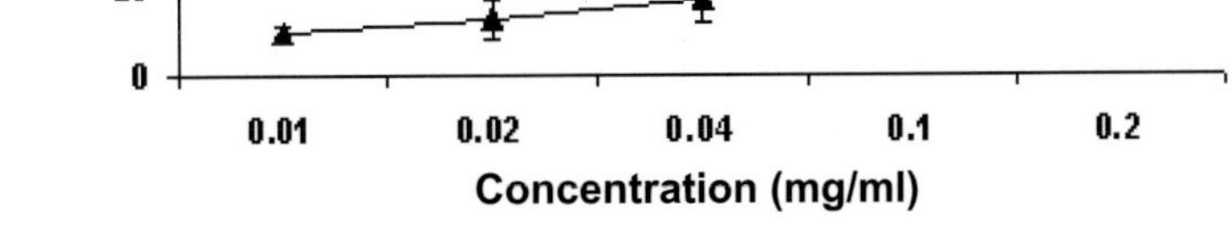

Fig. 7.2. Scavenging effect of different concentration (0.01–0.2mg/ml) of *Crataegus microphylla* (▲) and butylated hydroxytoluene (BHT) (●) on the 1,1-diphenyl-2-picrylhydrazyl (DPPH) free radical. The absorbance was recorded at 517nm after 30 min at room temperature. The experiment was performed in triplicate.

The frequency of micronuclei was increased in all groups of γ-irradiated mice in comparison to the control group. The frequency of MnPCE found in the extract-treated groups was significantly lower than that of the group treated only with radiation. In comparison to the irradiation-control group, the total MnPCE values were 2.1, 1.8 and 2.5-fold less in the citrus group after being exposed to 1.5Gy respectively, than those in the respective irradiated control (Table 7.1).

These values were 1.71, 2, 1.93 and 5.7 times less than those in the treatment group receiving hawthorn extract after being exposed to 2Gy. In this experiment, administration of amifostine at a dose 100mg/kg reduced the frequency of MnPCE up to 4.8 fold (Table 7.2). The data showed a suppressive effect of citrus and hawthorn extracts on radiation-induced clastogenic effects. Treatment of mice with citrus and hawthorn extract arrested radiation-induced decline in the PCE/PCE+NCE ratio, and the increase in the PCE. PCE+NCE ratio in the extract + irradiated groups was higher than that of the irradiated-alone group, though the difference was not statistically significant.

Antioxidant activity

DPPH free radical scavenging method was used to evaluate the antioxidant activity of specific compound. Excellent scavenging effect was observed with *Citrus* and *Crataegus* extract. The scavenging effects were obtained more 90% for these extracts (Figs. 7.1. and 7. 2).

Phenol and flavonoid analysis

The flavonoid content of *Citrus* extract was determined spectrophotometrically (Hosseinimehr, 2004). The flavonoid content was found to be 9.2 ± 0.17g/100g in extract powder, utilizing hesperidin as a standard, indicating that citrus peel extract contains high amounts of flavonoids. The phenolics contained in hawthorn fruits were analyzed by HPLC method (Hosseinimehr *et al.,* 2007). Three analytes *viz.,* chlorogenic acid, epicatechin and hyperoside, were identified in the hawthorn extract (Hosseinimehr *et al.,* 2007). The epicatechin content was 15.96mg/g of extract powder.

Discussion

We report that *Citrus* and *Crataegus* extracts could protect mice bone marrow cells against gamma irradiation (Hosseinimehr, 2003; Hosseinimehr *et al.*, 2007). Exposure to γ-irradiation increased DNA damage, and an elevation in the frequency of MnPCE was observed after irradiation in irradiated animal groups (Hosseinimehr, 2003; Hosseinimehr and Nemati, 2006; Hosseinimehr *et al.*, 2007). In previous reports we showed that synthetic compounds containing thiol protect mice against a lethal dose of γ-irradiation (Hosseinimehr *et al.*, 2001, 2002). Although synthetic compounds, mainly thiol compounds, have good radioprotective effects, they are limited in their use by side effects (Turrisi *et al.*, 1988). Natural compounds, including phenolic and flavonoids, may affect scavenging free radicals, such as hydroxyl radicals, generated by γ-rays in cells. There is a possibility that pre-treatment with herbal medicine could induce protection against gamma irradiation. Our studies showed that citrus and hawthorn extract contain high amounts of phenolic acid and flavonoids (Hosseinimehr *et al.,* 2003). Flavonoids have strong antioxidant activity (Hosseinimehr, 2004; Hosseinimehr and Nemati, 2006). These extracts also showed good antioxidant properties which could be attributed mainly to the presence of flavonoids and phenolic acid in these plants.

Maximum protection against the side effects of γ-irradiation was observed with citrus extract at a dose 250mg/kg (almost 2.1 fold) and hawthorn extract at a dose 200mg/kg (almost 5.7 fold). The results of these experiments showed that the percentage of MnPCE significantly reduced upon pre-treatment with these extracts. The percentage of PCE/PCE+NCE ratio declined in irradiated mice, since this ratio gave a direct index of cell division. Citrus and hawthorn protected mice against radiation-induced decline in cell proliferation, as was indicated by the increased PCE/PCE+NCE ratio. We have shown that hawthorn extract contains phenolic compounds (Hosseinimehr *et al.*, 2007). All of these compounds have antioxidant activity and exhibit protective effects against oxidative stress induced by hazardous chemicals.

In our study, we showed that single i.p. administration of hesperidin, a flavonone glucoside, at a dose of 80mg/kg 45 min prior to gamma irradiation (2Gy) reduced the frequencies of MnPCEs. The total MnPCE values were 2.85 times lower in the hesperidin group as compared to those in respective irradiated control (Hosseinimehr and Nemati, 2006). The mitigating effects of other flavonoids, orientin, vicenin, naringin, quercetin and rutin have been previously investigated against genotoxicity induced by radiation in the mice bone marrow cells (Jagetia and Reddy, 2002; Devi *et al.*, 1998; Shimoi *et al.*, 1994; Ganasoundari *et al.* 1998). These flavonoids significantly protected mice bone marrow when administrated at low doses before exposure to irradiation. The protective effect of flavonoids in mice may be attributed to the hydroxyl radical scavenging activity, which acts in a direct way or behaves like an endogenous enzyme (Shimoi *et al.*, 1994). The dose used for protective effects of herbal preparation was lower than the toxic dose, and is the biggest advantage of using these agents as compared to synthetic compounds. Further studies are necessary to identify the bioactive compounds responsible for the radioprotective efficacy, and to extend the time window (e.g., 24h prior irradiation). In conclusion, we report that citrus and hawthorn extracts contain high amounts of phenolic compounds with antioxidant activity and reduced genotoxicity induced by gamma irradiation in mice bone marrow cells.

Although many plants have been evaluated for reducing the radiation-induced sickness in animals, there have not been enough clinical trials for the potential use of these plants in patients during radiotherapy. Since citrus and hawthorn are edible plants, they hold promise for evaluation in clinical trials.

References

Devi, P.U., Bisht, K.S. and Vinitha, M. (1998) A comparative study of radioprotection by *Ocimum* flavonoids and synthetic aminothiol protectors. *British Journal of Radiology* 71, 782–784.

Ganasoundari, A., Uma Devi, P. and Rao, B.S.S. (1998) Enhancement of bone marrow radioprotection and reduction of WR-2721 toxicity by *Ocimum sanctum. Mutation Research* 397, 303–312.

Hosseinimehr, S.J. (2004) Determination of flavanone content and antioxidant activity of *Citrus aurantium. Chemistry and Industry Journal* 1, 635–638.

Hosseinimehr, S.J. and Nemati, A. (2006) Radioprotective effects of hesperidin against gamma irradiation in mouse bone marrow cells. *British Journal of Radiology* 79, 415–418.

Hosseinimehr, S.J., Azadbakht, M., Mousavi, S.M., Mahmoudzadeh, A. and Akhlaghpoor, S. (2007) Radioprotective effects of hawthorn fruit extract against gamma irradiation in mouse bone marrow cells. *Journal of Radiation Research* 48, 63–68.

Hosseinimehr, S.J., Shafiee, A., Mozdarani, H. and Akhlagpour, S. (2001) Radioprotective effects of 2-iminothiazolidine derivatives against lethal doses of gamma radiation in mice. *Journal of Radiation Research* 42(4), 401–408.

Hosseinimehr, S.J., Shafiee, A., Mozdarani, H., Akhlagpour, S. and Froughizadeh, M. (2002) Radioprotective effects of 2-iminothiazolidine-3-[(chromone-2-yl)carbonyl] thiazolidines against gamma-irradiation in mice. *Journal of Radiation Research* 43(3), 293–300.

Hosseinimehr, S.J., Tavakoli, H., Pourheidari, G.R., Sobhani, A. and Shafiee, A. (2003) Radioprotective effects of citrus extract against gamma irradiation in mouse bone marrow cell. *Journal of Radiation Research* 44, 237–241.

Hosseinimehr, S.J., Tavakoli, H., Pourheidari, G.R., Sobhani, A. and Shafiee, A. (2003) Radioprotective effects of citrus extract against gamma irradiation in mouse bone marrow cells. *Journal of Radiation Research* 44, 237–241.

Jagetia, G.C. and Reddy, T.K. (2002) The grapefruit flavanone naringin protects against the radiation-induced genomic instability in the mice bone marrow: a micronucleus study. *Mutation Research* 519, 37–48.

Karbownik, M. and Reiter, R.J. (2000) Antioxidative effects of melatonin in protection against cellular damage caused by ionizing radiation. *Experimental Biology and Medicine* 225, 9–22

Koleva, I., Van Beek, T.A., Linssen, J.P.H., De Groot, A. and Evstatieva, L.N. (2002) Screening of plant extracts for antioxidant activity: a comparative study on three testing methods. *Phytochemical Analysis* 13, 8–17.

Mozdarani, H. and Gharabi, A. (1993) Radioprotective effects of cimetidine in mouse bone marrow cells exposed to γ-rays as assayed by micronucleus test. *International Journal of Radiation Biology* 64, 189–194.

Schmid, W. (1975) The micronucleus test. *Mutation Research* 31, 9–15.

Shimoi, K., Masuda, S., Furugori, M., Esaki, S. and Kinae, N. (1994) Radioprotective effect of antioxidative flavonoids in γ-ray irradiated mice. *Carcinogenesis* 15, 2669–2672.

Turrisi, A.T., Glover, D.G. and Hurwitz, S. (1988) The final report of the phase I trial of single dose WR-2721, s-2-(3-aminopropylamino)ethyl phosphorothioic acid, intravenous injection. *Cancer Treatment Reports* 70, 1389–1393.

Zhang, Z., Chang, Q., Zhu, M., Huang, Y., Ho, W.K. and Chen, Z. (2001) Characterization of antioxidants present in hawthorn fruits. *Journal of Nutritional Biochemistry* 12, 144–152.

Chapter 8

The Healing Potential of Indigenous Essential Oils from New Zealand in the Prevention and Management of Radiation-induced Mucositis

W. Maddocks-Jennings

Introduction

There are two main indigenous essential oils produced commercially from plants in New Zealand for aromatherapy and medicinal purposes. The first is Manuka, *Leptospermum scoparium* J.R et G. Forst. The other is Kanuka, *Kunzea ericoides* A. Rich Thompson, (formerly *Leptospermum ericoides*). Manuka is also known as Kahikatoa, Red Manuka and teatree. Kanuka is also known as White or Tree Manuka as it is larger than Manuka, with smaller leaves, flowers and fruit with a white wood (Brooker *et al.,* 1987).

Of the 79 *Leptospermum* species, Manuka is the only one that is native to New Zealand (Perry *et al.,* 1997). The common name of tea tree relates to the uses that Captain Cook had for the dried leaves. Historically the leaves and twigs of both Manuka and Kanuka have been used by both the indigenous Maori and early European settlers. These uses include infusions for urinary and internal complaints, as a febrifuge, sucking the gum for coughs, bark infusion is also used for diarrhoea, vapour inhalations for colds, poultices for back pain and skin conditions, inflamed breasts, burns and scalds, mouthwashes and gargles, gum diseases etc. (Brooker *et al.,* 1987; Riley, 1994). The wood was also used for canoe structures, as fishing tools, gardening tools and war weapons. As it burned easily, it was used for firewood. The gum is called Manna, which has mannitol, a diuretic, as a main ingredient (Riley, 1994). Recent interest has focussed on the use of honey produced from Manuka flowers in the healing of chronic skin ulcers and other wounds. This use has been extensively researched and further information can be obtained from The Waikato Honey Research Unit at the University of Waikato, New Zealand (http://bio.waikato.ac.nz/honey/).

The areas where Manuka oil is harvested in New Zealand for distillation range in latitude from 35°24’ in Auckland, in the North Island, to 45°53’ in Otago in the South Island (see Fig. 8.1). Recently there has been interest in the possible therapeutic uses of the distilled essential oils of both Manuka and Kanuka, which predominantly centres on the anti-microbial attributes. This chapter explores the potential to use a combination of these oils to reduce the effects of radiation on the oropharyngeal mucosa during radiotherapy for head and neck (HAN) cancers.

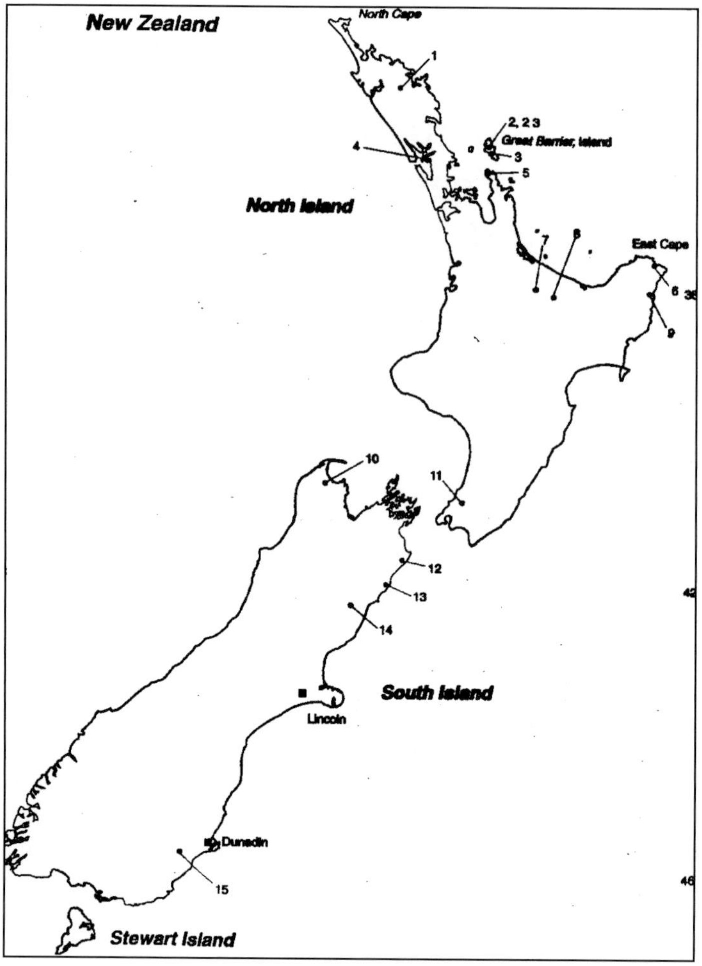

Figure 8.1. Locations in New Zealand from where Manuka essential oil is harvested (Perry *et al.,* 1997).

Mucositis

Radiation-induced mucositis is an expected and unpleasant side effect of radiotherapy, which includes any area of mucosal tissue. Exact figures vary, but it is generally accepted that 80% of patients will develop this, and of these between 11–16% will require either cessation of radiotherapy or hospitalization for complications (Scully *et al.,* 2003; Trotti *et al.,* 2003). Clearly these have an impact on the overall patient well-being and their long-term recovery from illness (Bhatia *et al.,* 1996). There are significant efforts worldwide in understanding the aetiology and management of radiation-induced mucositis, but it appears that there still is no definitive oral care solution available. The main symptoms experienced include erythema, ulceration and pain (Mueller *et al.,* 1995). Irradiation of the salivary glands removes the protective physical and bioactive layer, which can intensify the appearance of mucositis (Duncan and Grant, 2003). Any or all of these subsequently affect the ability to maintain adequate nutrition and hydration through oral intake.

The oral cavity is lined with stratified squamous epithelium, with an average cell turn over of 4–8 days. Underneath the mucosal epithelium, the lamina propria, a network of connective tissue and blood vessels, contains fibroblasts, macrophages and mast cells along with other cells of the immune system. Along with saliva, other substances are secreted which assist in protecting the mucosa. These include lactoferrin and lysozymes as well as epidermal growth factor (EGF) (Duncan and Grant, 2003).

Basal keratinocytes, the cells responsible for epithelial regeneration, have a four-day life cycle and in the oral mucosa the epithelium is 3–4 cells thick. It is thought though that there is a difference in the radiosensitivity of the basal layer, which may determine the degree of response to radiotherapy (Scully *et al.,* 2003, 2004). Radiation therapy for HAN cancer is usually given in fractionated doses of between 180–200cGy a day for 25–30 days. Radiation has a cumulative effect on the body and typically symptoms commence after the first week or 1500–2000cGy, due to the disruption of the rapidly dividing epithelial cells. Some degree of ulceration is expected by the third week or 3000cGy (Makkonen *et al.,* 2000; Scully *et al.,* 2003, 2004). Apart from the direct effects of radiation, epithelial loss and sub-mucosal damage is also due to physical, chemical and microbial factors involved in complex interactions during cellular death occurring via apoptosis (Epstein and Wong, 1994; Pevlová, 1999; Scully *et al.,* 2003).

The severity of mucositis in radiotherapy patients is determined by the radiation dose, the field size and fractionation schedules and can be negatively affected by smoking tobacco and concomitant chemotherapy (Scully *et al.,* 2003). It has been estimated that smoking during radiotherapy of the head and neck can increase the severity of the reaction by up to 50% as well as prolonging the recovery time (Rugg *et al.,* 1990).

Studies to investigate whether the presence or absence of certain microbes contributes to the development of mucositis remain inconclusive. It is, however, believed that there is some link. The presence of Gram-negative bacteria may potentiate the immune response, which may contribute in some way to the aetiology of mucositis. However, it is not clear how significant this role may be (Scully *et al.,* 2003). Recently it has become more apparent that Gram-positive organisms may have a role in the development of mucositis (Duncan and Grant, 2003). The normal oral flora may be altered by radiotherapy, which opens the way for other organisms to enter the tissue such as *Candida albicans*, α–haemolytic streptococci and herpes viruses (Scully *et al.,* 2003). Smoking, antibiotics, dental prosthetics and alcohol exacerbate oral candidiasis during radiotherapy (Scully *et al.,* 2004). Various combinations of antibacterial and anti-fungal mouthwashes have been trialled with mixed outcomes (Scully *et al.,* 2003). The addition of antifungal or antibacterial preparations may be routine in some centres (Gava *et al.,* 1996; Wijers *et al.,* 2001; Sutherland and Browman, 2001; Koç and Atkas, 2002; Stokman *et al.,* 2003; El-Sayed *et al.,* 2002a,b; Trotti *et al.,* 2004).

The disruption of mucosa causes pain, which is usually continuous and exacerbated by eating, drinking, performing mouth care and when sleeping. This is often combined with fatigue and a dry mouth. Pain affects up to 90% of patients and increases dramatically in the presence of any ulceration, impacting on daily life (McGuire *et al.,* 1993). Once mucositis has appeared it is necessary to protect the area from further damage and to reduce the experience of pain. Mild analgesic and anti-inflammatory agents are used, unless the pain is severe when opioids are the preferred agent.

Table 8.1. Geographical locations of Manuka oil in New Zealand with variations of composition of monoterpenes.

S. No.	Geographical location	% of monoterpenes
1	Ngawha, Auckland	40
2	Great Barrier Island	10
3	Great Barrier Island	11
4	Kaipara Harbour, Auckland	30
5	Coromandel Peninsula	6
6	East Cape	3
7	Rotorua	14
8	Mt. Tarawera	3
9	Gisborne	6
10	Nelson	19
11	Wellington	5
12	Marlborough	18
13	Marlborough	10
14	Canterbury	8
15	Otago	9

Manuka and Kanuka

Manuka is known to have approximately 100 different constituents, with around 50% being identified with certainty (Christoph *et al.*, 1998).

Essential oil yield ranges from 0.17% to 0.57%, depending on the location grown. Until recently it hasn't been appreciated that there are significant variations in the composition of Manuka and Kanuka oil depending on the geographical location of the plants. The main commercial interest has been in oil produced on the East Cape of the North Island where there are high levels of triketones in Manuka.

This constituent is attributed to having the main antibacterial and anti-infective properties (Porter and Wilkins, 1999). Oils from plants grown in the North Island, especially around the East Coast area have higher (>30%) triketone levels (Leptospermone and flaveone), whereas South Island oils contain more sesquiterpene hydrocarbons and oxygenated hydrocarbons (up to 65%) (Perry *et al.*, 1997). Variances in composition have also been noted depending on the age of the plant. For example the amount of monoterpenes (alpha pinene, beta pinene and myrcene) increases from less than 1% in young trees to between 17–34% from trees that are three years older. These variations are also seasonal, with pinene levels at their highest in the spring and summer time when the foliage is growing. These factors are all important when considering commercial production (Porter *et al.*, 1999).

Further classification has been provided by Porter and Wilkins (1999), who have identified four main groups depending on the presence or absence of triketones.

i) Group I Manuka are rich in triketones and occur primarily in the East Coast of the North Island of New Zealand.

(ii) Group II are high in linalool and eudesamol, and mainly occur in the Nelson region, Northern South Island.

(iii) Group III are rich in pinenes from Canterbury in the Central South Island

(iv) All others.

Despite the obvious differences in constitution, the geographical location is rarely noted on price lists.

One constituent of potential therapeutic interest is the presence of (-)-trans-calamenene, a sesquiterpene hydrocarbon (9.6–18.5%) in South Island samples. Sesquiterpene hydrocarbons contain 15 carbon atoms, and are not very common in essential oils, compared to the more plentiful monoterpene hydrocarbons (Tisserand and Balacs, 1995). As a group, sesquiterpene hydrocarbons are believed to be antiseptic, bactericidal, analgesic and anti-inflammatory (Sheppard-Hangar, 1994). Total sesquiterpene content can range from 60–70%, with about 30 individual sesquiterpenes noted (Christoph *et al.*, 1998). Table 8.2 summarizes the main chemical constituents from the essential oils of the Myrtaceae family. It can be clearly seen that there are chemical differences amongst them all and that it is not appropriate to believe they have interchangeable actions.

Table 8.2. Comparisons of main constituents (%) of Myrtaceae family (Christoph *et al.,* 2000; Perry *et al.,* 1997; Analysis of Oils from Brooklyn Valley Essential Oils, New Zealand).

	M. alternifolia	*M. cajeputi*	*M. quinquenervia*	S. Island *L. scopiarum*	N. Island *L. scopiarum*	S. Island *K. ericoides*
Alpha pinene	0	0	0	1.3	2.9	61.6
Beta pinene	0.9	0	20	0.5	0.5	0
Limonene	0	0	0.4	<1%	?	1.4
Monoterpenes	11.8	0.4	0	33	?	7
1,8,cineol	5.1	0	0	2.4%	?	5.1
Terpinene-4-ol	>10	60	0	0	0	0
Other alcohols	0	0.3	0	>7%	2.4	4.9
Sesquiterpenes	0	0	0	19	64	>1%
Gamma terpinene	0.5	0	0	0	-	1.4
Alpha terpinene	1.4	0	0	0	-	0
Leptospermone	0	0	0	17.6	1.7	0
Viridiflorol	0	0	3.2	0	0	3.2
Alpha terpineol	0	20	5.7	0	-	1.1
Beta triketones	0	0	0	27.8	2.4	0

Antibacterial potential

Most of the present interest in Manuka and Kanuka is related to the potential antimicrobial effects of these oils in comparison to tea tree (*M. alternifolia*) and other essential oils from the Myrtaceae family (Cooke and Cooke, 1991). Tea tree is more effective against Gram-negative organisms than both Manuka and Kanuka. However, Manuka was at least 20 times stronger on Gram-positive organisms than both tea tree and Kanuka. Manuka also had some activity against skin and ringworm fungi (Cooke and Cooke, 1991). Other researchers have found variable antimicrobial activity with some suggestion that combining Manuka and Kanuka is more effective than using either oil individually (Lis-Balchin *et al.,* 1996; Lis-Balchin and Hart, 1998). Kanuka oil has the additional potential to be used against Herpes Simplex Type 1 and Polio Type 1 viruses. The active compounds have been identified as isomers of isobutyryl methoxyresoorcinol and unnamed beta triketones (Bloor, 1992).

Other researchers have found that Kanuka was effective against Gram-positive organisms at concentrations of 0.2–0.4%, but was not effective against yeasts (Christoph *et al.,* 2000). Manuka had mixed effects on all microorganisms, with greater potency against the Gram-negative organisms at concentrations of 0.05%–0.15%. It was the most effective of all the whole oils against *P. aeruginosa* at 0.85%, compared to 1–2% for the rest of the whole oils. Manuka had virtually no effect against the moulds and yeast, but was most effective against the dermatophytes (at 0.3%), compared to 0.6%–1.1% for others. The lack of effect of Kanuka oil on *C. albicans* may be due also to the high % of terpene hydrocarbons (85%), whereas the other oils, which are effective, contain lower amounts and also sesquiterpene hydrocarbons (Christoph *et al.,* 2000).

Evaluating the benefits of Manuka and Kanuka on the oral mucosa

Based on the previously available information about Manuka and Kanuka possessing potential anti-inflammatory and analgesic effects, dependent on the sesquiterpene and alpha pinene content, a pilot study was conducted for evaluating the effects of a gargle of South Island grown Manuka and Kanuka essential oils on radiation-induced mucositis (RIM), following treatment for head and neck cancers (HAN). This study was conducted in Palmerston North, New Zealand, and is believed to be the first such study exploring the use of these essential oils in this way.

In this pilot placebo-controlled, randomized, blind study, patients gargled with a mix of the essential oils in water for up to five times per day every day of their radiotherapy and for two weeks afterwards. Control group patients did not gargle and the placebo group used a sterile water mouthwash. Besides evaluating the development of mucositis, other clinical outcomes were also measured. These included the total pain experience, nutritional status and the effects of fatigue and memory loss. Of the twenty-six patients who were originally recruited, nineteen completed all stages of the trial and sufficient data was available for analysis. Due to this, smaller than the anticipated number, it has not been possible to infer effects across all patient groups for all the variables measured. The key outcomes have been reproduced here, which are supportive of a positive outcome. Based on this study there is a clear need for larger studies. Of note though is that the gargle was generally well tolerated with a high degree of patient compliancy (over 80%, compared to 60% compliance with hospital prescribed gargles). No side effects were reported, which could be attributed to the gargle and/or swallow solution.

Quality of life

At the completion of their radiotherapy treatment, patients were asked to reflect back over their treatment as to how much they felt they were affected by their treatment and symptoms experienced. The following scoring system was used (refer Table 8.3). Clusters of symptoms, as reported by patients, were grouped together and are represented in Table 8.4.

Table 8.3. Measuring tool to record subjective experience of radiotherapy at the completion of treatment.

Score	Descriptors
0	Did not feel at all
1–3	Felt symptom mildly and infrequently, not affecting day-to-day activities
4–6	Felt symptom often, moderately intensely and starting to affect day-to-day activities
7–9	Feeling symptom every day, could barely function through the day or perform daily activities
10	Symptom overwhelmed the patient

Table 8.4. Subjective symptoms of patients reflected at the end of radiotherapy.

	Group of symptoms
Eating	All food tastes same, foul taste of food, inability to eat solid food, lack of taste, loss of appetite
Pain and discomfort	Pain when eating, sore throat, pain when not eating
Fatigue and memory loss	Loss of concentration, memory loss and fatigue
Lack of secretions	Dry cough, dry mouth, not enough mucous or saliva
Excess secretions	Too much saliva or mucous, nausea + vomiting

Clinical outcomes

The development of radiation-induced mucositis

Patients gargled up to five times per day with either the active gargle solution mixed with warm water or the placebo solution mixed with warm water. Gargles were timed to be either 30 minutes before or after eating, drinking or smoking.

On radiation days they were also asked to gargle before and after their radiotherapy, if they could. The hypothesis behind the use of the gargle was that the severity of their reactions experienced would be reduced, as well as the length of time the patient actually had mucositis. The actual radiation dose (centigray) that induced mucositis was the point of measure, along with the highest mucositis score obtained. The scoring system used was a 0–4 scale, with radiation ceasing at a grade 3/4 (Table 8.5). For the nineteen patients who completed this trial, their first mucosal reactions occurred from 400cGy to 4400cGy (Table

8.6). The usual daily dose was 200cGy for five days a week. This means that for one patient their mucositis developed after only two days of radiotherapy. As expected, all patients in this pilot study experienced some mucositis, with 42% (8/19) experiencing a grade 1/4 and 53% (10/19) experiencing a grade 2/4. One patient experienced a 3/4 reaction and had her treatment halted on two separate occasions to allow her oral area to heal.

Table 8.7 presents the mean results obtained relating to the development of RIM presented according to the three research groups. This shows the different radiation doses (cGy) where reactions were first experienced, as well as the highest mucositis scores experienced during treatment. The active gargle group went the longest time through treatment until mucositis first developed at a dose of 2667cGy, which is 13.3 treatment days at the usual dose of 200cGy per day. The placebo group experienced mucositis earliest at 1600cGy, or after eight treatment days.

Table 8.5. MidCentral health assessment tool for oral mucositis.

Grade	Reactions of the oropharynx	Reactions of the oral cavity	Patient effects
0	Odynophagia (no pain on swallowing)	Normal	No pain or discomfort, normal mucosa
1	Mild dysphagia-soft diet and/or local anaesthetic required	Erythema-mild pain, no analgesia required	Soft diet, mild pain, no analgesics required, may need topical anaesthetic for eating
2	Moderate dysphagia-liquid diet and narcotic analgesia required	Patchy mucositis-inflammation, moderate pain, serosanginous ooze	Liquid diet, moderate pain, requiring narcotic analgesia
3	Severe dysphagia dehydration, >15% weight loss, artificial feeding	Confluent fibrinous mucositis-severe pain requiring narcotic analgesia	Dehydration, weight loss of >15%, artificial feeding required, severe pain requiring regular narcotics
4	Complete obstruction, ulceration, perforation or fistula	Ulcer, haemorrhage, tissue necrosis	Patient extremely unwell. With careful management this would not be seen as treatment is paused or discontinued at a grade 3
5	Death		

Table 8.6. Development of mucositis for individual patients (expressed in cGy).

Patient	cGy at no mucositis (treatment days)	Highest reaction (0–4)
1	2400 (12)	2
2	2520 (14)	2
3	3000 (15)	2
4	3000 (15)	1
5	2000 (10)	1
6	800 (4)	1
7	1200 (6)	2
8	400 (2)	2
9	1800 (9)	2
10	1200 (6)	1
11	1600 (8)	3
12	1400 (7)	2
13	3200 (16)	1
14	1600 (8)	2
15	2000 (10)	2
16	2600 (13)	1
17	1800 (9)	1
18	4400 (22)	1
19	3600 (18)	2

Table 8.7. Development of mucositis across treatment according to research group allocation.

	Active (6)	Control (7)	Placebo (6)
cGy at no mucositis	2667 (1114)	2131 (855)	1600 (921)
cGy at 1/4	3120 (1136)	2163 (907)	1450 (661)
cGy at 2/4	4366 (1916)	3833 (382)	3350 (2311)
Max. RIM score	1.7 (0.5)	1.7(0.5)	1.8 (0.7)
Mean RIM score 1st half RT	0.4 (0.4)	0.5 (0.6)	1.0 (0.7)
Mean RIM score 2nd half RT	1.4 (0.3)	1.6 (0.5)	1.3 (0.7)

The mean cGy for each mucositis score is presented along with the standard deviation in brackets.

The experience of pain

Patients were asked to rate their oral pain scores at various times during the day using a 0–10 visual analogue scale, with 0 meaning no pain and 10 meaning the worst possible pain imaginable. At the conclusion of the trial, data was evaluated rating the total pain experienced over the day rather than at different points during the day. Any pain of 3/10 or more was considered, as it is at this point that some form of pain relief would be required such as mild analgesia or soothing mouthwashes. Eleven of the nineteen patients in this pilot study experienced pain of 3/10 or higher.

Only two patients who were using the active gargle experienced pain of 3/10 later in their treatment at a mean of 3600cGy, compared to 5/7 patients in the control group (3000cGy) and 4/6 patients in the placebo group (2100cGy). Of note for all patients, who developed mucositis, pain was not an early sign as the development of pain occurred after the development of mucositis. Daily medication use (other gargles and analgesia) was also recorded to see if this altered as the pain levels increased. The number of doses was averaged for each day rather the size of each dose. As shown in Table 8.8, daily medication use was lower over treatment time for the patients in the active gargle group once they stopped taking the gargle though their medication use increased in the post- radiotherapy period.

Table 8.8. Medication use across radiotherapy and in the post-radiotherapy period.

	1st half RT			2nd half RT			Post RT		
	Active	*Control*	*Placebo*	*Active*	*Control*	*Placebo*	*Active*	*Control*	*Placebo*
Mean daily medication use	0.2 (0.2)	0.4 (0.3)	0.4 (0.3)	0.5 (0.5)	0.8 (0.5)	0.6 (0.7)	0.5 (0.5)	0.3 (0.4)	0.6 (0.9)

Nutritional status

The effect of radiotherapy treatment on nutritional status was evaluated through monitoring of patient weight change during treatment, self reporting of taste changes, changes in oral secretions, eating difficulties and grading of impact of these changes on quality of life.

Weight change varied across all patients, from a weight gain of 4.35% to a loss of 11% over baseline body weight. Several other variables have the potential to impact on the nutritional status of patients. These include having a support person present who usually paid extra attention to preparing appetizing and nutritious food. Others found the communal eating environment (for out-of-town patients) a supportive experience as others in the group would “bolster” or “jolly” patients along. Some patients or relatives would prepare meals for others within the communal kitchen. The active group experienced a mean weight change of 1% loss, the control group 5.2% loss and the placebo 4.1% loss. Indeed the active group had two patients who gained weight over the course of their radiotherapy and one patient whose weight did not change at all.

Patients also reflected back on their radiotherapy and post-radiotherapy period as to how much each symptom affected their quality of life using a 0–10 scale. Table 8.9 presents the mean scores for each group in each treatment segment with the standard deviations in brackets. The placebo and control groups were most affected by a dry mouth and altered taste and appetite across treatment, including afterwards. Whilst nausea and vomiting and excess secretions were not a major concern for all patients, the active gargle group were affected the least of the three groups. This also shows that the active gargle did not increase any gastro-intestinal effects even though it was swallowed, although some patients had made the comment that they felt it "repeated" on them once they swallowed.

Table 8.9. Recollection of how much the patients were affected by symptoms.

Symptom	1st half RT			2nd half RT			Post RT		
	Active	*Control*	*Placebo*	*Active*	*Control*	*Placebo*	*Active*	*Control*	*Placebo*
Dry oral area	0.5 (0.5)	2.5 (1.9)	2.2 (1.0)	3.4 (1.7)	4.1 (2.3)	5.3 (1.8)	3.3 (2.4)	4.0 (2.1)	4.2 (2.0)
Altered taste and appetite	1.0 (0.9)	2.4 (2.0)	27 (1.6)	2.9 (3.4)	4.6 (2.0)	5.5 (2.2)	2.7 (3.5)	3.4 (2.6)	4.9 (3.3)
Excess secretions, nausea and vomiting	0	1.2 (1.2)	17 (2.4)	0.3 (0.5)	2.0 (1.7)	2.1 (2.2)	0.2 (0.4)	1.2 (1.2)	0.5 (0.8)

Conclusions

The potential benefits of using the essential oils of Manuka (*Leptospermum scoparium*) and Kanuka (*Kunzea ericoides*) have been briefly presented here. There is still much work to be done in the area of using Manuka and Kanuka for radiation-induced mucositis. Further comparative studies are required to see if other essential oils grown in different parts of the country could be used. There is also the potential to explore the value of the hydrosol or water produced as a part of the distillation process. Even though the number of patients in this study was small, the most significant result was that obtained with the development of mucositis, which was delayed in the active gargle group. Whether this had an effect on other outcomes such as pain relief and nutritional status could not be inferred from this study, the remarkable observation was that the gargle/swallow was well tolerated with a high degree of patient compliancy. There were no side effects, which could be attributed to the gargle. Even though both essential oils have a distinctive taste, patients reported that the mix was soothing and they were not put off by the aroma or taste.

References

Bhatia, V., Harari, P.M., Thompson, S.P., Ford, C.N. and Hartig, G.K. (1996) Head and neck cancer patient perspectives regarding pain, cosmesis, swallow function, voice quality and overall quality of life. *International Journal of Radiation Oncology, Biology, and Physics* 36(1042), 265.

Bloor, S.J. (1992) Antiviral phloroglucinols from New Zealand Kunzea species. *Journal of Natural Products* 55(1), 43–47.

Brooker, S.G., Cambie, R.C. and Cooper, R.C. (1987) *New Zealand Medicinal Plants*. Reed Books, Auckland, New Zealand.

Christoph, F., Kaulfers, P.M. and Stahl-Biskup, E. (2001) *In vitro* evaluation of the antibacterial activity of β-triketones admixed to *Melaleuca* oils. *Planta Medica* 67(8), 768–771.

Christoph, F., Kubeczka, K.-H. and Stahl-Biskup, E. (1998) The composition of commercial Manuka oils from New Zealand. *Journal of Essential Oil Research* 11, 705–710.

Cooke, A. and Cooke, M.D. (1991) An investigation into the anti-microbial properties of Manuka and Kanuka oil. *Cawthron Institute Report*, February, 1991.

Duncan, M. and Grant, G. (2003) Review article: oral and intestinal mucositis–causes and possible treatments. *Alimentary Pharmacology Therapy* 18, 853–874.

El-Sayed, S., Epstein, J., Minish, E., Burns, P., Hay, J. and Laukkanen, E. (2002a) A pilot study evaluating the safety and microbiologic efficacy of an economically viable antimicrobial lozenge in patients with head and neck cancer receiving radiation therapy. *Head and Neck* 24(1), 6–15.

El-Sayed, S., Nabid, A., Shelley, W., Hay, J., Balogh, J., Gelinas, M., MacKenzie, R., Read, N., Berthlet, E., Lau, H., Epstein, J., Delvecchio, P., Ganguly, P.K., Wong, F., Burns, P., Tu, D. and Pater, J. (2002b) Prophylaxis of radiation-associated mucositis in conventionally treated patients with head and neck cancer: a double-blind, Phase III, randomized, controlled trial evaluating the clinical efficacy of an anti-microbial lozenge using a validated mucositis scoring system. *Journal of Clinical Oncology* 20(19), 3956–3963.

Epstein, J.B. and Wong, F.L. (1994) The efficacy of sucralfate suspension in the prevention of oral mucositis due to radiation therapy. *International Journal of Radiation Oncology Biology and Physics* 28(3), 693–698.

Gava, A., Ferrarese, F., Tonetto, V., Coghetto, F., Marazzato, G. and Zorat, P.L. (1996) Can the prophylactic treatment of mycotic mucositis improve the time of performing radiotherapy in head and neck tumours. *Radiological Medicine* 91(4), 452–455.

Koç, M. and Atkas, E. (2002) Prophylactic treatment of mycotic mucositis in radiotherapy of patients with head and neck cancers. *European Journal of Clinical Oncology* 33(2), 57–60.

Lis-Balchin, M. and Hart, S.L. (1998) An investigation of the actions of the essential oils of Manuka (*Leptospermum scopiarum*) and Kanuka (*Kunzea ericoides*), Myrtaceae on guinea pig smooth muscle. *Journal of Pharmacy and Pharmacology* 50(7), 809–811.

Lis-Balchin, M. Dr, Hart, S. Deans, S.G. and Eaglesham, E. (1996) Comparison of the pharmacological and antimicrobial action of commercial plant essential oils. *Journal of Herbs, Spices and Medicinal Plants* 4(2), 69–86.

Makkonen, T.A., Minn, H., Jekunen, A., Vilja, P., Tuominen, J. & Joensuu, H. (2000). Granulocyte macrophage-colony stimulation factor (GM-CSF) and sucralfate in prevention of radiation-induced mucositis: a prospective randomised study, *International Journal radiation Oncology, Biology, Physics,* 46(3), 525-534.

McGuire, D.B., Altomonte, V., Peterson, D.E., Wingard, J.R., Jones, R.J. and Grochow, L.B. (1993) Patterns of mucositis and pain in patients receiving preparative chemotherapy and bone marrow transplantation. *Oncology Nursing Forum* 20(10), 1493–502.

Mueller, B., Millheim, E., Farrington, E., Brusko, C. and Wiser, T. (1995) Mucositis management for hospitalised patients: national survey results. *Journal of Pain and Symptom Management* 10(7), 510–519.

Perry, N.B., Brennan, N.J., van Klink, J.W., Harris, W., Douglas, M.H., McGimpsey, J., Smallfield, B.M. and Anderson, R.E. (1997) Essential oils from New Zealand Manuka and Kanuka. *Chemotaxonomy of Leptospermum* 44, 1485–1494.

Porter, N.G. and Wilkins, A.L. (1999) Chemical, physical and antimicrobial properties of essential oils of *Leptospermum scoparium* and *Kunzea ericoides. Phytochemistry* 50(3), 407-415.

Porter, N.G., Smale, P.E., Nelson, M.A., Hay, A.J., van Klink, J.W. and Dean, C.M. (1998) Variability in essential oil chemistry and plant morphology within *Leptospermum scopiarum* population. *The New Zealand Journal of Botany* 36, 125–133.

Riley, M. (1994) *Maori Healing and Herbal New Zealand Ethnobotanical Sourcebook,* Viking Seven Seas, New Zealand.

Rugg, T., Saunders, M.I. and Dische, S. (1990) Smoking and mucosal reactions to radiotherapy. *British Journal of Radiology* 63(751), 554–556.

Scully, C., Epstein, J. and Sonis, S. (2003) Oral mucositis: a challenging complication of radiotherapy, chemotherapy, and radiochemotherapy: Part 1: Pathogenesis and prophylaxis of mucositis. *Head and Neck* December, 1057–1070.

Scully, C., Epstein, J. and Sonis, S. (2004) Oral mucositis: a challenging complication of radiotherapy, chemotherapy, and radiochemotherapy: Part 2: Diagnosis and management of mucositis. *Head and Neck* January, 77–84.

Sheppard-Hangar, S. (1994) *The Aromatherapy Practitioner's Reference Manual*, Vol I and II, Atlantic Institute of Aromatherapy, Tampa, Florida, USA.

Stokman, M.A., Spijkervet, F.K., Burlage, F.R., Dijkstra, P.U., Manson, W.L., de Vries, E.G. and Roodenburg, J.L. (2003) Oral mucositis and selective elimination of oral flora in head and neck cancer patients receiving radiotherapy: a double-blind randomized clinical trial. *British Journal of Cancer* 88(7), 1012–1016.

Sutherland, S.E. and Browman, G.P. (2001) Prophylaxis of oral mucositis in irradiated head and neck cancer patients: a proposed classification scheme of interventions and meta-analysis of randomized controlled trials. *International Journal of Radiation Oncology, Biology and Physics* 49(4), 917–930.

Tisserand, R. and Balacs, T. (1995) *Essential Oil Safety, A Guide for Health Professionals*, Churchill Livingstone, United Kingdom.

Trotti, A., Bellm, L.A., Epstein, J.B., Frame, D., Fuchs, H.J., Gwede, C.K., Komaroff, E., Nalysnyk, L. and Silberberg, M.D. (2003) Mucositis incidence, severity, and associated outcomes in patients with head and neck cancer receiving radiotherapy with or without chemotherapy: a systematic literature review. *Radiotherapy and Oncology* 66(3), 253–262.

Trotti, A., Garden, A., Warde, P., Symonds, P., Langer, C., Redman, R., Pajak, T.F., Fleming T.R., Henke, M., Bourhis, J., Rosenthal, D.I., Elizabeth, J., Cmelak, A., Sheehan, F., Pulliam, J., Devitt-Risse, P., Fuchs, H., Chambers, M., O'Sullivan, B. and Ang, K.K. (2004) A multinational, randomized phase III trial of iseganan HCl oral solution for reducing the severity of oral mucositis in patients receiving radiotherapy for head-and-neck malignancy. *International Journal of Radiation Oncology, Biology, and Physics* 58(3), 674–681.

Wijers, O.B., Levendag, P.C., Harms, E.R., Gan-Teng, M., Schmitz, P.I., Hendriks, W.D., Wilims, E.B., van der Est, H. and Visch, L.L. (2001) Mucositis reduction by selective elimination of oral flora in irradiated cancers of the head and neck: a placebo-controlled double-blind randomized study. *International Journal of Radiation Oncology, Biology, and Physics* 50(2), 343-352.

SECTION III

BIOLOGICAL AND BEHAVIOURAL RADIOPROTECTION BY DIETARY INGREDIENTS/NUTRACEUTICALS

Chapter 9

Piper betel Leaves : A Potential Gold Mine of Radioprotective and Photoprotective Compounds

Debashish Banerjee and Subrata Chattopadhyay

Introduction

Traditional medicine all over the world is nowadays revalued by an extensive activity of research on different plant species and their therapeutic principles (Arora *et al.,* 2004, 2005, 2006a,b). A herbal drug is a natural alternative for other, more conventional drugs and stimulants. Because they are culled straight from nature, they are mostly unregulated and can have astonishing effects. Recently, the World Health Organization (WHO) estimated that 80% of people worldwide rely on herbal medicines for some aspect of their primary healthcare. In the last twenty years in the United States, increasing public dissatisfaction with the cost of prescription medications, combined with an interest in returning to natural or organic remedies, has led to an increased use of herbal medicines. Approximately 25 percent of Americans who consult their physician about a serious health problem use unconventional therapy (expenditure was nearly $4 billion in 1998 alone on herbal products), but only 70 percent of these patients inform their physician of such use. In Germany, roughly 600 to 700 plant-based medicines are available and are prescribed by approximately 70% of German physicians. The popularity of herbal medicines can be attributed to the perceived notion about their lesser side effects and also low cost.

Ayurveda, the Indian traditional health care system (*ayus* = life, *veda* = knowledge, meaning science of life), is the oldest medical system in the world, and is believed to have originated from the Vedas and other ancient religious scripts. Ayurveda has been an integral part of Indian culture and Materia Medica, and is being revived in its complete form by several organizations/institutions, including Maharishi Ayurved (Glaser, 1988), while its efficacy has been approved by WHO. Long before the discovery of the allopathic system by Hippocrates or homeopathic system by Hannemahn, and the practise of the Unani System in Greece, Ayurveda was practised exclusively. It provides an approach to prevent and treat different diseases by a large number of medical procedures and pharmaceuticals. One of the clinical specialities of Ayurveda is Rasayana, which is not only a drug therapy, but is a specialized procedure practised in the form of rejuvenating recipes and dietary regimen to promote health. The purpose of rasayana is two-fold: prevention of disease and counteraction of aging processes, which result from optimization of homeostasis. Rasayana (*rasa* = essence, water; *ayana* = going) essentially refers to nutrition and its acquisition, movement, circulation and perfusion in the body tissues. This magic spell even now strongly and vibrantly persists in India, despite the tall claims of scientific discoveries, brought about by the Western systems of medicine.

Experimental evidence suggests that free radicals and reactive oxygen species (ROS) are involved in a large number of diseases (Richards and Sharma, 1991). As plants produce a lot of antioxidants to control the oxidative stress caused by sunbeams and oxygen, they represent a source of new compounds with antioxidant activity. The antioxidant activity of rasayana compounds has been reported to be 1000 times more potent than ascorbic acid, α-tocopherol and probucol. From the rich Indian biodiversity, Ayurveda has identified various plants/herbs that have potential therapeutic efficacy. *Piper betel* is one such indigenous medicinal plant, which has a folk (besides medicinal use in Siddha and Ayurveda) reputation in rural India. The present deliberation is restricted to its antioxidant, radioprotective and free radical scavenging activities. In addition, factors such as the active principles, differential activity of its different varieties as well as mammalian toxicity are also dealt with.

Piper betel: Uses in Traditional/Modern Medicine

The *Piper betel* plant (Family: Piperaceae) is found widely growing in the tropical humid climate of Southeast Asia, and its leaves (commonly known as pan), with a strong pungent and aromatic flavour, are widely consumed as a mouth freshener. Use of this plant against various diseases is mentioned even in the ancient Vedic literature. The Aryans called it "tambula" and the Arabs "tambul". People chew it to sweeten the breath, and colour the lip and tongue. Usually the people of South Asia, Gulf states, Southeast Asia and Pacific islands consume pan. In Bengal, pan is traditionally chewed by all classes of people not only as a habit but also as an item of rituals, etiquette and manners. During the aristocratic age, pan preparation and the style of garnishing it on a plate was a recognized folk art. The production and marketing of pan led to the rise of an occupational caste called Barui. In the Muslim period, they were the richest people among the cultivating classes due to the huge demand for pan at that time. At present, pan has a worldwide market due to the global dispersion of the pan-chewing South Asians, and scientific recognition of its medicinal values.

Different varieties of *P. betel* are available in India, of which the leaf of the "Magahi" variety (literally from the Magadha region; grown near Patna in Bihar, India) is the most expensive, and a connoisseur's delight due to its tender nature. Besides, the most commonly available varieties are *P. betel* Bangla, *P. betel* Sweet and *P. betel* Mysore. The leaves of the Bangla variety (name derived from its cultivation in West Bengal, India) are moderately green, most pungent and consumed widely in India. The leaves of the sweet variety, commonly known as "meetha pan" are dark green and most expensive, and valued for their least pungency (explaining its sweet name; meetha (Hindi or Urdu for sweet)) and exotic flavour. The leaves of the Mysore variety are light green and moderately pungent, with a rather unusual flavour liked by the people of Southern India.

In traditional medicine, *P. betel* leaves are credited with wound healing, digestive and pancreatic lipase stimulant activities (Chatterjee and Pakrashi, 1995), which has also been proved with experimental animals (Prabhu *et al.,* 1995). Earlier, we have reported cytoprotective and healing properties of the leaf extract against indomethacin-induced gastric lesions in experimental animals (Majumdar *et al.,* 2003). In addition, its antimicrobial (Ramji *et al.,* 2002), antifungal and anti-inflammatory (Ambastha, 1986) activities are also reported. Some of the beneficial roles of *P. betel* leaf such as anti-diabetic (Santhakumari *et al.*, 2006), hepatoprotective (Saravanan *et al.*, 2002) and anti-ulcerogenic activities are correlated with its antioxidant property.

P. betel as a Non-toxic Radioprotector

The potential of antioxidants to reduce the cellular damage induced by ionizing radiations has been studied in animal models for more than 50 years (Dale *et al.*, 1949; Patt *et al.*, 1949; Bacq *et al.*, 1951; Yuhas *et al.*, 1980; Bump and Malakar, 1997). Some antioxidant nutrients and phytochemicals have the advantage of low toxicity, although they are generally protective when administered at pharmacological doses (Weiss and Landauer, 2003; Venkatachalam and Chattopadhyay, 2004; Arora *et al.*, 2005). Many naturally occurring antioxidants exhibit a long window of protection, including post-irradiation protection against lethality and mutagenesis. A number of phytochemicals have multiple physiological effects, as well as antioxidant activity, which result in radioprotection *in vivo* (Weiss and Landauer, 2003; Uma Devi, 2006; Chattopadhyay, 2006). The potential application of many of these antioxidants shows promise either prophylactically for anticipated exposures in emergency situations or therapeutically after radiation accidents/incidents. Many antioxidant nutrients and phytochemicals have antimutagenic properties also, and their modulation of long-term radiation effects, such as cancer, needs further examination.

In recent years, many studies have shown that diets containing high content of phytochemicals can provide protection against various diseases. Approximately 90% of all cancer cases correlate with environmental factors, including one's dietary habits, and one-third of all cancer deaths are avoidable by changing dietary habits only (Willett, 1995). These discoveries have rapidly amplified the consumer's awareness of the potential benefits of naturally occurring compounds from plants in health promotion and maintenance, and research in nutraceuticals and functional foods and natural health products have become the hot topic in recent years (Goldberg, 1998; Nice, 1997; Swanson, 1998). The protective effects of fruits, vegetables and spices and herbs were found not only for cancer (Swanson, 1998; Toniolo *et al.*, 2001), but also other chronic diseases such as cardiovascular diseases. From this perspective, *P. betel* seems well qualified for exploration as a non-toxic and inexpensive radioprotectant. To this end, our group and others have assessed the antioxidant property and also evaluated the preventive potential of *P. betel* leaf extracts against ionizing and non-ionizing radiation-induced cellular damages.

Protective property of *P. betel* leaves against ionizing radiation

Exposure of mammalian systems to ionizing radiation induces damaging effects leading to cell death and an increased risk of diseases, particularly cancer (Pierce and Preston, 2000). A dose of 4–8Gy is considered fatal for humans and other mammals. Due to the high concentration of water in metabolizing cells, radiation exposure of biological systems primarily leads to its radiolysis furnishing e^-_{aq}, •OH and H•. In aerobic cells, the e^-_{aq}, in turn, also generates the $O_2^{-\bullet}$ radicals. The biological damage induced by the low linear energy transfer (LET) radiations is mostly indirect, and mediated by the ROS mentioned above (Von Sonntag, 1987; Breen and Murphy, 1995).

These reactive species are known to cause degradation of important macromolecules including DNA and membranes (Von Sonntag, 1987). Thus, the high level of unsaturated lipids is most susceptible to oxidative damage, resulting in disruption of cellular integrity, inactivation of cellular components etc. that lead to cytotoxicity and cause several diseases and aging (Packer and Ong, 1998; Rice-Evans and Burdon, 1993). Likewise, the DNA molecules are also prone to radiation-induced lesions due to the presence of various reactive sites (base and sugar) in them (O'Neill and Fieden, 1993). For a variety of tissues, the pathophysiological importance of ROS-mediated oxidative injury caused by the exposure to radiation is widely evaluated.

With regard to the radioprotective property, the efficacy of *P. betel* in inhibiting the γ-radiation induced lipid peroxidation (LPO) has been reported. Choudhary and Kale (2002) found that oral supplementation with the *P. betel* leaf extract (1, 5 and 10mg/kg/day × 14 days) to Swiss albino mice increased the hepatic GSH content and specific activity of SOD in a dose-dependent manner. The radioprotective effect of the extract was attributed to its capacity to enhance the antioxidant status and scavenge free radicals. The extract, however, did not modify the stress conditions as revealed from its insignificant effect on the glyoxalase system.

For our studies, we used the *P. betel* Bangla variety ethanol extract (designated as PBE). Its protective capacity against γ-ray (total dose 450Gy) induced peroxidation of rat liver mitochondria was studied by assaying the thiobarbituric acid reactive substances (TBARS) formed during lipid peroxidation. The TBARS assay is fairly non-specific since thiobarbituric acid (TBA) reacts with various chemicals, including sugars, to give a pink-red colour. Hence, the protection induced by PBE was also assayed by measuring lipid hydroperoxides (LOOH), as well as conjugated dienes (CD), two relatively unstable products of lipid peroxidation. Representative results are summarized in Fig. 9.1.

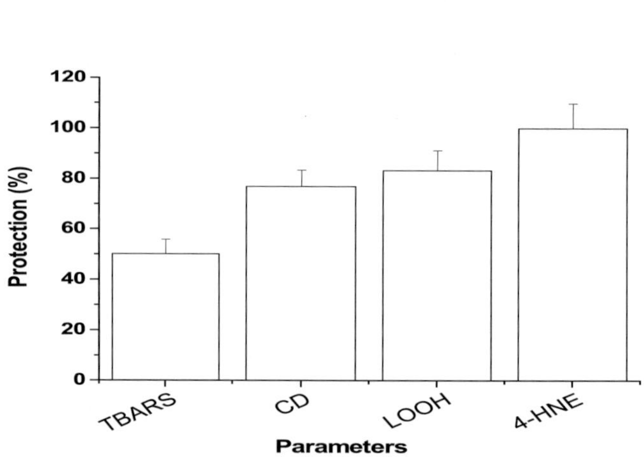

Fig. 9.1. Protective activity of PBE (50mg/ml) against γ-ray-induced lipid peroxidation. Mitochondria (2mg protein/ml) were exposed to γ-radiation (total dose 450Gy) and the formation of TBARS, LOOH, CD and 4-HNE in the presence and absence of PBE was measured. The percentages of protection were calculated by comparing the values with the control and treated samples. The values are mean ± SEM (n = 5). $p < 0.001$ compared to radiation alone.

The percent protection offered by 5, 10, 50 and 100μg/ml of PBE was 30, 36, 50 and 56 respectively. The positive control, ascorbic acid (50μg/ml), showed ~30% protection under similar experimental conditions. Likewise, PBE prevented the formation of LOOH and CD by 77 and 83% respectively. All these results established significant protective activity of PBE against the γ-ray induced lipid peroxidation.

LOOH produced during lipid peroxidation is known to furnish various reactive aldehydes on subsequent cleavage. Amongst these, 4-hydroxynonenal (4-HNE) has been associated with human diseases and experimental models (Esterbauer *et al.,* 1991). It reacts with glutathione, hampering the antioxidant defense. In addition, it inactivates various key proteins by reacting with the amino acids (Bosch-Morell *et al.,* 1999). Consequently, the preventive activity of PBE against γ-ray-induced 4-HNE formation was also assessed. Exposure of rat liver mitochondria to the γ-rays increased the 4-HNE levels by 385%, compared to those in the unirradiated samples. This was brought back to the basal level by PBE (50μg/ml).

It is well known that exposure of humans even to a considerably lower dose of γ-radiation causes extensive lipid peroxidation. From that perspective, the chosen radiation dose was admittedly too high. However, our *in vitro* experiments carried out by exposing rat liver mitochondria with lower doses (75, 150 and 300Gy) of γ-irradiation did not produce significant amounts of TBARS. Hence, a high radiation dose was necessary to generate reproducible and consistent results. Thus, the choice of the radiation dose was primarily dictated to demonstrate the capacity of PBE in preventing the γ-ray induced lipid peroxidation. It was found that PBE could inhibit even extensive lipid peroxidation caused by a very high dose of γ-ray. Thus, it is anticipated that it can prevent similar lipid damage in cells.

DNA scission is apparently the main factor contributing to radiation-induced mutation and carcinogenesis (Von Sonntag, 1987; Breen and Murphy, 1995). The oxidative damage to DNA is caused by the reactions of the ROS at base and sugar moiety. Hence, the protective capacity of PBE against γ-ray induced DNA single strand breaks (SSB) was also assessed. Addition of PBE (0.25–0.75mg/ml) prior to γ-irradiation (total dose: 22.5Gy) progressively reduced (81–96%) the intensity of the band due to the open circular form (Fig. 9.2.).

Amongst the ROS produced due to cellular exposure to ionizing radiations, $^{\bullet}$OH radical is the most toxic and contributes maximum in DNA and lipid damages. Exposure of aerobic cells to ionizing radiations can also generate the superoxide ($O_2^{-\bullet}$) radical, which has been implicated to play crucial roles in ischaemia-reperfusion injury and promoting human gout (Hatano *et al.,* 1991). Further, although the radical by itself is not very reactive, it can generate the toxic hydroxyl radical via a superoxide-driven Fenton process. Hence, for a better understanding of the radioprotective mechanism of PBE, its scavenging abilities against these ROS were also assayed. Under similar conditions, the hydroxyl radicals scavenging activity of PBE (1mg/ml) and mannitol (10mM) were 52% and 75% respectively. The scavenging activity of PBE (1mg/ml) for the $O_2^{-\bullet}$ radicals was similar (72%) to that of SOD (3U). These results revealed that PBE could scavenge both $^{\bullet}$OH and $O_2^{-\bullet}$ radicals efficiently, accounting for its impressive radioprotective potential.

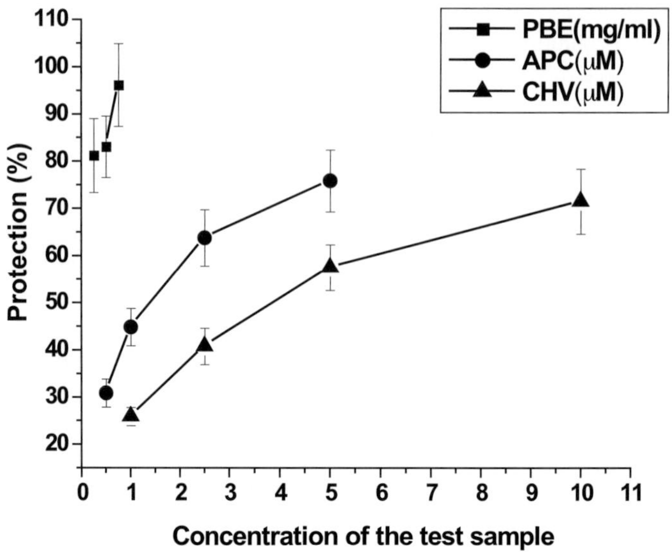

Fig. 9.2. Concentration-dependent protective activity of PBE, APC and CHV against γ-ray-induced DNA damage. Plasmid DNA (200ng) samples in 10mM potassium phosphate buffer, pH 7.4, in a 14μl volume were irradiated at 25°C up to a dose of 12.25Gy using a ^{60}Co source (dose rate: 5Gy/min) in the presence and absence of different concentrations of the test samples. The open-circular (OC) and super-coiled (SC) forms of DNA were separated by gel electrophoresis; the respective DNA bands were stained with ethidium bromide and visualized under UV light. The relative intensities of the bands were quantified with a Kodak Gel Logic 200 imaging system to determine the percentage of protection. The values are mean ± SEM (n = 5). $p < 0.01$ compared to radiation alone.

Besides the above antioxidant mechanism, many of the radioprotective compounds operate via their immunomodulatory activity. Our studies revealed a concentration-dependent lymphoproliferative activity of PBE against the peripheral blood mononuclear cells, as revealed by the MTT assay. The activity of PBE was ten times that of the positive control, phytohemaglutinnin (PHA) at the same concentration (0.25mg/ml). Overall PBE was found to be a potential radioprotectant, exerting the activity through its superior radical scavenging and immunomodulatory properties. It is worth noting that as compared to an earlier study (Choudhary and Kale, 2002) that used considerably higher doses of PBE, our investigation established its radioprotecting capacity at a much lower dose (Bhattacharya *et al.,* 2005).

Protective property of *P. betel* against photosensitization

Photosensitization involving light, photosensitizer and oxygen is a potentially damaging event in biological systems. It is a widely occurring phenomenon in biological systems due to the ubiquitous nature of visible light and a number of pigments and related compounds, which can act as photosensitizers (Black, 1987). Even endogenous compounds have been reported to act as photosensitizers. This generates several types of ROS that are capable of damaging various sub-cellular structures and molecules. Among these, singlet oxygen ($^1O_2$) has been identified as one of the major species responsible for the biological damage caused by photosensitization (Weishaput *et al.*, 1976). Singlet oxygen is an electronically excited species of oxygen produced in mammalian cells under various normal and pathophysiological conditions. Due to its relatively long half-life (10–50μs), it is capable of travelling appreciable distance in the cellular milieu causing damage to membrane lipids, DNA etc. (Epe, 1991).

Membrane damage has been considered as a crucial event contributing to the cytotoxicity of a photosensitizer. Besides resulting in drastic alteration of membranes, the ROS-mediated lipid peroxidation generates toxic metabolites, which are capable of inducing indirect damages at sites far away from the sites of their generation. The induced alterations in membrane permeability, transport systems, loss of membrane-bound enzymes etc. can eventually lead to cell lysis and death under specific conditions (Valenzeno and Tarr, 1991). Adverse effect of photosensitization to specific carriers for succinate, citrate and oxalacetate has been reported (Salet and Moreno, 1990). The generation of free radicals in skin by solar ultraviolet light accelerates skin cancer, photoaging and other light-related skin pathogenesis (Witt *et al.*, 1993). Cellular exposure to UV light also leads to iron release, resulting in excessive production of ROS and ultimately to pathogenesis (Halliwell and Gutterridge, 1989).

Against the above backdrop, the need to develop suitable formulations against photo-induced biological damages assumes significance. To this end, the protective capacity of PBE against photo-induced peroxidation of rat liver mitochondria was studied by assaying the TBARS, LOOH and conjugated diene (CD) formed during the lipid peroxidation. Photosensitization was carried out with methylene blue plus light (a 100W tungsten lamp, light intensity 32.45W m^{-2}). It was found that PBE could effectively inhibit the TBARS formation in a concentration-dependent manner. The percent protection offered by 50, 75 and 100μg/ml of PBE was 17, 24 and 40% respectively, while sodium azide (10mM), a $^1O_2$ quencher, offered 76% protection against TBARS formation.

Likewise, mitochondrial photosensitization led to the formation of LOOH and CD, which, however, were prevented by PBE. Addition of PBE (50μg/ml) to the mitochondria prior to photo-irradiation could significantly inhibit the formation of LOOH and CD by 70% and 36% respectively. The results were better than that of the well-known $^1O_2$ quencher, sodium azide (10mM), that offered 52% and 16% protection against the formation of LOOH and CD respectively.

Protein oxidation is inherent to aerobic life. Oxidation of membrane proteins by ROS, a process independent of lipid peroxidation, is also a highly damaging event. The amino acids of proteins are easily modified by photosensitization-induced oxidation resulting in loss of protein structure and function.

We observed that photosensitization for 30 and 60 min significantly deactivated the mitochondrial SOD level by 74% and 88% respectively. Addition of PBE (50μg/ml) to the mitochondria prior to its photo-irradiation could restore the SOD activity substantially (66% and 55% restoration). Conversion of proteins into oxidized species (e.g., oxyacids) is one of the earliest observable events during the radical-mediated oxidation of proteins. The extent of oxidative protein damage can be easily assayed by measuring the amount of protein carbonyls formed due to the oxidation. The preventive effect of PBE against protein oxidation was also assessed by the protein carbonyl assay. It was found that photosensitization of rat liver mitochondria for 0.5h increased the amounts of protein carbonyls 5-fold compared to the control value. PBE could reduce it by 78%, while ascorbic acid (10mM) showed 57% prevention of protein carbonyls formation. The results of protective activity of PBE against the photosensitization-induced SOD deactivation and protein carbonyl formation are shown in Figs. 9.3 and 9.4 respectively.

The combination of methylene blue, light and oxygen, as employed in our studies, is a widely used biologically relevant system. It mainly generates $^1O_2$, which can induce oxidative damage in membrane (Valenzeno and Tarr, 1991). Our investigations revealed that PBE could scavenge $^1O_2$ significantly, albeit at a higher concentration (>47μg/ml). Overall, PBE was found to be a good photoprotector (Bhattacharya *et al.,* 2007). However, the protective effect of PBE against ionizing radiation is significantly higher (Bhattacharya *et al.,* 2005) than that against photosensitization. This might be attributed to its lower reactivity with $^1O_2$.

Quality control of *P. betel*-based herbal radioprotector

For most herbs, the specific ingredient that causes a therapeutic effect is unknown. Whole herbs contain many ingredients, and it is likely that they work together to produce the desired medicinal effect. Unlike conventional drugs, herbal products are not regulated for purity and potency. Thus, some of the adverse effects and drug interactions reported for herbal products could be caused by impurities (e.g., allergens, pollen and spores) or batch-to-batch variability.

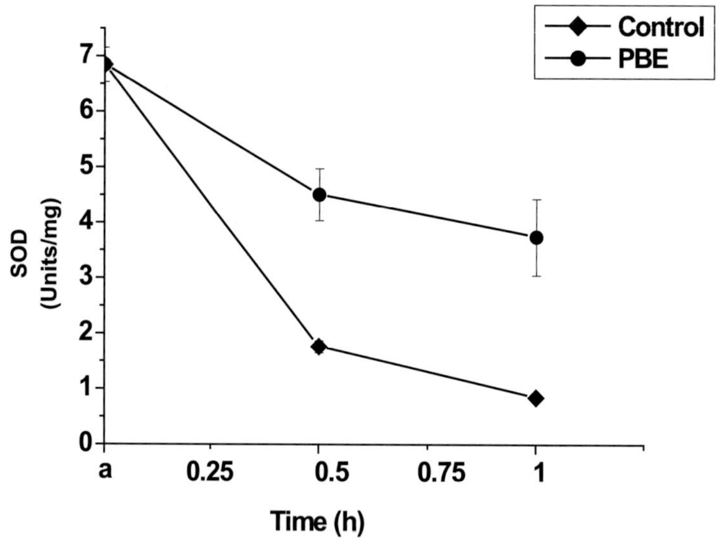

Fig. 9.3. Photosensitization-induced depletion of superoxide dismutase (SOD) in rat liver mitochondria and its restoration by PBE. Mitochondria (0.5mg protein/ml) were exposed to light (100W) in the presence of methylene blue and oxygen for 30 and 60 mins without and with PBE (50 μg/ml), and the activity of SOD was measured. The values are mean ± SEM (n = 5). $p < 0.001$ as compared to photosensitization at 30 min. $p < 0.01$ as compared to photosensitization at 60 min.

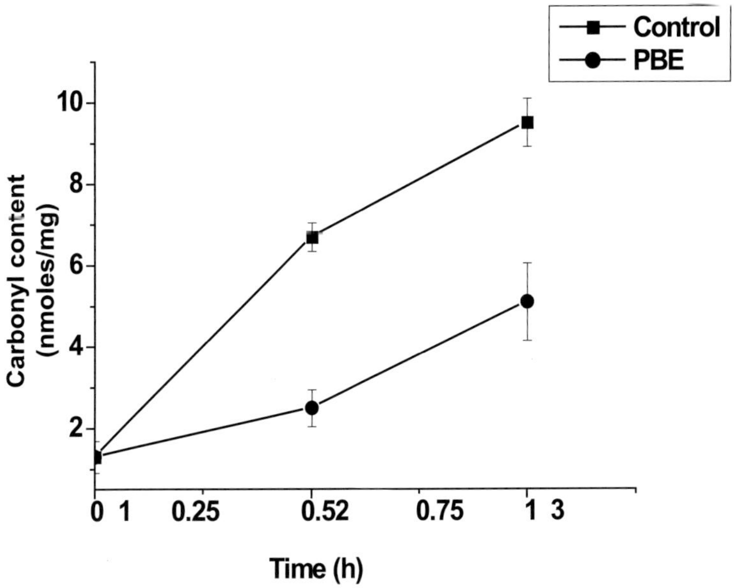

Fig. 9.4. Photosensitization-induced protein oxidation in rat liver mitochondria and its prevention by PBE. Mitochondria (0.5mg protein/ml) were exposed to light (100W) in the presence of methylene blue and oxygen for 30 and 60 mins without and with PBE (50 µg/ml) and formation of protein carbonyls was measured. The values are mean ± SEM (n = 5). $p < 0.001$ as compared to photosensitization at 30 and 60 min.

In addition, the potency of any herbal product may increase the possibility of adverse effects. Herbal medicines are being manufactured on a large scale in mechanical units, where manufacturers come across many problems such as availability of good quality raw materials, authentication of raw materials, availability of standards, proper standardization methodology of single drugs and formulations, quality control parameters, etc.

In many respects, the mechanism of action of the herbal drugs differs from that of the synthetic drugs or pure compounds. This can be characterized as a polyvalent action and interpreted as additive or, in some cases, potentiating. However, one or more active principles in the plant-based or herbal preparations are primarily responsible for their health benefits. Consequently, their presence in optimum amounts is essential to get the desired effects of these drugs. Due to the lack of proper quality control, this aspect is often neglected leading to the availability of spurious drugs. It is well established that factors such as habitation, time of collection, maturity of the plants etc. affect the concentrations of their bioactive chemical constituents.

Previous studies on the radioprotective and antioxidative activities of *P. betel* leaves were carried out with non-standardized crude preparations, without identifying the active principles. For a proper quality control of the plant-based drugs, it is essential to identify the bioactive constituents, and analyze their contents. To address this aspect, we identified the active principles, and also quantified them in three varieties of *P. betel* leaves that are commonly available and consumed in India.

Evaluation of chemical constituents and quality control of P. betel

For our quality control studies, we used ethanol extracts of *P. betel* Bangla, *P. betel* sweet and *P. betel* Mysore varieties (designated as PBE, PSE and PME respectively). Preliminary TLC analyses of PBE, PSE and PME showed similar profiles of the constituent chemicals. Subsequently, the constituents in the extracts were quantified by HPTLC analyses that were standardized using various mobile phases. All the extracts contained four major compounds, although in different concentrations. A detailed phytochemical investigation of PBE by conventional chromatography furnished chevibetol (CHV) and allylpyrocatechol (APC), as well as their glucosides, **1** and **2** (Fig. 9.5).

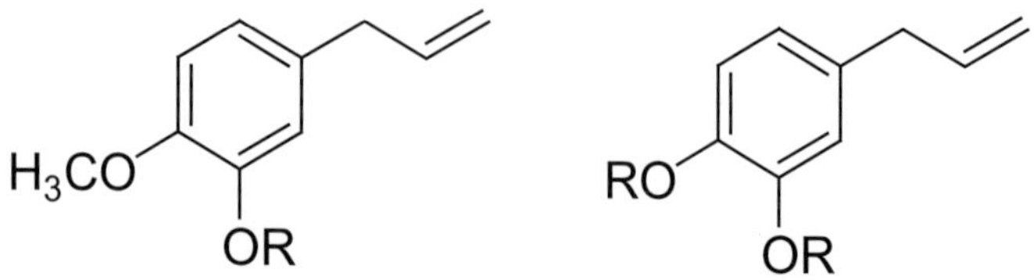

CHV: R = H
1: R = Glu

APC: R = H
2: R = Glu

Fig. 9.5. Chemical structures of chevibetol (CHV), allylpyrocatechol (APC) and their glucosides.

PBE was found to be most enriched with CHV and APC. Although we did not quantify the concentrations of the glucosides PSE and PME, TLC analysis revealed these glucosides to be the major constituents of the extracts. The lesser concentrations of the free phenolics in the extracts might account for less pungency. *P. betel* leaf oil is known (Evans *et al.*, 1984) to contain primarily a class of allylbenzene compounds such as CHV, chavicol, estragole, eugenol, methyl eugenol and APC, along with monoterpenes, para-cymene and terpinene, monoterpenoids, eucalyptol and carvacrol, as well as sesquiterpenes, cadinene and caryophyllene. However, our extracts did not show any appreciable amount of many of these compounds. Comparison of our results with different varieties of *Piper*, as well as the previous reports from other laboratories, clearly established significant variation in the chemical compositions of the herbal extracts. This emphasized the need for proper analyses of the active principles in order to ensure efficacy as well as the non-toxicity of the herbal drugs.

Among the extracts of the different varieties of *Piper betel*, PBE showed the best DPPH radical scavenging activity, 57% and 71% scavenging at 6 and 7.5μg/ml concentrations respectively. PSE and PME were effective only at higher concentrations (>20μg/mL). Their respective scavenging activities were 33% and 16%, at a concentration of 50μg/ml. PBE also showed significantly higher reducing potential than PSE, PME and even BHT used at ten-fold higher concentrations. The reducing power of BHT was better than those of PSE and PME at comparable concentrations. These factors are likely to contribute significantly towards the observed radical scavenging properties of PBE > PME > PSE.

It is well known that plant phenolics, in general, are highly effective free radical scavengers and antioxidants. The antioxidant activities of plant/herb extracts often show good correlation with their respective total phenolic contents. The total phenolic content of PBE (820.5μg GAE/mg) was significantly higher than both PSE and PME (72.3 and 106.7μg GAE/mg). Overall, the order of the phenolic contents of the different *Piper betel* extracts matched with their respective DPPH radical scavenging abilities and reducing powers.

Identification of the active antioxidant constituents of P. betel

It is well known that the radioprotecting ability of a test sample can often be correlated with its antioxidant or free radical scavenging property. Consequently, we assessed the antioxidant activity of CHV and APC through a series of *in vitro* experiments, which confirmed their status as the active principles. Earlier, the DPPH scavenging activities of APC and α-tocopherol were similar, and significantly higher than that of CHV. The anti-lipid peroxidation (LPO) activities of CHV and APC against Fe(II)-ascorbic acid-mediated peroxidation of liposome was assessed by measuring the TBARS and LOOH. APC was a very strong antioxidant, preventing lipid peroxidation by 46% even at a very low concentration (2μM). Overall, CHV, APC and α-tocopherol prevented lipid peroxidation with IC_{50}-values of 25, 2 and 10μM respectively. With rat brain homogenate as the lipid source also, APC was found to be the best antioxidant, the IC_{50} values of CHV and APC being 20 ± 2 and 0.44 ± 0.1μM respectively. The data correlated well with their respective DPPH scavenging activities. Likewise, APC (2μM) inhibited the iron-stimulated formation of LOOH (from liposomes) by 54%, while the positive control, α-tocopherol (10μM), showed 35% protection. Thus, APC was the best antioxidant amongst the *P. betel* constituents, its activity being significantly better than that of α-tocopherol.

The vastly enhanced activity of APC was due to its better radical scavenging potential arising out of the constituent-unprotected catechol moiety. It is well known that the presence of an electron-donating substituent in a phenol lowers the O-H bond dissociation enthalpy of the phenolic group and increases the rate of H-atom transfer to peroxyl radicals. The electron-donating effect of the substituent depends on its nature and the position with respect to the phenol moiety. Thus, the activity of a strong electron donating group, such as the hydroxyl group, at the ortho and para positions is much higher than that at the meta position, explaining the relative order of the antioxidant potencies of catechols, quinols and resorcinols. Between CHV and APC, the latter contains a free catechol moiety. Hence, its antioxidant capacity is anticipated to be better than various other catechols.

Both APC and CHV prevented the γ-ray induced DNA cleavage (total dose, 12.5Gy) concentration-dependently, the former being significantly more potent than CHV. For example, APC (0.5, 1.0, 2.5 and 5.0μM) offered 31, 45, 64 and 76% protection against the radiation-induced single strand break (ssb) formation. In comparison, the protecting efficiency of CHV even at higher concentrations (1, 2.5, 5.0 and 10μM) were significantly less (26, 41, 58 and 72% respectively). Fig. 9.2 shows the comparative protective activities of PBE and its constituent phenolics, CHV and APC.

As already mentioned, the cellular damage due to exposure of ionizing radiations are primarily mediated through ROS such as $^{\bullet}OH$, $O_2^{-\bullet}$ and H_2O_2. Hence, the scavenging abilities of CHV and APC against these ROS were also assayed. Both CHV and APC showed almost equal $^{\bullet}OH$ scavenging efficiency (IC_{50}: 67 ± 2 and 67 ± 1μM respectively) as revealed by the 2-deoxyribose assay. Under similar conditions, mannitol (1.0mM) showed ~19%

scavenging of the hydroxyl radical. However, APC showed superior $O_2^{\cdot-}$ scavenging ability (26%), compared to CHV (6% scavenging), at the same concentration (each 50μg/ml) accounting for its better potency against the γ-ray-mediated DNA cleavage.

Although not a radical species, H_2O_2 is an important ROS contributing to oxidative stress. Naturally occurring iron complexes are believed to react with H_2O_2 *in vivo* to generate the highly reactive ${}^{\cdot}OH$ radicals via the Fenton reaction. CHV and APC depleted H_2O_2 in a concentration-dependent manner with the IC_{50} values of 76 ± 0.5 and 62 ± 0.2μM respectively. In comparison, the positive control α-tocopherol was marginally less effective (IC_{50}: 81 ± 0.5μM) in scavenging H_2O_2.

Quality control assessment and recommendation

Many of the therapeutic properties of *P. betel* are attributable to its antioxidant action. However, the antioxidant principles have so far not been established. This would be important for developing plant-based drugs, since chemical constituents of plants are dependent on various factors. Our studies utilizing liposome, rat brain homogenate and plasmid DNA as *in vitro* models explicitly showed that the antioxidant and radioprotective activities of *P. betel* ethanol extract can be attributed to one of its constituent phenolics, APC; while the contribution of the other partially methylated phenol (CHV) is insignificant.

The studies also revealed that among the three most widely consumed varieties (Bangla, Sweet and Mysore) of *P. betel*, the Bangla variety is rich in phenolics, especially APC, and hence is the most effective antioxidant. Similar differential antioxidant activities in three varieties of *P. betel* have also been reported, wherein the anti-lipid peroxidation activity of the plant was found to be better than that of tea (Dasgupta and De, 2004). Thus, in order to promote *P. betel* as an effective antioxidant, it is essential to assay the content of the total phenolics and CHV and APC in the samples (Rathee *et al.*, 2006). More recently, the free radical scavenging activity of *P. betel* (Sri Lankan variety) ethanol extract was found (Arambewela *et al.*, 2006) to be higher than that of butylated hydroxyl toluene (BHT). Although the antioxidant activity was not reduced in storage up to a year, it was adversely affected at elevated temperatures. This indicated the importance of the processing methodology in developing a herbal drug from the plant.

Acute toxicity of *P. betel* extract and its active principle

Our own evaluation of the acute toxicity of PBE (up to a dose of 500mg/kg body wt.) and APC (up to a dose of 25mg/kg body wt.) on mice did not reveal any observable physical sign change, and the animals had normal food and water, as well as stool, during the experimental period. After one month of the administration of the test samples, there were no signs of abnormality in the liver of the mice as revealed by histology (data not shown). It is worth mentioning that in spite of the reported (Evans *et al.*, 1984) fungicidal and nematocidal activity of *P. betel* and allergic reaction of eugenol, we did not observe any side effect of even the whole extract (PBE) in our previous studies (Bhattacharya *et al.*, 2005). All these findings suggested that APC or PBE, administered even at a sufficiently high dose, do not have any potential side effect in the animals. In the light of the above, PBE and APC appear to be promising candidates for further *in vivo* bioevaluation.

Conclusion

Scientific research and perfection have struck gold for the economic benefits and the prosperity of these industries. The most modern and latest investigations in molecular biology and biotechnology also help to find new solutions for problems of health and eradication of hitherto unconquerable diseases. The magic of herbs and plants is there all around us waiting to be discovered, understood and used. They are now definitely recognized, and accepted as perennial storehouses of infinite, limitless benefits to man. People are buying herbal remedies for everything from migraine to memory preservation to depression.

India is singularly blessed with inexhaustible natural resources of plants and herbs, which mankind is hitherto not aware of. In the past few decades, an exponential growth of herbal drug research has provided an expanded international market where India is anticipated to play a big role by contributing in innovative research using its vast repertoire of biodiversity and wealth of knowledge in traditional medicines. But these opportunities come with a warning. Mixing herbal remedies and prescription drugs could be harmful due to herb-drug interactions. Some herbal medicines may cancel the effect of a prescription drug; others may reduce it, or even exaggerate it. Further, no organization or government body regulates the manufacture or certifies the labelling of herbal preparations. Currently, only a few herbal preparations are standardized, that is guaranteed to contain a specific amount of the active ingredients of the herb. Another important aspect in herbal medicine originates from the ambiguity present in Ayurveda itself. This is reflected in the interpretation of names and description of drugs given in the ancient books like Charaka Samhita and Sushruta Samhita, etc. Due to lack of scientific names in the original texts, under one name, different plants are known in different parts of the country as per the description, which makes the drug controversial.

All these may have serious consequences, both from the point of view of toxicity as well as efficacy. The importance and challenges of conducting clinical research with herbal drugs, developing simple bioassays for standardization, pharmacological, toxicological and safety evaluation with various animal models is warranted to realize the full potential of the herbal drugs. In the present deliberation, we have provided a huge databank with *P. betel* extract to confirm and check its antioxidative and radioprotective efficacy. In addition, keeping in view of the variations in the phytoconstituents, different varieties of *P. betel* were also assessed for the designated activity and the most potent constituents have been identified. These results, along with the results on the acute toxicity studies, revealed *P. betel* Bangla variety as a promising herbal antioxidant/radioprotectant where quantification of its active principle, allylpyrocatechol, would serve the purpose of proper quality control.

Acknowledgements

The authors thank Dr. S. K. Bandyopadhyay and Dr. Sayanti Bhattacharya, Department of Biochemistry, Dr. B.C. Roy Post-Graduate Institute of Basic Medical Sciences, and IPGMERR, Kolkata, India, for their encouragement in preparing the manuscript. One of the authors (DB) thanks the Board of Research in Nuclear Science (BRNS), Department of Atomic Energy, Government of India, for generous support in providing financial grants.

References

Ambastha, S.P. (1986) *The Useful Plants of India*, Council of Scientific and Industrial Research (CSIR) Publications, New Delhi, vol. 4, p. 59.

Arambewela, L., Arawwawala, M. and Rajapaksa, D. (2006) *Piper betel*: a potential natural antioxidant. *International Journal of Food Science and Technology* 41, 10–14.

Arora, R. and Sharma, R.K. (2006a) *Herbal Drug Research: A Twenty First Century Perspective*. Jaypee Brothers Medical Publishers Private Limited, Delhi, India, p. 688.

Arora, R., Chawla, R., Singh, S., Sagar, R.K., Kumar, R., Sharma, A., Prasad, J., Singh, S., Gurudatta, G.U. and Sharma, R.K. (2006b) Bioprospection for Radioprotective Molecules from Indigenous Flora. In: Govil, J.N. (ed) *Recent Progress in Medicinal Plants*. Studium Press, USA, Vol. 16, pp. 179–219.

Arora, R., Gupta, D., Chawla, R., Sagar, R., Sharma, A., Kumar, R., Prasad, J., Singh, S., Samanta, N. and Sharma, R.K. (2005) Radioprotection by plant products: Present status and future prospects. *Phytotherapy Research* 19, 1–22.

Arora, R., Gupta, D., Sharma, A.K., Sagar, R.K., Prasad, J., Singh, S., Samanta, N., Kumar, R., Goel, H.C., Sinha, A.K., Chawla, R. and Sharma, R.K. (2004) Modification of radiation–induced damage in mammalian model systems by natural plant products: implications for radiation protection. *Proceedings of the IUPAC Sponsored International Conference on Biodiversity and Natural Products*. University of Delhi, Delhi, India, p. 56.

Bacq, Z.M., Herve, A., Lecomte, J., Fischer, P. and Blavier, J. (1951) Protection against x–rays by beta–mercaptoethylamine. *Archives Internationales de Physiologie* 59, 442–447.

Bhattacharya, S., Mula, S., Gamre, S., Kamat, J.P., Bandyopadhyay, S.K. and Chattopadhyay, S. (2007) Inhibitory property of *Piper betel* extract against photosensitization–induced damages to lipids and proteins. *Food Chemistry* 100, 1474–1480.

Bhattacharya, S., Subramanian, M., Roychowdhury, S., Bauri, A.K., Kamat, J.P., Chattopadhyay, S. and Bandyopadhyay, S.K. (2005) Radioprotective property of the ethanolic extract of *Piper betel* leaf. *Journal of Radiation Research* 46, 165–171.

Black, H.S. (1987) Potential involvement of free radical reaction in ultraviolet light–mediated cutaneous damage. *Photochemistry and Photobiology* 46, 213–221.

Bosch-Morell, F., Flohe, L., Marin, N. and Romero, F.J. (1999) 4–Hydroxynoneal inhibits glutathione peroxidase: Protection by glutathione. *Free Radical Biology and Medicine* 26, 1383–1387.

Breen, A.P. and Murphy, J.A. (1995) Reactions of oxyl radicals with DNA. *Free Radical Biology and Medicine* 18, 1033–1077.

Bump, E.A. and Malakar, K. (1997) (eds) *Radioprotectors: Chemical, Biological and Clinical Perspectives*, CRC Press, Boca Raton, Florida, USA, p. 431.

Chatterjee, A. and Pakrashi, S.C. (1995) *Treatise of Indian Medicinal Plants*, Vol. 1, Publications and Information Directorate, Council of Scientific and Industrial Research (CSIR), New Delhi, India, p. 26.

Chattopadhyay, S. (2006) Biological Effects of Ionizing Radiations: Modulation by Phytochemicals. In: Sharma, R.K. and Arora, R. (eds) *Herbal Drugs Research: A Twenty First Century Perspective*. Jaypee Brothers Medical Publishers Pvt. Ltd., New Delhi, India, pp. 246–256.

Choudhary, D. and Kale, R.K. (2002) Antioxidant and non–toxic properties of *Piper betel* leaf extract: *in vitro* and *in vivo* studies. *Phytotherapy Research* 16, 461–466.

Dale, W.M., Gray, L.H. and Meredith, W.J. (1949) The inactivation of an enzyme (carboxypeptidase) by X-and γ-radiation. *Philosophical Transactions of the Royal Society, London* A Math Phy Sci 242, 33–62.

Dasgupta, N. and De, B. (2004) Antioxidant activity of *Piper betel* L. leaf extract *in vitro*. *Food Chemistry* 88, 219–224.

Epe, B. (1991) Genotoxicity of singlet oxygen. *Chemico-Biological Interactions* 80, 239–260.

Esterbauer, H., Schaur, R.J. and Zollner, H. (1991) Chemistry and biochemistry of 4-hydroxynoneal, malonaldehyde and related aldehydes. *Free Radical Biology and Medicine* 11, 81–128.

Evans, P.H., Bowers, W.S. and Funk, E.J. (1984) Identification of fungicidal and nematocidal components in the leaves of *Piper betel* (Piperaceae). *Journal of Agricultural and Food Chemistry* 32, 1254–1256.

Glaser, J.L. (1988) *Maharishi Ayurveda: An Introduction to Recent Research. Modern Science and Vedic Science* 2, 89–108.

Goldberg, I. (1998) In: Goldberg, I. (ed) *Functional Foods-Designer Foods, Pharma Foods, Nutraceuticals*, Chapman and Hall, New York, USA, pp. 3–16.

Halliwell, B. and Gutterridge, J.M.C. (1989) Free radicals, ageing, and disease. In: *Free Radicals in Biology and Medicine*, 2nd edn., Clarendon Press, Oxford, pp. 416–508.

Hatano, T., Yasuhara, T., Yoshihara, R., Ikegami, Y., Matsuda, M., Yazaki, K., Agata, I., Nishibe, S., Noro, T., Yoshizaki, M. (1991) Inhibitory effects of galloylated flavonoids on xanthine oxidase. *Planta Medica* 57, 83–84.

Majumdar, B., RoyChaudhury, S., Roy, A. and Bandyopadhyay, S.K. (2003) Effect of ethanol extract of *Piper betel* Linn. leaf on healing of NSAID-induced experimental ulcer- a novel role of free radical scavenging action. *Indian Journal of Experimental Biology* 41, 311–315.

Nice, D. (1997) In: Yalpani, M. (ed) *New Technologies for Healthy Foods and Nutraceuticals*, ATL Press Inc., Shrewsbury, MA, pp. 105–123.

O'Neill, P. and Fieden, E.M. (1993) Primary Free Radical Processes. In: *DNA: Advances in Radiation Biology*, Academic Press, New York, USA, pp. 53–120.

Packer, L.S. and Ong, A.S.H. (1998) *Biological Oxidants and Antioxidants: Molecular Mechanisms and Health Effects*, AOCS Press, Champaign, Illinois, USA.

Patt, H.M., Turee, E.B., Straube, R.L. and Smith, D.E. (1949) Cysteine protection against X-irradiation. *Science* 110, 213–214.

Pierce, D.A. and Preston, D.L. (2000) Radiation-related cancer risks at low doses among atomic bomb survivors. *Radiation Research* 154, 178–186.

Prabhu, M.S, Patel, K., Saraawathi, G. and Srinivasan, K. (1995) Effect of orally administered betel leaf (*Piper betel* leaf Linn.) on digestive enzymes of pancreas and intestinal mucosa and on bile production in rats. *Indian Journal of Experimental Biology* 33, 752–756.

Ramji, N., Iyer. R. and Chandrasekaran, S. (2002) Phenolic antibacterials from *Piper betel* in the prevention of halitosis. *Journal of Ethnopharmacology* 83, 149–152.

Rathee, J.S., Patro, B.S., Mula, S., Gamre, S. and Chattopadhyay, S. (2006) Antioxidant activity of *Piper betel* leaf extract and its constituents. *Journal of Agricultural and Food Chemistry* 54, 9046–9054.

Rice-Evans, C. and Burdon, R. (1993) Free radical-lipid interactions and their pathological consequences. *Progress in Lipid Research* 32, 71–110.

Richards, R.T. and Sharma, H.M. (1991) Free radicals in health and disease. *Indian Journal of Clinical Practice* 2, 15–26.

Salet, C. and Moreno, G. (1990) Photosensitization of mitochondria. Molecular and cellular aspects. *Journal of Photochemistry and Photobiology* B 51, 133–150.

Santhakumari, P., Prakasam, A. and Pugalendi, K.V. (2006) Antihyperglycemic activity of *Piper betel* leaf on streptozotocin-induced diabetic rats. *Journal of Medicinal Food* 9, 108–112.

Saravanan, R., Prakasam, A., Ramesh, B. and Pugalendi, K.V. (2002) Influence of *Piper betel* on hepatic marker enzymes and tissue antioxidant status in ethanol-treated Wistar rats. *Journal of Medicinal Food* 5, 197–204.

Swanson, C.A. (1998) In: Bidlack, W.R., Omaye, S.T., Meskin, M.S. and Johner, D. (eds) *Phytochemicals- A New Paradigm*, Technomic Publishing Company Inc., Pennsylvania, USA, pp. 1–12.

Toniolo, P., Van Kappel, A.L., Akhmedkhanov, A., Ferrari, P., Kato, I., Shore, R.E. and Riboli, E. (2001) Serum carotenoids and breast cancer. *American Journal of Epidemiology* 153, 1142–1147.

Uma Devi, P. (2006) Development of a radioprotective drug from *Ocimum sanctum*. In: Sharma, R.K. and Arora, R. (eds) *Herbal Drugs: A Twenty First Century Perspective*. Jaypee Brothers Medical Publishers Pvt. Ltd., New Delhi, India, pp. 291–300.

Valenzeno, D.P. and Tarr, M. (1991) Membrane photomodification and its use to study reactive oxygen species. In: Rabek, J.F. (ed) *Photochemistry and Photophysics*, CRC Press, Boca Raton, Florida, USA, vol. III, pp. 139–191.

Venkatachalam, S.R. and Chattopadhyay, S. (2003) Natural radioprotectors: An Overview. *Current Organic Chemistry* 9, 389–404.

Von Sonntag, C. (1987) *The Chemical Basis of Radiation Biology*, Taylor and Francis, London, UK.

Weishaput, K.R., Goomer, C.J. and Dougherty, T.J. (1976) Identification of singlet oxygen as the cytoxic agent in photo-inactivation of a murine tumour. *Cancer Research* 36, 2326–2329.

Weiss, J.F. and Landauer, M.R. (2003) Protection against ionizing radiation by antioxidant nutrients and phytochemicals. *Toxicology* 189, 1–20.

Willett, W.C. (1995) Diet, nutrition, and avoidable cancer. *Environmental Health Perspective* 103 (suppl 8), 165–170.

Witt, E.H., Motchink, P. and Packer, L. (1993) Evidence for UV light as an oxidative stressor in skin. In: Fuchs, J. and Packer, L. (eds) *Oxidative Stress in Dermatology*, Mercel Dekker, New York, USA, pp. 29–47.

Yuhas, J.M., Spellman, J.M. and Culo, F. (1980) The role of WR-2721 in radiotherapy and/or chemotherapy. *Cancer Clinical Trials* 3, 211–216.

Chapter 10

Dietary Antioxidants and Phytochemicals in Radioprotection and Therapy

Carmia Borek

Introduction

Radiotherapy has been used in cancer treatment for many decades, to eradicate cancer and as a palliative to relieve pain associated with metastasis. During the course of treatment, radiation produces many biological perturbations in the exposed cells. Because the danger of toxicity to normal cells must limit the dose for effective cancer treatment, approaches are designed to strike a balance between eliminating cancer cells and protecting normal tissues. The primary focus in radiotherapy is to increase DNA damage in tumour cells, as double strand breaks are important in cell death. Another approach is to alter cellular homeostasis, by modifying signal transduction pathways, the redox state and sensitivity to programmed cell death (apoptosis). The ideal situation is to enhance the killing of cancer cells while protecting normal cells and minimizing the probability of their death. This approach has highlighted the efficacy of antioxidants and phytochemicals with antioxidant activity in protecting normal cells from radiation damage and death by apoptosis, while at the same time, in some cases, having the potential to increase the death of cancer cells.

The Effect of Intracellular Free Radicals

Ionizing radiation consists of electromagnetic radiation (photons), including X-rays and gamma rays, and particulate radiation, such as electrons, protons and neutrons. Radiation oncologists treating cancer use largely electromagnetic radiation and particulate radiation, mostly electrons and to a lesser extent neutrons and protons (Coia and Moyland, 1998).

Radiation damages cells by direct ionization of DNA and other targets and by an indirect effect through free radicals produced in the tissue environment; these include hydroxyl radicals (the most damaging), superoxide and other oxidants such as hydrogen peroxide. Additional destructive radicals are formed through various chemical interactions. Approximately two-thirds of X-ray and gamma ray damage is caused by indirect action while heavy particles, such as neutrons, act mostly by direct ionization. While highly effective in killing cancer cells, free radicals and other reactive oxygen species that are produced in radiotherapy threaten the survival of surrounding normal cells. The response of cells to radiation depends on the type and dose of radiation, tissue sensitivity and potential for repair. Other factors that come into play are the position in the cell cycle, oxygen concentration and levels of thiols and other antioxidants that quench free radicals. Intracellular oxygen determines the extent of DNA damage by X-rays and gamma rays. Without oxygen, the indirect radiation damage can be repaired. Oxygen binds to (oxidation) free radical sites in

DNA, "fixing" the damage (Coia and Moyland, 1998).

Antioxidants, including internal thiols such as glutathione, chemically reduce the free radicals, and repair the damage. Reduction of transient free radicals is one mechanism by which antioxidants influence the indirect action of radiation. The extent of danger to normal tissues in radiation therapy depends on the dose, tissue sensitivity and repair capacity, the specific affected organs and prevailing endogenous antioxidant defences.

Mitotic Death and Apoptosis

Radiation induces mitotic cell death in dividing cells and activates pathways that lead to death by apoptosis in interphase cells and differentiated cells. Apoptosis is regulated by the Bcl-2 family of proteins that include the pro-apoptosis proteins Bax and others and the apoptosis-inhibitors Bcl-2, Bcl-X_L, Bag and others. Apoptosis is associated with characteristic morphological changes in cells and their DNA (Pardo *et al.*, 1995) that result largely from the action of activated cysteine proteinases (caspases) (Zimmermann *et al.*, 2001).

Influence of Radiation on Tissue Antioxidants

Radiation at doses used in therapy depletes cellular antioxidants in normal cells, thereby increasing their risk of damage; animal studies show that whole-body exposure to X-ray irradiation decreases tissue concentrations of vitamins C and E (Umegaku *et al.*, 1995). A decline in tissue vitamins E and Se during radiation therapy for breast cancer and a fall in vitamins A, C, E and Se during breast cancer treatment with free radicals – producing adriamycin – may increase normal tissue sensitivity to radiation damage (Borek, 2004a).

Antioxidants and Radiation–A Conundrum

Antioxidant supplementation during radiation therapy, aimed at protecting normal cells, poses a conundrum for the radiation oncologist, as antioxidants that protect normal cells from reactive oxygen species may provide the same benefits to cancer cells and reduce treatment efficacy. Antioxidants can prevent short- and long-term injury to healthy cells, including tissue damage and oncogenic transformation, as seen experimentally (Borek *et al.*, 1986). Clinicians treating cancer patients are faced with this conundrum, and while rarely restricting the intake of antioxidant-rich food, they may advise patients to refrain from taking antioxidant supplements, lest they interfere with the course of treatment. New data, however, show that some dietary antioxidants may have potential as adjuvants in cancer therapy by their ability to induce apoptosis (Borek, 2004a). Studies in cell cultures show that vitamin E, vitamin C, selenium and some phytochemicals selectively induce apoptosis in cancer cells while sparing normal cells (Borek, 2004a; Borek *et al.*, 1986; Sageness *et al.*, 1997; Borek and Pardo, 2002; Rodemann *et al.*, 1999); studies using a model of metastatic growth show that vitamin C is angiostatic and may have potential in aiding host resistance to tumour growth and preventing cancer invasiveness by stopping angiogenesis (Ashino *et al.*, 2003). Antioxidants also show promise in cancer therapy by their palliative action, reducing painful side effects associated with treatment (Kennedy *et al.*, 2001).

Antioxidants differ in modulating radiation effects

Antioxidant action depends on the oxygen partial pressure in a tissue and the nature of the antioxidant. Beta-carotene is an effective chain-breaking antioxidant and thus a potential radioprotector at low PO_2; at high oxygen pressure it is less efficient and may even act as a prooxidant due to autooxidation. By contrast, alpha tocopherol is an efficient antioxidant and potentially a radioprotector, in cells with a high PO_2, for example, in the lung. This is a point of consideration in antioxidant supplementation; radiation therapy aims to increase the oxygen content of tumours to enhance cell killing. As oxygen partial pressure differs among tissues, so does the modulating effect of radiation treatment by antioxidants, which varies with the antioxidant used and the tissue PO_2 (Young and Lowe, 2001).

The potential of antioxidants and a wide range of radioprotectors to reduce cellular damage by ionizing radiation has been studied in animal models as well as *in vitro,* in cell culture (Umegaku *et al.,* 1995; Borek, 2004a; Weiss and Landauer, 2003). There are limited data on the use of antioxidants as radioprotectors in humans, although endogenous antioxidants, such as thiols, and antioxidant enzymes, such as superoxide dismutase, can provide some protection (Coia and Moyland, 1998; Borek, 2004a; Borek *et al.,* 1986; Weiss and Landauer, 2003).

Antioxidant protection and apoptosis

Selenium

The micronutrient selenium is an integral part of glutathione peroxidase, which destroys peroxides, toxic molecules that are produced in exposure to radiation. Treatment of cells with selenium protects them against radiation induced transformation and cell death (Borek, 2004a; Borek *et al.,* 1986; Weiss and Landauer, 2003); while there may be several mechanisms involved, studies show that selenium treatment of cells prior to radiation exposure significantly increases cellular levels of internal antioxidants, glutathione peroxidase, catalase and superoxide dismutase (Borek *et al.,* 1986); selenium has been shown to act selectively to protect normal cells. Selenium stimulates DNA repair in cells with functional p53 (normal cells) (Borek, 2004a) and protected normal human fibroblasts from single and multiple doses of radiation though not squamous cell carcinoma cells (Rodemann *et al.,* 1999).

Vitamin E

Similar to selenium, with which it can in synergistic fashion protect cells against radiation-induced transformation (Borek, 2004a; Borek *et al.,* 1986). Other studies have shown that vitamin E (alpha-tocopherol succinate) inhibits the growth of human cervical, ovarian and lung carcinoma cells in a dose-dependent manner, but has no effect on normal human fibroblasts (Jha *et al.,* 1999). Vitamin E decreased mitotic accumulation in all three-tumour cell lines but did not produce such an effect in any of the normal fibroblasts. Irradiation of both normal and tumour cells with 1Gy decreased mitotic accumulation, but pre-treatment with vitamin E 24 hours before, during and after irradiation further decreased mitotic accumulation in the human tumour cells, but not in the normal cells, suggesting by the authors of the study that "vitamin E alone or in combination with gamma-irradiation, are selective for tumour cells. Therefore, existing fear that antioxidants such as vitamin E may protect cancer cells from free radical damage during radiation therapy is not justified" (Jha *et al.,* 1999).

Another study found that combined treatment with vitamins E and C protects human endothelial cells against apoptosis more effectively than each alone, while increasing Bcl-2 and down regulating the pro-apoptotic Bax (Haendeler *et al.,* 1996). By contrast, vitamin E induces apoptosis in human breast and prostate cancer cells as well as leukemia (Sageness *et al.,* 1997) and glioblastoma cells (Rodemann *et al.,* 1999). Pre-treatment of cells with vitamin E and selenium increases the levels of glutathione, glutathione peroxidase and catalase, while doubling the breakdown of toxic peroxide and reducing transformation (Borek, 2004a; Borek *et al.,* 1986).

Vitamin E has been shown to protect against radiation-induced lethality in experimental animals (Weiss and Landauer, 2003); vitamin E has been shown to protect the intestine from physiological damage by radiation, although it does not protect against high radiation doses that result in death of gastrointestinal cells. Vitamin E has been shown to prevent the loss of jejunal, ileal and colonic fluid absorption in irradiated rats (Weiss and Landauer, 2003).

Vitamin E, neurons and glioblastoma

Brain tissue is highly sensitive to free-radical damage because of its low level of endogenous antioxidants, notably vitamin E and superoxide dismutase. In addition, aging neuronal mitochondria are prone to oxidative damage. Vitamin E protects the integrity of acetyl choline receptors in normal neurons and prevents toxicity and apoptosis induced by free radical-producing amyloid beta-peptides that increase in the brain with age and in dementia (Behl and Moosmann, 2002). Glioblastoma multiforme is the most common and aggressive brain cancer in humans and resists all forms of therapy. Vitamin E (alpha- tocopherol succinate) induces apoptosis in glioblastoma cells but not in normal cells, in a dose-related manner, thus serving to selectively protect the normal cells. A 48-hour exposure to 50-micromol/l vitamin E results in a 15% increase in apoptosis in the glioblastoma cells over control (Pardo *et al.,* 1995). Pre-treatment with vitamin E may have a potential role in sensitizing glioblastoma to radiotherapy while protecting normal cells.

Vitamin E decreases painful side effects

Vitamin supplementation may help treat side effects of radiation therapy. Vitamin E (400 IU) and vitamin C (500mg) have been shown to offer protection against proctitis, a painful chronic injury that affects 5–20% of people receiving radiation therapy for cervical and prostate cancer (Kennedy *et al.,* 2001); a striking regression of chronic radiation-induced fibrosis was seen in a clinical trial that combined radiation treatment of head and neck cancer with vitamin E (1000IU) and pentoxyfylline (0.8g/d) supplementation (Delanian *et al.,* 1999).

Other antioxidants

Radioprotection by dietary vitamin A and beta-carotene in mice exposed to partial-body irradiation or total-body irradiation has been reported, reducing mortality, morbidity and radiation-induced damage in the adrenals and thymus as well as reducing lymphocytopenia (reviewed in Weiss and Landauer, 2003).

Vitamin C protects against radiation-induced chromosomal damage in mice even when administered after irradiation. In studies of mice transplanted with fibrosarcomas, ascorbic acid (4.5g/kg IP) administered 50 min before whole-body irradiation protected against lethality and skin damage (Weiss and Landauer, 2003).

Phytochemicals

Fruits and vegetables and a variety of herbs and spices contain a wide range of antioxidant phytochemicals that are radioprotective in experimental systems (Borek, 2004a; Weiss and Landauer, 2003; Borek, 2001). These include flavonoids, polyphenols, organosulfur compounds, carotenoids, soy products, compounds in cruciferous vegetables (e.g., cabbage and broccoli), tea, *Gingko biloba* extract (flavone glycosides and terpene lactones), milk thistle (silymarin), curcumin and garlic (Weiss and Landauer, 2003; Borek, 2001; Arora *et al.,* 2005). Experimental data are limited and human studies for the most part point to the potential cancer preventive effects of many of these phytochemicals by virtue of their presence in diets that are rich in fruits and vegetables (Borek, 2004a). Cancer preventive effects of phytochemicals that have antioxidant activity are potentially radioprotective (Borek, 2004b).

Phytochemicals: Role in radioprotection

Phytochemicals may act as antioxidants and protect normal cells from free radicals and other reactive oxygen species that are produced during exposure to ionizing radiation. In addition, many of these phytochemicals have the ability to induce apoptosis in cancer cells but not in normal cells (Borek, 2004a).

Garlic

Garlic has been used for centuries as a remedy and cure for many conditions (Borek, 2001). Garlic contains a wide variety of antioxidants, largely organosulfur compounds, that are amplified by aging the cloves in a standardized process of extraction and aging, that yields the odourless supplement aged garlic extract (kyolic) (Borek, 2001). Aged garlic extract is rich in water soluble organosulfur antioxidant compounds, the main one being S-allyl cysteine, which is highly effective in protecting cells from oxidant damage by radiation (Borek, 2001; Lau, 1989); S-allyl mercaptocysteine, an antioxidant organosulfur compound that is unique to aged garlic extract, has apoptotic effects in cancer cells and, as suggested experimentally, may serve potentially as an adjuvant in radiation therapy (Xiao *et al.,* 2003). Other antioxidants present in aged garlic extract include allixin and selenium (Borek, 2001, 2004b; Ide and Lau, 1999, 2001; Kodera *et al.,* 2002). Radioprotection by aged garlic extract (the most studied form of garlic), and its antioxidant compounds, can be achieved in several ways (Borek, 2001, 2004b): by scavenging free radicals and hydrogen peroxides (Ide and Lau, 1999, 2001) that are formed in cells exposed to radiation (Coia and Moyland, 1998; Borek, 2004b) and by enhancing cellular glutathione, an internal radioprotector (Borek, 2001) and reducing its depletion by oxidation (Ide and Lau, 1999, 2001).

Studies *in vivo* also show the protection of garlic compounds against the deleterious effects of radiation and radiomimetics. Garlic has been shown to increase the survival in rats exposed to 4Gy radiation (Jaiswal and Bordia, 1996); aged garlic extract has been found to prevent the toxic effects of the radiomimetic doxorubicin on normal tissue (Kojima *et al.*, 1994). Doxorubicin, also known as adriamycin, is an anthracyclin glycoside, used in treatment of solid tumours, including breast cancer, ovarian carcinoma, small lung cell carcinoma, gastric carcinoma and lymphomas (Borek, 2001). However, therapeutic treatment is limited by cardiotoxicity related directly to the cumulative dose of doxorubicin administered. The cardiotoxic effects of doxorubicin are related to oxidant stress caused by the semiquinone radical of doxorubicin and by superoxides, singlet oxygen and peroxyl radicals, which are generated by the interaction of doxorubicin with mitochondrial membranes, causing lipid peroxidation. The resulting structural and functional mitochondrial damage impairs myocardial function and may result in arrhythmia and congestive heart failure.

Garlic, in the form of aged garlic extract and some of its components, protects mice and cardiac cells *in vitro* against the cardiotoxic effects of doxorubicin, preventing doxorubicin-induced lipid peroxidation (Kojima *et al.*, 1994). The protective effects by aged garlic extract may have potential applications in therapy, reducing the risk of cardiotoxicity in cancer patients receiving treatment with doxorubicin.

Recent studies suggest that aged garlic extract may have an additional potential role as an adjuvant in cancer therapy; the water-soluble organosulfur compound, S-allyl mercaptocysteine, that is unique to aged garlic extract, induces apoptosis in human prostate, breast and colon cancer cells as well as leukemia cells (Borek, 2001, 2004a; Xiao *et al.*, 2003). S-allyl mercaptocysteine activates caspase 3, the executioner enzyme in apoptosis, inhibits anti-apoptotic protein Bcl-2 and disrupts microtubules in the cancer cells, preventing further growth (Xiao *et al.*, 2003).

Curcumin

Curcumin, a yellow pigment present in turmeric (and curry) and used widely in Asian countries, is a powerful antioxidant with radioprotective action. Curcumin has been reported in model systems to protect against damage induced by ionizing radiation in normal cells and enhance the effect of radiation in producing apoptosis in cancer cells. It has been suggested that the radioprotective effect might be mainly due to its ability to reduce oxidative stress and inhibit transcription of genes related to oxidative stress and inflammatory responses, whereas the radiosensitive activity might be due to the upregulation of genes responsible for cell death (Jagetia, 2007).

Recent experimental studies show that curcumin may have a dual action in radiotherapy, on one hand protecting normal cells from damage and on the other hand inducing apoptosis, for example in colon cancer and ovarian cancer cells, one mechanism being the suppression of the anti-apoptotic protein, Bcl-2, in the cancer cells, enhancing their death (Chauhan, 2002; Zheng *et al.*, 2004), though not inducing apoptosis in normal cells (Chauhan, 2002; Zheng *et al.*, 2004).

Resveratrol

An important component of grapes, wine and other grape products, as well as peanuts, resveratrol has gained much interest as a health promoting substance, because of its antioxidant and anti-inflammatory properties and its cancer preventive activity, in experimental models. Thus, resveratrol is a potential radioprotective agent and, as some studies show it has potential in radiation therapy, by inducing apoptosis in cancer cells while protecting the normal tissue by virtue of its antioxidant activity. In a study on cancer cell response to treatments with X-rays and resveratrol, alone or in combination, in terms of DNA damage, cell cycle delays and induction of apoptosis, the results showed that resveratrol induced apoptosis and a block of cell cycle progression at an early step of S-phase. Furthermore, resveratrol mitigated the apoptotic clearance of irradiated cells and prevented the G2 phase cell cycle arrest induced by X-rays (Fiore *et al.*, 2005).

Genistein

The isoflavone genistein that is a major component of soy and has antioxidant activity was found to exert radioprotection (Weiss and Landauer, 2003; Landauer *et al.*, 2003; also see Chapter 12 of this book). When administered as a single subcutaneous dose of genistein 24h before a lethal dose of gamma radiation (9.5Gy of Cobalt-60 at 0.6Gy/min), the mice showed a significant increase in 30-day survival. If given genistein only one hour before irradiation, the 30-day survival rates were not significantly different from those of control mice (Landauer *et al.*, 2003).

Caffeine

Caffeine is a commonly consumed methylxanthine that was found to be a radioprotector in a human study. A retrospective study was undertaken to determine an association between caffeine consumption and a lower incidence of late radiation toxicity in humans. Patients with cervical cancer and endometrial cancer were treated with primary or adjuvant radiation therapy at the University of Washington, USA. Patients reported their average daily caffeine consumption during the time of radiotherapy. Acute radiation toxicity was not associated with caffeine consumption for cervical or endometrial cancer. There was a non-statistically significant trend toward a decrease in overall late radiation toxicity with increased caffeine intake for cervical cancer patients. Subgroup analysis showed that this trend could be attributed to a decreased incidence of severe late radiation injury in cervical cancer patients who consumed higher levels of caffeine at the time of their radiotherapy ($p = 0.02$). This relationship was not seen in the endometrial cancer patients due to the low incidence of severe late injury following radiation for that disease (Stelzer *et al.*, 1994).

Tea polyphenols

Green tea is consumed as a popular beverage worldwide, particularly in Asian countries like China, Korea, Japan and India. Black tea consumption is more common in Western countries. Tea contains polyphenols, also known as epicatechins, which are antioxidant in nature. A study aimed at determining whether tea polyphenols protect human cells against radiation-induced DNA damage found that the addition of green or black tea extracts, or their polyphenols to cultures of human skin fibroblasts significantly reduced the frequencies of radiation-induced chromatid breaks. The protective action of tea polyphenols results from their ability to scavenge hydroxyl free radicals, the most DNA-damaging free radical, produced during radiation (Borek, 2004b). Frequencies of chromatid breaks in cells arrested immediately after irradiation or 0.5 to 1.5 hours post-irradiation, in the presence or absence of a DNA repair inhibitor, to provide a measure of DNA damage. Thus, experimentally, tea can be seen as a radioprotector to human cells against radiation-induced cellular damage (Parshad *et al.,* 1998).

Green tea extract and more effectively, its major compound (-) epigallocatechin-3 gallate (EGCG), was also found to protect normal cells from apoptosis induced by free radical producing agents (Nie *et al.,* 2002), while inducing apoptosis in tumour cells, by activating caspases (Hsu *et al.,* 2003).

Conclusion

This short review focused on the radioprotective efficacy of some antioxidant nutrients and phytochemicals, in their potential to reduce various forms of radiation damage. At the same time, these radioprotective agents offer potential action as adjuvants in cancer therapy by selectively killing cancer cells by apoptosis. Results from experimental studies *in vivo* and *in vitro* show that antioxidant nutrients, such as vitamin E, C and selenium protect against damaging and lethal effects of radiation effects. Some antioxidant nutrients and phytochemicals have the advantage of low toxicity and are generally protective when administered at pharmacological doses (Weiss and Landauer, 2003). Some antioxidants may provide an extended protection against low-dose, low-dose-rate irradiation, even when administered after irradiation. A number of phytochemicals, including caffeine, genistein, garlic compounds, resveratrol, curcumin and the supplement aged garlic extract have the ability to induce apoptosis in cancer cells while protecting the normal cells from free radical damage, including malignant transformation.

Epidemiological studies link the intake of antioxidant-rich foods with a reduced risk of certain cancers. Many people consume antioxidant supplements, sometimes in large doses, to protect against excess free radicals and oxidant stress and these may help in radioprotection. Yet many questions remain unanswered and require additional research. Some of the ones that stand out are: Which antioxidants are most protective? Are isolated antioxidants available as supplements as effective? What would be an optimal dose, under different circumstances of risk, in health and disease and in different populations? What is the role of antioxidants once cancer has been diagnosed? What is the risk/benefit relationship in taking antioxidants during radiation therapy? Antioxidants do protect against radiation-induced oncogenic transformation in experimental systems; however, we do not have comparable human studies that show the same association. Antioxidants do reduce the painful side effects of radiation therapy, thus supporting the beneficial effects of antioxidants in protecting normal cells in radiation therapy and in being used in conjunction with treatment for certain cancers. When considering antioxidant supplementation during treatment, it is doubtful whether high doses of radiation given in certain treatments would be rendered less effective if patients took a daily supplement of antioxidants – for example, at RDA levels – yet, we do not know and more research is needed (Borek, 2004b).

More studies are needed when considering adjuvant therapy of antioxidants and phytochemicals with radiation. At present, with limited available data, many radiation oncologists counsel their patients to refrain from taking antioxidant supplements during radiation therapy. Others, however, consider the data and suggest that a cautious and judicious use of antioxidants that helps the patient maintain a good quality of life may be helpful in cancer treatment and protect against normal tissue damage. Additional research is needed to establish the role of phytochemicals in conjunction with cancer treatment by radiation.

References

Arora, R., Gupta, D., Chawla, R., Sagar, R., Sharma, A., Kumar, R., Prasad, J., Singh, S., Samanta, N. and Sharma, R.K. (2005) Radioprotection by plant products: Present status and future prospects. *Phytotherapy Research* 19, 1–22.

Ashino, H., Shimamura, M. and Nakashima, H. (2003) Novel function of ascorbic acid as an angiostatic factor. *Angiogenesis* 6, 259–269.

Behl, C. and Moosmann, R. (2002) Oxidative nerve cell death in Alzheimer's disease and stroke, antioxidants as neuroprotective compounds. *Biological Chemistry* 383, 521–536.

Borek, C. (2001) Antioxidant health effects of aged garlic extract. *Journal of Nutrition* 131, 1010S–1015S.

Borek, C. (2004a) Dietary antioxidants and human cancer. *Integrative Cancer Therapies* 3, 333–341.

Borek, C. (2004b) Antioxidants and radiotherapy. *Journal of Nutrition* 134, 3207S–3209S.

Borek, C. and Pardo, F. (2002) Vitamin E and apoptosis, a dual role. In: Pasquier, C. (ed.) *Biennial Meeting of the Society for Free Radicals Research International*, Paris, France, July 16–20, 2002.

Borek, C., On, A., Mason, H., Donahue, L. and Beagle, J.E. (1986) Selenium and vitamin E inhibit radiogenic and chemically induced transformation *in vitro* via different mechanisms. *Proceedings of the National Academy of Sciences of the United States of America*, 83, 1490–1494.

Chauhan, D.P. (2002) Chemotherapeutic potential of curcumin for colorectal cancer. *Current Pharmaceutical Design* 8, 1695–1706.

Coia, L.R. and Moyland, D.J. (1998) *Introduction to Clinical Radiation Oncology*. Medical Physics Publishing, Madison, WI, USA.

Delanian, S., Balla-Mekias, S. and Lefaix, J.L. (1999) Striking regression of chronic radiotherapy damage in a clinical trial of combined pentoxyfilline and tocopherol. *Journal of Clinical Oncology* 17, 3283–3290.

Fiore, M., Festa, F., Cornetta, T., Ricordy, R. and Cozzi, R. (2005) Resveratrol affects X-ray induced apoptosis and cell cycle delay in human cells *in vitro*. *International Journal of Molecular Medicine* 15, 1005–12.

Haendeler, J., Zeiher, A.M. and Dimmler, S. (1996) Vitamin C and vitamin E prevent lipopolysaccharide-induced apoptosis in human endothelial cells by modulating Bcl-2 and Bax. *European Journal of Pharmacology* 17, 407–411.

Hsu, S., Lewis, J., Sigh, B. and Schoenlein, P. (2003) Green tea polyphenol target the mitochondria in tumour cells inducing caspase-3 dependent apoptosis. *Anticancer Research* 23, 1533–1539.

Ide, N. and Lau, B.H. (1999) Aged garlic extract attenuates intracellular oxidative stress. *Phytomedicine* 6, 125–131.

Ide, N. and Lau, B.H. (2001) Garlic compounds minimize intracellular oxidative stress and inhibit nuclear factor-kappa β activation. *Journal of Nutrition* 131, 1020S–1026S.

Jagetia, G.C. (2007) Radioprotection and radiosensitization by curcumin. *Advances in Experimental Medicine and Biology* 595, 301–320.

Jaiswal, S.K. and Bordia, A. (1996) Radioprotective effect of garlic *Allium sativum* Linn. in albino rats. *Indian Journal of Medical Sciences* 50, 231–233.

Jha, M.N., Bedford, J.S., Cole, W.C., Edward-Prasad, J. and Prasad, K.N. (1999) Vitamin E(d-tocopheryl succinate) decreases mitotic accumulation in gamma-irradiated human tumor, but not in normal cells. *Nutrition and Cancer* 35(2), 189–194.

Kennedy, M., Brunuinga, K., Mutlu, E.A., Losurdo, J., Choudhary, S. and Keshavarian, A. (2001) Successful and sustained treatment of chronic radiation proctitis with antioxidant vitamins E and C. *American Journal of Gastroenterology* 96, 1080–1084.

Kodera, Y., Ichikawa, M., Yoshida, J., Kashimoto, N., Uda, N., Sumioka, I., Ide, N. and Ono, K. (2002) Pharmacokinetic study of allixin, a phytoalexin produced by garlic. *Chemical and Pharmaceutical Bulletin* (Tokyo) 50(3), 354–363.

Kojima, R., Toyama, Y. and Ohnishi, S.T. (1994) Protective effects of an aged garlic extract on doxorubicin-induced cardiotoxicity in the mouse. *Nutrition and Cancer* 22(2), 163–173.

Landauer, M.R., Srinivasan, V. and Seed, T.M. (2003) Genistein treatment protects mice from ionizing radiation injury. *Journal of Applied Toxicolgy* 23, 379–385.

Lau, B.H.S. (1989) Detoxifying, radioprotective and phagocyte-enhancing effect of garlic. *International Nutrition Review* 9, 27–31.

Nie, G., Cao, Y. and Zhan, B. (2002) Protective effects of green tea polyphenols and their major component (-) epigallocatechin-3-gallate (EGCG), on 6- hydroxydopamine induced apoptosis in PC12 cells. *Redox Reports* 7, 171–177.

Pardo, F. S., Mei, S. and Borek, C. (1995) Cyclin D1 induced apoptosis maintains the integrity of the G1/S checkpoint following ionizing radiation. *Somatic Cell and Molecular Genetics* 22, 135–144.

Parshad, R., Sanford, K.K., Price, F.M., Steele, V.E., Tarone, R.E., Kelloff, G.J. and Boone, C.W. (1998) Protective action of plant polyphenols on radiation-induced chromatid breaks in cultured human cells. *Anticancer Research* 18, 3263–3266.

Rodemann, H.P., Hehr, T. and Bamberg, M. (1999) Relevance of the radioprotective effect of sodium selenite. *Medizinische Klinik* 94, 39–41.

Sageness, G., Anagnostu, A. and Steiner, M. (1997) Dl-alpha tocopherol induces apoptosis in erythroleukemia, prostate and breast cancer cells. *Nutrition and Cancer* 28, 30–35.

Stelzer, K.J., Koh, W.J., Kurtz, H., Greer, B.E. and Griffin, T.W. (1994) Caffeine consumption is associated with decreased severe late toxicity after radiation to the pelvis. *International Journal of Radiation Oncology, Biology and Physics* 30, 411–417.

Umegaku, K., Aoki, S. and Esashi, T. (1995) Whole-body irradiation to mice decreases ascorbic acid concentrations in bone marrow, comparison with vitamin E. *Free Radical Biology and Medicine* 19, 493–497.

Weiss, J.F. and Landauer, M.R. (2003) Protection against ionizing radiation by antioxidant nutrients and phytochemicals. *Toxicology* 189, 1–20.

Xiao, D., Pinto, J.T., Soh, J.T., Deguchi, A., Gunderson, G.G., Palazzo, A.F., Yoon, J.T., Shirin, H. and Weinstein, I.B. (2003) Induction of apoptosis by the garlic-derived compound S-allyl mercaptocysteine (SAMC) is associated with microtubule depolymerization and c-Jun NH (2) -terminal kinase 1 activation. *Cancer Research* 15, 6825–6842.

Young, A.J. and Lowe, G.M. (2001) Antioxidant and prooxidant properties of carotenoids. *Archives of Biochemistry and Biophysics* 385, 20–27.

Zheng, L., Tong, Q. and Wu, C. (2004) Growth inhibitory effects of curcumin on ovary cells and its mechanism. *Journal of Huazhong University of Science and Technology. Medical Sciences* 24, 55–58.

Zimmermann, K.C., Bonzon, C. and Green, D.R. (2001) The machinery of programmed cell death. Pharmacology and Therapeutics 92, 57–70.

Chapter 11

Effects of Berry Fruits on Neurocognitive Deficits Produced by Exposure to Space Radiation

Bernard M. Rabin, James A. Joseph and Barbara Shukitt-Hale

Introduction to Space Radiation and its Biological Effects

Future manned missions in space will involve long-term travel outside the protection provided by the magnetic field of the earth. Astronauts who participate in these missions will be exposed to types of radiation that are not experienced in low earth orbit, where the space shuttle and the International Space Station operate (Badhwar, 1997; Letaw *et al.*, 1989; Schimmerling *et al.*, 2003; Townsend *et al.*, 1992). These types of radiation, galactic cosmic rays, are composed of alpha particles, protons and particles of high energy and charge (HZE particles). The primary source of high-energy protons is solar particle events ("solar flares"). HZE particles are of celestial origin and, while some may be given off as a consequence of solar particle events, most are free particles in space remaining from the formation of the universe.

HZE particles, such as ^{56}Fe, can be characterized by the energy of the particle and by the amount of energy (Linear Energy Transfer [LET]) that is deposited in the tissue through which the particle passes. In general, the higher the LET of the particle the greater the amount of energy that is deposited in the tissue and the greater amount of tissue destruction. In terms of their physical characteristics, HZE heavy particles differ from lower LET types of radiation such as X-rays or gamma rays on several dimensions (e.g., Choudhury *et al.*, 1998; Nelson, 2003). In contrast to conventional types of radiation, HZE particles do not exert diffuse effects on the target tissue, but rather deposit their energy along a well-defined track. The length of the track is determined by the energy of the particle. If the particle stops in tissue there is a significant increase in energy deposition at the point at which the particle stops. This differs from conventional electromagnetic types of radiation which show an exponential loss of energy as a function of depth in tissue.

Despite differences in physical characteristics, both electromagnetic radiation and HZE particles will lead to the development of cancer (Cucinotta *et al.*, 2001; Cucinotta and Durante, 2006; Barcellos-Hoff *et al.*, 2005; Hellweg and Baumstark-Khan, 2007). These types of radiation differ, however, in their effects on neurocognitive performance. In contrast to exposure to low doses of gamma- or X-rays, which do not affect central nervous system function in mature organisms, low dose, non-lethal exposures to ^{56}Fe particles produce changes in neuronal functioning (Joseph *et al.*, 2000) and a significant disruption of cognitive/behavioural performance (Rabin *et al.*, 1998, 2000; Shukitt-Hale *et al.*, 2004, 2007).

Neurocognitive Effects of Exposure to Heavy Particles: Accelerated Aging

Exposing rats to HZE particles can affect performance on a variety of neurobiological and behavioural endpoints. Following exposure to ^{56}Fe particles rats show a reduction in potassium-stimulated dopamine release (Joseph *et al.,* 1992). The neurochemical deficits include changes in signalling molecules that regulate signal transduction processes in the striatum (Joseph *et al.,* 1993, 1994; Denisova *et al.,* 2002). Irradiation with ^{56}Fe particles also produces deficits in hippocampal neurogenesis (Casadesus *et al.,* 2005; Rola *et al.,* 2005).

Coincident with the changes in neuronal function, deficits have been observed in both motor and cognitive behaviours. Exposure to low doses of ^{56}Fe particles produces a decrease in upper body strength, measured by the length of time rats can maintain their grip on a wire suspended above the ground (Joseph *et al.,* 1992). Similarly, exposure to low doses of HZE particles will prevent the acquisition of a conditioned taste aversion (CTA) produced by the dopamine agonist amphetamine (Rabin *et al.,* 1998). A CTA is produced by pairing a novel taste solution (10% sucrose) with an unconditioned stimulus (amphetamine). As a result of this pairing the rat will avoid ingestion of the solution at a subsequent presentation. Because amphetamine is a dopamine agonist, the development of an amphetamine-induced CTA requires an intact dopamine system. Similarly, treating rats with the dopamine antagonist haloperidol also produces a disruption of amphetamine-induced CTA learning (Rabin *et al.,* 1998). Both exposure to ^{56}Fe particles and injection of haloperidol disrupt dopaminergic function and interfere with the acquisition of an amphetamine-induced CTA.

Aging (Shukitt-Hale *et al.*, 1998, 2001) and exposure to ^{56}Fe particles also affect spatial learning and memory, measured using the Morris water maze (Shukitt-Hale *et al.,* 2000) and the radial-arm maze (Denisova *et al.,* 2002). Overall, the results of these experiments indicate that irradiated rats are deficient in their ability to perform a task requiring the use of spatial cues. The Morris water maze is a test of cognitive ability in which rats are required to use spatial cues to locate a platform placed just below the surface of the water. There are no differences in performance between the non-irradiated controls and the irradiated rats in the initial acquisition of the task. However, when the platform is moved to a different location in the maze, rats exposed to 1.5Gy of ^{56}Fe particles show significantly poorer performance than the non-irradiated control rats. Similarly, when the platform is absent during probe trials, the irradiated rats spend significantly less time in the quadrant in which the platform had been located than do the control rats, indicating impaired spatial memory (Shukitt-Hale *et al.,* 2000).

Performance on the radial-arm maze is also disrupted following exposure to ^{56}Fe particles (Denisova *et al.,* 2002). The radial-arm maze consists of eight equally spaced arms radiating from a central area. Each maze arm contains a small dish at its end that can be baited with a small food pellet or saccharin water reward. The task involves the successive selection of arms in order to obtain the reward. An optimal strategy involves using spatial memory to visit each arm only once, since the arms are not baited again after the rats have obtained the reward on that particular trial. Various control experiments strongly suggest that the rat performs the radial arm maze test by utilizing extra-maze cues. The radial arm maze provides a test of both working and reference memory. Working memory is required while the rat is selecting the next arm to visit, apparently "keeping track of" which arms were previously selected during the specific trial. Reference memory is tested on trials in which only some of the arms contain reward because performance on this version of the test requires that the rats remember which arms have been baited from trial to trial. Rats exposed to ^{56}Fe particles had no difficulty in performing the task when all eight arms are baited (i.e., no working memory deficit) because they were using kinesthetic strategies, or chaining, which is simply moving from one arm to the next adjacent arm, in order, without attending to spatial cues. However, when only four arms were baited on subsequent days, the irradiated rats continued to visit each arm in order and committed more reference memory errors than control rats, who quickly learned to enter only the baited arms (Denisova *et al.*, 2002). The performance deficit in reference memory is correlated with alterations in signalling molecules (such as protein kinases) and synaptic vessicle proteins (synaptophysin and synaptobrevin) in the stratum and frontal cortex (Denisova *et al.*, 2002).

Rats that have been exposed to ^{56}Fe particles also show deficits on operant conditioning tasks (Rabin *et al.*, 2002a). Operant conditioning, in which the organism learns to make a response in order to obtain reward or to avoid punishment, is broadly construed to include all forms of complex learning. The specific task requires that the rats respond on an ascending fixed-ratio (FR) reinforcement schedule. On an FR schedule, a rat is required to make a fixed number of responses (lever presses) in order to obtain reinforcement (45 mg food pellet). On an FR-1 schedule, the rat is rewarded for every lever press, while on an FR-20 schedule the rat makes 20 lever presses for a single food pellet, and on an FR-35 reinforcement schedule the rat is required to make 35 responses in order to be rewarded with a single food pellet. When tested 3 months following exposure to ^{56}Fe particles, only the rats exposed to 2.0Gy (but not 1.0 or 1.5Gy) showed a disruption of responding at schedules of reinforcement of FR-25 or greater. There were no effects of irradiation at schedules less that FR-20. When tested 7 months after exposure all irradiated groups showed significantly decreased performance at reinforcement schedules of FR-20 or greater compared to the non-irradiated controls.

Another cognitive task that is affected by exposure to low doses of ^{56}Fe particles is anxiety, which is typically measured by performance on the elevated plus-maze (Rabin *et al.*, 2007a). Anxiety levels are typically inferred by measuring the amount of time a rat will spend exploring the normally aversive open arms of the maze. In general, the greater the level of anxiety, the less time spent in the open arms of the maze. Concordant with the results of previous research (Andrade *et al.*, 2003; Baguszewski and Zagrodzka, 2002; Bessa *et al.*, 2005; Rabin *et al.*, 2007a), old rats spent less time exploring the open arms of the maze. Exposure to ^{56}Fe particles also produced decreased exploration of the open arms of the plus-maze, indicating that radiation produced an increase in baseline anxiety. The dose needed to produce increased levels of anxiety was a function of age at the time of irradiation, such that lower doses were needed to produce reduced exploration time in the rats radiated at 7 and 12 months of age than were needed in the 2-month-old rats (Rabin *et al.*, 2007a).

The neurochemical and behavioural deficits detailed above are also observed in old rats. Old rats show deficits in potassium-stimulated dopamine release and related deficits in motor behaviour. Similarly, there are age-related deficits in signal transduction processes and in signalling molecules in the striatum. Coincident with the changes in dopaminergic function and neuronal signalling, there are significant decreases in neurocognitive performance of old rats. Old rats show significant decreases in performance in the Morris water maze, indicating decreased ability to utilize spatial cues in a learning task (Shukitt-Hale *et al.*, 1998). In addition, the partial loss of dopaminergic neurons produced by treatment with the neurotoxin 6-hydroxydopamine, which does not affect the performance of young rats on an ascending fixed-ratio schedule, does cause a significant impairment in the performance of older rats (Lindner *et al.,* 1999).

In sum, the changes in neuronal function and in behaviour seen in rats that have been exposed to ^{56}Fe particles parallel those seen in aged animals. As such, it has been proposed that exposing rats to HZE particles produces "accelerated aging" (Joseph *et al.,* 1992, 1993).

Role of Oxidative Stress

Current theories of aging stress the role of free radicals and oxidative stress in the process (Barja, 2004; Bokov *et al.,* 2004; Floyd and Hensley, 2002). Oxidative stress occurs when endogenous and exogenous sources of reactive oxygen species (ROS) exceed the capacity of the endogenous antioxidant systems to remove them. Endogenous sources of ROS include the aerobic metabolism of mitochondria and the destruction of dopamine by monoamine oxidase. Acting to mitigate the effects of oxidative stress are a variety of endogenous antioxidant defense systems (superoxide dismutase and glutathione peroxidase) and exogenous sources of antioxidants (vitamins and flavonoid antioxidants). The consequences of oxidative stress include aging (Finkel and Holbrook, 2000), carcinogenesis (Oberly, 2002) and a variety of neurodegenerative disorders such as Parkinson's and Alzheimer's diseases (Halliwell, 2001).

Radiation, like other toxic stimuli, produces oxidative stress (Riley, 1994; Miura, 2004; Denisova *et al.,* 2002; Fang *et al.,* 2002), which may mediate changes in neuronal function that affect performance. Following exposure to ^{56}Fe particles there is evidence of increased oxidative stress in hippocampus, striatum and frontal cortex and changes in signalling molecules which are correlated with decreased performance in the radial arm maze (Denisova *et al.,* 2002).

The neurocognitive effects of both the aging process and exposure to HZE particles are mediated by oxidative stress. It may be this common factor, oxidative stress, that is responsible for another effect of exposure to HZE particles: "accelerated aging" (Joseph *et al.,* 1992, 1993). As indicated above, exposure to ^{56}Fe particles produces many of the same changes in neurochemical function and behaviour that are observed in aged animals. Additional research has shown that there is an interaction between age and exposure to HZE particles such that doses of ^{56}Fe particles that do not affect the performance of younger rats do produce a significant disruption in performance as the animals age (Rabin *et al.,* 2005a). Also, doses of ^{56}Fe particles that do not affect the performance of younger organisms do affect the performance of older rats (Rabin *et al.,* 2007a).

Berry Fruits as Compounds to Prevent Age-Related Decrements in Behaviour

One treatment that has been effective in ameliorating the neurobehavioural effects of aging has been the use of dietary antioxidants, such as are found in fruits, particularly berries. Measured as Oxygen Radical Absorbance Capacity, the free radical scavenging capacity of blueberries and strawberries is much higher than that of vitamin E (Prior *et al.,* 1998; Wang *et al.,* 1996). Research using old animals has shown that maintaining rats on diets containing 2% blueberry or strawberry extract for eight weeks prevents and/or reverses the age-related changes in neurochemical functioning and in the behaviours which show age-related decrements, including spatial learning and memory (Bickford *et al.,* 2000; Joseph *et al.,* 1998, 1999) and in object recognition memory (Goyarzu *et al.,* 2004). This observation is consistent with current theories, which suggest that oxidative stress and the production of ROS are key factors in the aging process (Finkel and Holbrook, 2000).

Berry Fruits as Radioprotectants

To the extent that oxidative stress mediates the neurobehavioural effects of exposure to HZE particles, then antioxidant treatments that are effective in ameliorating the neurochemical and neurobehavioural effects that accompany the aging process may also function to ameliorate the effects of irradiation. Specifically, given the similarity in the neurobehavioural effects of aging and irradiation, it is possible that maintaining rats on diets containing flavonoid antioxidants may also counteract the effects of exposure to HZE particles.

Blueberries and strawberries contain a variety of compounds that may function as antioxidants. The polyphenolics contained in fruits include the hydroxycinnamates and the flavonoids such as the anthocyanins and flavonols. The relative amounts of these compounds in different fruits vary, which may account for differences in the antioxidant capacity. For instance, blueberries have higher levels of the proanthocyanins whereas strawberries tend to have higher levels of hydrolyzable tannins such as ellagitannins (Seeram, 2005). Additional work will be needed in order to determine the active compounds and their effects on specific neurobehavioural endpoints. Nonetheless, as summarized below, maintaining rats on diets containing either blueberry or strawberry extract can ameliorate the effects of exposure to HZE particles on specific neurochemical and behavioural endpoints.

Effects on Neurocognitive Endpoints

When rats are maintained on diets containing 2% blueberry or strawberry extract for two months prior to exposure to ^{56}Fe particles (1.5Gy, 1GeV/n), the radiation-induced decreases in potassium-stimulated dopamine release in the striatum are prevented (Shukitt-Hale *et al.,* 2007a). These results are similar to those obtained with aged rats maintained on identical diets (Joseph *et al.,* 1998, 1999; Bickford *et al.,* 2000).

Concordant with the neurochemical effects, antioxidant diets also ameliorate the cognitive/behavioural deficits produced by exposure to HZE particles, although the effectiveness of the blueberry and strawberry diet varies as a function of the specific endpoint. For CTA learning, the rats maintained on either the blueberry or strawberry diet failed to show the ^{56}Fe particle-induced disruption of an amphetamine-induced CTA (Rabin *et al.,* 2002b). That is, following exposure to either 1.5Gy or 2.0Gy the irradiated rats maintained on either diet for two months prior to exposure showed the acquisition of a CTA following injection of the dopamine agonist amphetamine. In contrast, the irradiated rats fed a control diet failed to acquire an amphetamine-induced taste aversion, which is consistent with previous research (Rabin *et al.,* 1998, 2000).

Similar results were obtained in a test of spatial learning and memory using the Morris water maze (Shukitt-Hale *et al.*, 2007a,b) with rats that had been exposed to 1.5Gy of ^{56}Fe particles. Although the irradiated rats maintained on either the blueberry or strawberry diet showed improved performance compared to radiated rats maintained on the control diet, there were diet-specific differences in the pattern of responding. The rats fed the strawberry diet were better able to retain place information, which is mediated by the hippocampus and, therefore, showed less of a deficit in spatial performance compared to the control animals. In contrast, the rats maintained on the blueberry diet showed better performance on the reversal task; a behaviour which is dependent upon the striatum.

The effects of antioxidant diets on operant responding also varied as a function of the specific diet. Seven months following exposure to 1.5Gy of ^{56}Fe particles, there was no effect of irradiation or diet on operant responding. When the rats were tested eleven months following irradiation, the animals fed either the control or blueberry diets showed significantly poorer performance on an ascending fixed-ratio reinforcement schedule than the non-irradiated rats (Rabin *et al.,* 2005a). The performance of the rats fed the strawberry diet was not significantly different from that of the non-irradiated controls and significantly better than that of the irradiated rats fed the blueberry or control diet. Similar results were obtained following exposure to 2.0Gy of ^{56}Fe particles, except that the effects of irradiation and diet were observed when the rats were first tested five months following irradiation (Rabin *et al.,* 2005b).

These results raise several issues with regard to the mechanism of action of the diets containing blueberry or strawberry extract. The results obtained by Shukitt-Hale *et al.,* (2007a,b) looking at spatial learning and memory suggest that diets containing strawberry extract are more effective in ameliorating the effects of exposure to HZE particles on hippocampally mediated behaviours, whereas diets containing blueberry extract are more effective with striatally mediated behaviours. This suggestion is not consistent with the results obtained by Rabin *et al.,* (2005a,b) using operant responding on an ascending fixed-ratio schedule. This behaviour is mediated by the nigrostriatal dopamine system (Lindner *et al.,* 1997, 1999; Salamone *et al.,* 1993). As such, the latter results would suggest that the antioxidants contained in strawberry are more effective in providing protection against the deleterious effects of exposure to ^{56}Fe particles on a striatally mediated behaviour. The currently available data does not permit a determination of the factors that might influence the differential effectiveness of the different diets as a function of the specific brain areas that might be involved in mediating the behaviour or the component of the behaviour.

A second issue relating to the differential effectiveness of the two diets on neurocognitive endpoints is the chemical composition of the blueberry and strawberry diets. All berries contain bioactive chemicals including phenolics, anthocyanins, hydroxycinnamates and flavonols. However, the relative amounts of these constituents varies as a function of the specific berry. Again, blueberries have more proanthocyanins whereas strawberries have more ellagitannins. In addition, strawberries and blueberries differ in terms of their antioxidant capacity and in terms of their ability to cross the blood-brain barrier (Youdim *et al.,* 2004). However, further research will be necessary to define the specific phytochemicals in berries, the nature of their interaction with specific brain regions and the specific endpoints under consideration.

Effects on Tumourigenesis

A recent development in radiation protection has been the use of phytochemicals following exposure to low LET gamma- or X-irradiation (Lee *et al.*, 2005; Fang *et al.*, 2002; Miura, 2004; Arora *et al.*, 2005; Hayes 2005). Because exposure to ionizing radiation causes oxidative stress, which can play a role in carcinogenesis (Oberly, 2002), it is logical to hypothesize that phytochemicals that have antioxidant properties will be able to provide some degree of protection against the carcinogenic consequences of irradiation.

However, as indicated in the introduction, the physical properties of HZE particles such as ^{56}Fe are very different from those of low LET gamma- and X-rays. As such, the mechanisms by which they interact with living tissue are also different. It is therefore possible that the use of diets containing antioxidant phytochemicals might not be as effective against HZE particle-induced carcinogenesis as they are against low LET types of radiation.

Nonetheless, to the extent that oxidative stress does play a role in tumour development, the use of antioxidant diets should reduce the development of tumours following irradiation. In this regard, Kennedy and her colleagues (Kennedy and Todd, 2003; Guan *et al.*, 2004) have shown that the use of the chemical compound L-selenomethionine, either alone or in combination with antioxidants such as vitamin E, can reduce the risk of carcinogenesis following exposure to HZE particles or protons. Similarly, in a preliminary study (Rabin *et al.*, 2007b) we have shown that dietary phytochemicals (strawberry or blueberry extract) also provide a significant degree of protection against HZE particle-induced tumourigenesis. As part of another study (Rabin *et al.*, 2005b) rats that had been exposed to 1.5Gy of ^{56}Fe particles were studied for up to 56 weeks post-irradiation. In addition to their performance on an operant conditioning task, data were also collected on the effects of diet on tumour development. Analyses indicated that both blueberry and strawberry diets significantly reduced the occurrence of HZE particle-induced tumours. In contrast to the irradiated rats maintained on the control diet, which showed significant tumour development over the one-year period that they were followed, the frequency of tumour development in the rats that had been maintained on the diets containing either blueberry or strawberry extract prior to irradiation did not differ from the non-irradiated rats maintained on these diets. Because the rats were maintained on the diets containing antioxidant phytochemicals for only 1 week following exposure to ^{56}Fe particles, these results suggest that the reduction of oxidative stress is necessary only at the time of irradiation in order to reduce the frequency of HZE particle-induced tumourigenesis.

Conclusions

During exploratory class missions to other planets astronauts will be exposed to types and doses of radiation that are not experienced in low earth orbit. In addition to the carcinogenic consequences of exposure, cosmic ray irradiation has the potential to produce neurocognitive deficits which can interfere with the ability of astronauts to successfully meet mission requirements. The neurocognitive deficits seen following exposure to ^{56}Fe particles, a ground-based model for exposure to cosmic rays, are similar to those that accompany the aging process. As such, there is the possibility that the deficits in neurochemical and behavioural function result from increased oxidative stress produced by irradiation. The data reviewed above is consistent with the hypothesis that oxidative stress plays a significant role in the disruption of neurocognitive function following exposure to space radiation and that short-term treatment with diets containing antioxidant phytochemicals provides a significant degree of protection against the deleterious effects of HZE particle irradiation. Despite the differences in the mechanisms by which exposure to space radiation affects neuronal function and despite the fact that the neurocognitive deficits observed following exposure to ^{56}Fe particles are not observed following exposure to gamma- or X-rays, the use of antioxidant phytochemicals are equally effective following both types of radiation. As such, proper diets containing antioxidant phytochemicals may function to prevent or ameliorate potential neurocognitive deficits for astronauts on long-term exploratory class missions.

Acknowledgements

Preparation of this review was supported in part by Grant NNJ06HD93G from the National Aeronautics and Space Administration (NASA). The research cited in the paper was supported by grants from NASA, with additional support from UMBC, ARS-USDA and from the Department of Defense, USA.

References

Andrade, M.M., Tome, M.F., Santiago, E.S., Lucia-Santos, A. and de Andrade, T.G. (2003) Longitudinal study of daily variation of rats' behaviour in the elevated plus-maze. *Physiology and Behavior* 78(1), 125–133.

Andrade, M.M.M., Tomé, M.F., Santiago, E.S., Lucia-Santos, A. and de Andrade, T.G. (1991) Longitudinal study of daily variation of rats behaviour in the elevated plus-maze. *Physiology & Behavior* 78, 125–133.

Arora, R., Gupta, D., Chawla, R., Sagar, R., Sharma, A., Kumar, R., Prasad, J., Singh, S., Samanta, N. and Sharma, R.K. (2005) Radioprotection by plant products: Present status and future prospects. *Phytotherapy Research* 19, 1–22.

Badhwar, G.D. (1997) The radiation environment in low earth orbit. *Radiation Research* 148, S3–S10.

Baguszewski, P. and Zagrodzka, J. (2002) Emotional changes related to age in rats– a behavioural analysis. *Behavioural Brain Research* 133, 323–332.

Barcellos-Hoff, M.H., Park, C. and Wright, E.G. (2005) Radiation and the microenvironment–tumourigenesis and therapy. *Nature Reviews Cancer* 5, 867–875.

Barja, G. (2004) Free radicals and aging. *Trends in Neurosciences* 27, 595–600.

Bessa, J.M., Olivera, M., Cerqueira, J.J., Almeida, O.F.X. and Sousa, N. (2005) Age–related qualitative shift in emotional behaviour, Paradoxical findings after re–exposure of rats in the elevated–plus maze. *Behavioural Brain Research* 162, 135–142.

Bickford, P., Gould, T., Briederick, L., Chadman, K., Pollock, A., Young, D., Shukitt-Hale, B. and Joseph, J.A. (2000) Antioxidant–rich diets improve cerebellar physiology and motor learning in aged rats. *Brain Research* 866, 211–217.

Bokov, A., Chaudhuri, A. and Richardson, A. (2004) The role of oxidative damage and stress in aging. *Mechanisms of Ageing and Development* 125, 811–826.

Choudhury, D., Srivastava, M., Sarma, A. and Kale, R.K. (1998) Effect of high linear energy transfer radiation on biological membranes. *Radiation and Environmental Biophysics* 37, 177–185.

Cucinotta, F.A. and Durante, M. (2006) Cancer risk from exposure to galactic cosmic rays, implications for space exploration by human beings. *The Lancet-Oncology* 7, 431–435.

Cucinotta, F.A., Schimmerling, W., Wilson, J.W., Peterson, L.E., Badhwar, G.D., Sagaanti, P.B. and Dicello, J.F. (2001) Space radiation cancer risks and uncertainties for Mars missions. *Radiation Research* 156, 682–688.

Denisova, N., Shukitt-Hale, B., Rabin, B.M. and Joseph, J.A. (2002) Brain signaling and behavioural responses induced by exposure to ^{56}Fe radiation. *Radiation Research* 158, 725–734.

Fang, Y–Z., Yang, S. and Wu, G. (2002) Free radicals, antioxidants, and nutrition. *Nutrition* 18, 872–879.

Finkel, T. and Holbrook, N.J. (2000) Oxidants, oxidative stress and the biology of aging. *Nature* 408, 239–247.

Floyd, R.A. and Hensley, K. (2002) Oxidative stress in brain aging, Implications for therapeutics of neurodegenerative diseases. *Neurobiology of Aging* 23, 795–807.

Goyarzu, P., Malin, D.H., Lau, F.C., Taglialatela, G., Moon, W.D., Jennings, R., Moy, E., Moy, D., Lippold, S., Shukitt-Hale, B. and Joseph, J.A. (2004) Blueberry supplemented diet. Effects on object recognition memory and nuclear factor-kappa B levels in aged rats. *Nutritional Neuroscience* 7, 75–83.

Guan, J., Wan, X.S., Zhou, Z., Ware, J., Donahue, J.J., Biaglow, J.E. and Kennedy, A.R. (2004) Effects of dietary supplements on space radiation-induced oxidative stress in Sprague-Dawley rats. *Radiation Research* 162, 572–579.

Halliwell, B. (2001) Role of free radicals in the neurodegenerative diseases; therapeutic implications for antioxidant treatment. *Drugs & Aging* 18, 685–716.

Hayes, D.F. (2005) The protective role of fruits and vegetables against radiation–induced cancer. *Nutrition Reviews* 63, 303–311.

Hellweg, C.E. and Baunstark-Kahn, C. (2007) Getting ready for the manned mission to Mars, the astronauts risks from space radiation. *Naturwissenschaften* 94, 517–526.

Joseph, J.A, Hunt, W.A., Rabin, B.M. and Dalton, T.K. (1992) Possible "accelerated aging" induced by ^{56}Fe heavy particle irradiation. Implications for manned space flights. *Radiation Research* 130, 88–93.

Joseph, J.A., Erat, S. and Rabin, B.M. (1998) Selective efficacy of space-like radiation effects (^{56}Fe particles) on muscarinic neurotransmitter sensitivity and motor behaviour. *Advances in Space Research* 22, 216–224.

Joseph, J.A., Hunt, W.A., Rabin, B.M., Dalton, T.K. and Harris, A.H. (1993) Deficits in striatal muscarinic receptor sensitivity induced by ^{56}Fe heavy particle irradiation, Further "age-radiation" parallels. *Radiation Research* 135, 257–261.

Joseph, J.A., Shukitt-Hale, B., Denisova, N.A., Bielinski, D., Martin, A., McEwen, J.J. and Bickford, P.C. (1999) Reversals of age-related declines in neuronal signal transduction, cognitive and motor behavioural deficits with diets supplemented with blueberry, spinach or strawberry dietary supplementation. *Journal of Neuroscience* 19, 8114–8121.

Joseph, J.A., Shukitt-Hale, B., Denisova, N.A., Prior, R.L., Cao, G., Martin, A., Taglialatela, G. and Bickford, P.C. (1998) Long-term dietary strawberry, spinach, or vitamin E supplementation retards the onset of age-related neuronal signal-transduction and cognitive behavioural deficits. *Journal of Neuroscience* 18, 8047–8055.

Joseph, J.A., Shukitt-Hale, B., McEwen, J. and Rabin, B.M. (2000) CNS–induced deficits of heavy particle irradiation in space, the aging connection. *Advances in Space Research* 25, 2057–2064.

Kennedy, A.R. and Todd, P. (2003) Biological countermeasures in space radiation health. *Gravitation Space Biology Bulletin* 16, 37–44.

Kovacic, P. and Jacintho, J.D. (2001) Mechanisms of carcinogenesis. Focus on oxidative stress and electron transfer. *Current Medicinal Chemistry* 8, 773–796.

Lee, T-K., Johnke, R.M., Allison, R.R., O'Brien, K.F. and Dobbs, Jr. L.J. (2005) Radioprotective potential of ginseng. *Mutagenesis* 20, 237–243.

Letaw, J.R., Silberberg, R. and Tsao, C.H. (1989) Radiation hazards on space missions outside the magnetosphere. *Advances in Space Research* 9, 285–291.

Lindner, M.D., Cain, C.K., Plone, M.A., Frydel, B.R., Blaney, T.J., Emerich, D.F. and Hoane, M.R. (1999) Incomplete nigrostriatal dopaminergic cell loss and partial reductions in striatal dopamine produce akinesia, rigidity, tremor and cognitive deficits in middle-aged rats. *Behavioural Brain Research* 102, 1–16.

Lindner, M.D., Plone, M.A., Francis, J.M., Blaney, T.J., Salamone, J.D. and Emerich, D.F. (1997) Rats with partial striatal dopamine depletions exhibit robust and long-lasting behavioural deficits in a simple fixed-ratio bar-pressing task. *Behavioural Brain Research* 86, 25–40.

Miura, Y. (2004) Oxidative stress, radiation–adaptive responses and aging. *Journal of Radiation Research* 45, 357–372.

Nelson, G.A. (2003) Fundamental space radiobiology. *Gravitation Space Biology Bulletin* 16, 29–36.

Oberley, T.D. (2002) Oxidative damage and cancer. *American Journal of Pathology* 160, 403–408.

Prior, R.L., Cao, G., Martin, A., Sofic, E., McEwen, J., O'Brien, C., Lischner, N., Ehlenfeldt, M., Kalt, W. and Krewer, M. (1998) Antioxidant capacity as influenced by total phenolic and anthocyanin content, maturity and variety of Vaccinium species. *Journal of Agricultural and Food Chemistry* 46, 2586–2593.

Rabin B.M., Joseph, J.A., Shukitt-Hale, B. and Carey, A.N. (2007b) Dietary modulation of the effects of exposure to ^{56}Fe particles. *Advances in Space Research* 40, 576–580.

Rabin, B.M., Buhler, L.L., Joseph, J.A., Shukitt-Hale, B. and Jenkins, D.G. (2002a) Effects of exposure to ^{56}Fe particles or protons on fixed-ratio operant responding in rats. *Journal of Radiation Research* 43 (Suppl.), S225–S228.

Rabin, B.M., Carrihill-Knoll, K.L., Carey, A., Shukitt-Hale, B. and Joseph, J.A. (2005b). Effect of diet on the disruption of operant responding at different ages following exposure to ^{56}Fe particles. *AGE* 27, 69–73.

Rabin, B.M., Carrihill-Knoll, K.L., Carey, A.N., Shukitt-Hale, B., Joseph, J.A. and Foster, B.C. (2007a) Elevated plus-maze performance of Fischer-344 rats as a function of age and of exposure to ^{56}Fe particles. *Advances in Space Research* 39, 981–986.

Rabin, B.M., Joseph, J.A. and Erat, S. (1998) Effects of exposure to different types of radiation on behaviours mediated by peripheral or central systems. *Advances in Space Research* 22, 217–225.

Rabin, B.M., Joseph, J.A. and Shukitt-Hale, B. (2005a) Effects of age and diet on the heavy particle-induced disruption of operant responding produced by a ground-based model for exposure to cosmic rays. *Brain Research* 1036, 122–129.

Rabin, B.M., Joseph, J.A., Shukitt-Hale, B. and McEwen, J. (2000) Effects of exposure to heavy particles on a behaviour mediated by the dopaminergic system. *Advances in Space Research* 25, 2065–2074.

Rabin, B.M., Shukitt-Hale, B., Szprengiel, A. and Joseph, J.A. (2002b) Effects of heavy particle irradiation and diet on amphetamine- and lithium chloride-induced taste avoidance learning in rats. *Brain Research* 953, 31–36.

Riley, P.A. (1994) Free radicals in biology. Oxidative stress and the effects of ionizing radiation. *International Journal of Radiation Biology* 65, 27–33.

Rola, R., Sarkissian, V., Obenaus, A., Nelson, G.A., Otsuka, S., Limoli, C.L. and Fike, J.R. (2005) High-LET radiation induces inflammation and persistent changes in markers of hippocampal neurogenesis. *Radiation Research* 164, 556–560.

Salamone, J.D., Kurth, P.A., McCullough, L.D., Sokolowski, J.D. and Cousins, M.S. (1993) The role of brain dopamine in response initiation. Effects of haloperidol and regionally specific dopamine depletions on the local rate of instrumental responding. *Brain Research* 628, 218–226.

Schimmerling, W., Cucinotta, F.A. and Wilson, J.W. (2003) Radiation risk and human space exploration. *Advances in Space Research* 31, 27–34.

Seeram, N.P. (2005) Berries and Cancer. In: Heber, D., Blackburn, G.L. and Go, V.L.W. (eds) *Nutritional Oncology*, Academic Press, London, UK.

Shukitt-Hale, B., Carey, A.N., Jenkins, D., Rabin, B.M. and Joseph, J.A. (2007a) Beneficial effects of fruit extracts on neuronal function and behaviour in a rodent model of accelerated aging. *Neurobiology of Aging* 28, 1187–1194.

Shukitt-Hale, B., Casadesus, G., Carey, A., Rabin, B.M. and Joseph, J.A. (2007b) Exposure to ^{56}Fe irradiation accelerates normal brain aging and produces deficits in learning and memory. *Advances in Space Research* 39, 1087–1092.
Shukitt-Hale, B., Casadesus, G., McEwen, J.J., Rabin, B.M. and Joseph, J.A. (2000) Spatial learning and memory deficits induced by ^{56}Fe radiation exposure. *Radiation Research* 154, 28–33.
Shukitt-Hale, B., Casadeus, G., Cantuti-Castelvetri, I. and Joseph, J.A. (2001) Effect of age on object exploration, habituation, and response to spatial and non-spatial change. *Behavioral Neuroscience* 115, 1059–1064.
Shukitt-Hale, B., Mouzakis, G. and Joseph, J.A. (1998) Psychomotor and spatial memory performance in aging male Fischer 344 rats. *Experimental Gerontology* 33, 615–624.
Shukitt-Hale, B., Szprengiel, A., Pluhar, J., Rabin, B.M. and Joseph, J.A. (2004) The effects of proton exposure on neurochemistry and behaviour. *Advances in Space Research* 33, 1334–1339.
Townsend, L.W., Cucinotta, F.A. and Wilson, J.W. (1992) Interplanetary crew exposure estimates for galactic cosmic rays. *Radiation Research* 129, 48–52.
Wang, H., Cao, G. and Prior, R. (1996) Total antioxidant capacity of fruits. *Journal of Agricultural and Food Chemistry* 44, 701–705.
Youdim, K.A. and Joseph, J.A. (2001) A possible emerging role of phytochemicals in improving age-related neurological dysfunctions. A multiplicity of effects. *Free Radical Biology and Medicine* 30, 583–594.
Youdim, K.A., Shukitt-Hale, B., Martin, A., Wang, H., Denisova, N., Bickford, P.C. and Joseph, J.A. (2004) Short-term dietary supplementation of blueberry polyphenolics, beneficial effects on aging brain performance and peripheral tissue function. *Nutritional Neuroscience* 3, 383–397.

Chapter 12

Radioprotection by the Soy Isoflavone Genistein

Michael R. Landauer

Introduction

Natural products have been the source of many new pharmaceutical agents (Newman and Cragg, 2007) and the flavonoids in particular have been the origin of a variety of phytomedicines (Montoro *et al.,* 2005). The flavonoids are phenolic compounds and have wide distribution in the plant kingdom. Flavonoids play a major role in the ecology of plants and function as stress protectants in plant cells by scavenging reactive oxygen species produced by the photosynthetic electron transport system (Pietta, 2000). They are also believed to protect the plant from ultraviolet radiation (Winkel-Shirley, 2002).

Isoflavones, a class of flavonoids, are plant derived, non-steroidal compounds that are weak estrogens. These phytoestrogens are most abundant in legumes, with soybeans being the major source of isoflavones. The most predominant isoflavone in soy is genistein (4',5,7-trihydroxyflavone), and is primarily found as the glycoside. Its molecular formula is $C_{15}H_{10}O_5$ and its molecular weight is 270.24 daltons (Fig. 12.1).

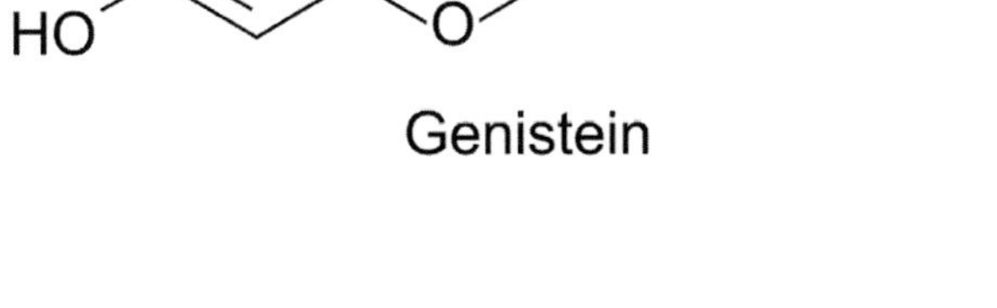

Fig. 12.1. Structure of Genistein (4',5,7-trihydroxyflavone).

Genistein has a number of biological properties that have been associated with radioprotection (Bump and Malaker, 1998). It has been shown to have antioxidant (see Arora *et al.,* 2005), free radical scavenging (Kruk *et al.,* 2005), estrogenic (Valachovicova *et al.,* 2004), anti-microbial (Hong *et al.,* 2006), anti-inflammatory (Verdrengh *et al.,* 2003) and protein tyrosine kinase inhibitory properties (Akiyama *et al.,* 1987).

Genistein has gained increasing attention because of its association with the beneficial effects for treatment of cardiovascular disease, high blood pressure, osteoporosis, breast cancer and prostate cancer (McCue and Shetty, 2004; Sugano, 2006; Valachovicova *et al.,* 2004). The effects of genistein's ability to protect against radiation-induced lethality will be the subject of this chapter.

The need for effective, non-toxic radiation countermeasures has become increasingly important. With the rise in terrorist activity and the dissemination of nuclear materials, an individual's chance of being exposed to dirty bombs or improvized nuclear devices has risen dramatically in recent years (Flynn and Goans, 2006). Radiation threat agents include radiation dispersal weapons, which scatter radioactive material with conventional explosives; improvized nuclear devices; and criticality devices, traditional nuclear weapons that release large prompt doses of radiation. Another possible threat is the destruction of nuclear power plants. In addition to these threats, radiation countermeasures also have applications in clinical oncology, space travel and radiation site cleanup.

The exposure of humans to high dose, total body irradiation results in a dose-dependent, severe and usually fatal illness known as the acute radiation syndrome (ARS). It is characterized by the haematopoietic (>1Gy), gastrointestinal (GI; >10Gy) and cerebrovascular syndromes (CVS; >20Gy) (Waselenko *et al.,* 2004). Great potential exists for developing treatments for the lower radiation doses that result in the haematopoietic syndrome. The primary cause of death from this syndrome is bone marrow destruction, which results in significant reductions in neutrophils and platelets. This can lead to infection, haemorrhage, and eventually death. It is more difficult to develop protective countermeasures against the higher radiation doses that lead to the GI and CVS syndromes.

Radiation countermeasure agents have been categorized into three main areas. These include 1) radioprotectors or prophylactic agents that are administered prior to irradiation, 2) mitigators that are given during or shortly after radiation exposure but before the appearance of overt signs of radiation injury, and 3) radiation therapeutics or treatments that are given after the manifestation of clinical symptoms (Stone *et al.,* 2004).

Currently, amifostine is the only radiation protector approved by the United States Food and Drug Administration (FDA). This drug is used in conjunction with radiation therapy for head and neck cancer to protect the parotid salivary glands (Buntzel *et al.,* 2007). Unfortunately, amifostine has severe dose-limiting side effects including nausea, vomiting and pronounced hypotension, making it unacceptable for emergency personnel, the military or civilian populations (Rades *et al.,* 2004). The need for a non-toxic radioprotectant for protection against ARS is therefore of great interest.

At present, several radiation mitigation agents are available. The FDA has approved the use of calcium-DTPA (diethylenetriamine pentaacetic acid) and zinc-DTPA for treating internal contamination from the transuranium elements plutonium, americium and curium (Food and Drug Administration, 2007). These drugs increase the rates of elimination of these substances from the body, thereby reducing the risk of radiation-induced cancer in the future. However, they have no effect on preventing the consequences of exposure to external radiation leading to ARS. Prussian blue, also approved by the FDA for treating internal radiation contamination, is limited for use with caesium-137 and thallium. In addition, the FDA approval of potassium iodide (KI) is limited to reducing the risk of thyroid cancer in emergencies involving the release of radioactive iodine (from power plants), but it has no protective effect against damage to the haematopoietic system (Food and Drug Administration, 2007).

Presently, there are no FDA-approved radiotherapeutic drugs for the treatment of ARS. However, there is an investigational new drug (IND) application for the hematopoietic growth factor, granulocyte-colony stimulating factor (G-CSF). G-CSF is currently available as a radiotherapeutic agent for emergency use only (Pellmar, 2006).

Because of the many biological properties of genistein and the need for radiation countermeasures, we investigated the ability of this isoflavone to prevent radiation-induced lethality. In the experiments discussed below, we evaluated 30-day survival in mice treated with genistein. Mortality from the doses of radiation used in these studies tends to occur later than acute lethality from most toxic substances. Therefore, mice were observed for 30-days post-irradiation in order to quantify protection from doses of radiation that could destroy the haematopoietic system (Storer *et al.*, 1982).

Experimental Studies

Animals

Male CD2F1 mice (Harlan Laboratories, Indianapolis, IN) weighing 24 to 30g were used in these studies. Mice were housed in a facility accredited by the Association for Assessment and Accreditation of Laboratory Animal Care International. Animal rooms were maintained at 21 ± 2°C with 50% ± 10% humidity on a 12-hr light/dark cycle. Commercial rodent ration (Harlan Teklad Rodent Diet 8604) was available freely as was acidified (pH=2.5-3.0) water to control opportunistic infections (McPherson, 1963). All animal handling procedures were performed in compliance with guidelines from the U.S. National Research Council and were approved by the Institutional Animal Care and Use Committee of the Armed Forces Radiobiology Research Institute (AFRRI), USA.

Isoflavone preparation

Genistein was solubilized in polyethylene glycol with a molecular weight of 400 (PEG) on the day of the experiment by 20 sec of sonication (Heat Systems-Ultrasonics Inc., Plainview, NY). Genistein and PEG were obtained from Sigma Chemical Company (St. Louis, MO). In some experiments, saline was used as a control excipient and was obtained from Abbott Laboratories (North Chicago, IL). All drugs were administered either subcutaneously (SC) in a volume of 0.1ml or by oral gavage (PO) in a volume of 0.2ml.

Experimental Design and Results

Toxicity studies

In the studies discussed below, separate groups of mice were used to assess the acute toxicity of genistein in non-irradiated mice. Radioprotection studies were performed once an acceptable non-toxic dose of genistein was determined. The behaviour of non-irradiated mice following administration of a single SC injection of saline, PEG vehicle or 100-, 200- or 400mg/kg genistein was observed. Locomotor activity, grip strength and motor coordination were evaluated over 14 days. At the termination of the study, tissues from the testes, liver, adrenal gland, mesenteric lymph node, spleen and bone marrow of the femur and sternum were histologically evaluated. The results indicated that there were no significant effects of genistein or the vehicle on behaviour when compared to the saline-treated control group. Moreover, no gross morphological changes or histopathological alterations were observed (Landauer *et al.*, 2003b).

Radiation studies

Mice received total body irradiation in a bilateral gamma radiation field at AFRRI's Cobalt-60 (^{60}Co) facility. The alanine/electron spin resonance (ESR) dosimetry system (American Society for Testing and Materials, Standard E 1607) was used to measure dose rates (to water) in the cores of acrylic mouse phantoms. To simulate a mouse, the phantoms were three inches in length and one inch in diameter. During the field mapping, all compartments in the exposure rack contained a phantom. Three alanine dosimeters were inserted into every other phantom. The ESR signals were measured using a calibration curve based on the standard calibration dosimeters provided by the National Institute of Standards and Technology (NIST; Gaithersburg, MD, USA). The overall uncertainty in the doses given to the calibration dosimeters at NIST was approximately 1.8% at 2σ. The accuracy of the calibration curve was verified by parallel measurements of doses to selected dosimeters at AFRRI and the National Physical Laboratory (Middlesex, UK). The only corrections applied to the dose rates in phantoms were for the decay of ^{60}Co and the small difference in the mass energy-absorption coefficients for water and soft tissue.

In the event that a radioprotective agent must be given to the general population, the three most appropriate routes of administration would be subcutaneous (SC), intramuscular (IM) or oral (PO). While intravenous administration may be acceptable in the clinic, it is not practical for use by emergency personnel, the military or the general population in a mass casualty scenario. Because drugs are not given to humans via the intraperitoneal (IP) route, we did not evaluate this route of administration. Using a murine model for our studies, we investigated both the SC and PO routes of administration. The administration of drugs by the IM route to mice is not recommended because of their small muscle mass (Shimizu, 2004).

Subcutaneous administration

For the initial radioprotection studies, mice received a single SC injection of genistein before 9.5Gy gamma irradiation at a dose rate of 0.6Gy/min. Mice were each administered a single SC injection of saline, vehicle or genistein at doses of 3 to 400mg/kg, 24 hr before irradiation and monitored for 30 days (N = 16–48/group).

A single SC injection of genistein administered 24 hr before irradiation was found to significantly ($p < 0.001$) enhance 30-day survival in groups of mice that received genistein doses of 25 to 400mg/kg (Fig 12.2). The optimal radioprotective dose of 200mg/kg genistein afforded significantly ($p < 0.01$) more protection (91% survival) than the lower doses (25–100mg/kg). The level of protection at 200mg/kg genistein was not significantly different from the 400mg/kg dose (Landauer *et al.*, 2003a).

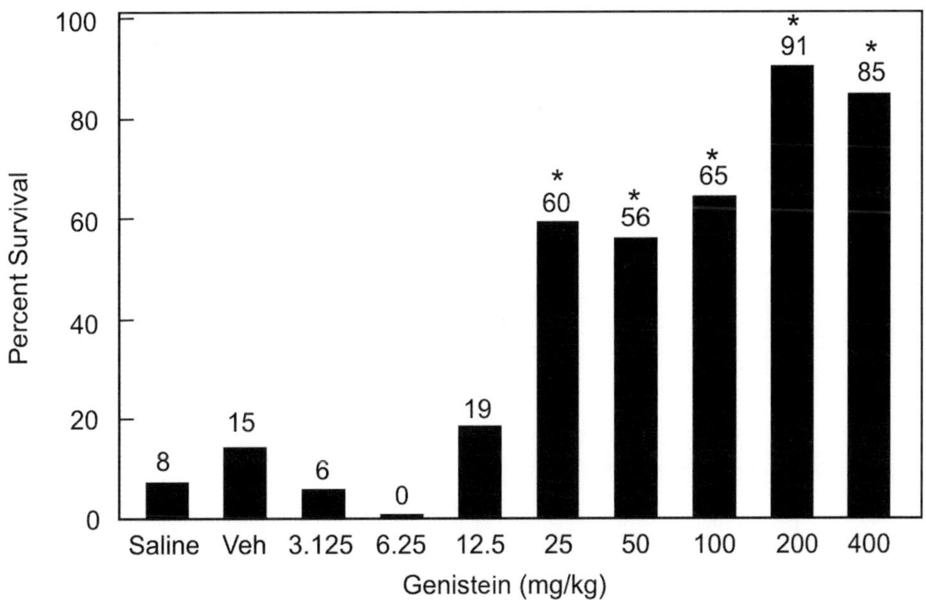

Fig. 12.2. Histogram illustrating 30-day survival of mice treated 24 hr before 9.5Gy ^{60}Co gamma irradiation with a single, subcutaneous injection of saline, vehicle (PEG) or genistein at a dose of 3.125 to 400mg/kg of body weight. Mice that received a genistein dose of 25, 50, 100, 200 or 400mg/kg exhibited a significant increase in survival 30 days after radiation exposure (*$p < 0.001$ from PEG vehicle control and saline groups).

Following this study, the radioprotective effect of genistein was evaluated when it was administered 30 min or 1 hr before irradiation, or 30 min or 1 hr after irradiation. When a single dose of genistein (200mg/kg) was given at these time points, there were no significant differences when compared to the vehicle control group (Landauer *et al.*, 2002). Thus, the optimal time for administration by the SC route was 24 hr before irradiation.

The dose reduction factor (DRF), defined as the ratio of the $LD_{50/30}$ for irradiated mice pretreated with genistein divided by the $LD_{50/30}$ for vehicle-treated mice, was determined when 200mg/kg genistein was administered SC 24 hr before irradiation. The $LD_{50/30}$s were obtained using probit analysis. The DRF was 1.16 and was significantly different from the vehicle control group ($p < 0.001$) (Landauer *et al.*, 2003b). While this may be considered a modest level of protection, it was obtained at a non-toxic drug dose that did not result in performance decrement. This is in contrast to many other drugs that yield a higher protective factor but also result in more severe side effects that would be unacceptable to first responders and other emergency personnel (Landauer, 2002; Landauer *et al.,* 2001; Landauer *et al.,* 1987; Landauer *et al.,* 1992). A modest level of protection may be enough to reduce the effective radiation dose to a level where treatment can be successfully initiated.

Oral administration

Genistein's protective effects when administered by the PO route were also evaluated. In this experiment, when genistein (400mg/kg) was administered PO by gavage as a single dose 24 hr or 1 hr before 9.5Gy radiation, there was no significant difference in protection between genistein and the vehicle-treated mice. In a subsequent experiment, mice received 100mg/kg or 400mg/kg genistein for 4 days before (pre), 4 days after (post), or 4 days before and 4 days after (pre + post) a lethal dose of gamma radiation (9.5Gy) (N = 16/group). Animals that received genistein pre-irradiation were given PEG vehicle post-irradiation, while mice that were administered genistein post-irradiation received PEG vehicle before irradiation. Thus, all animals received eight daily oral gavages of either vehicle or genistein.

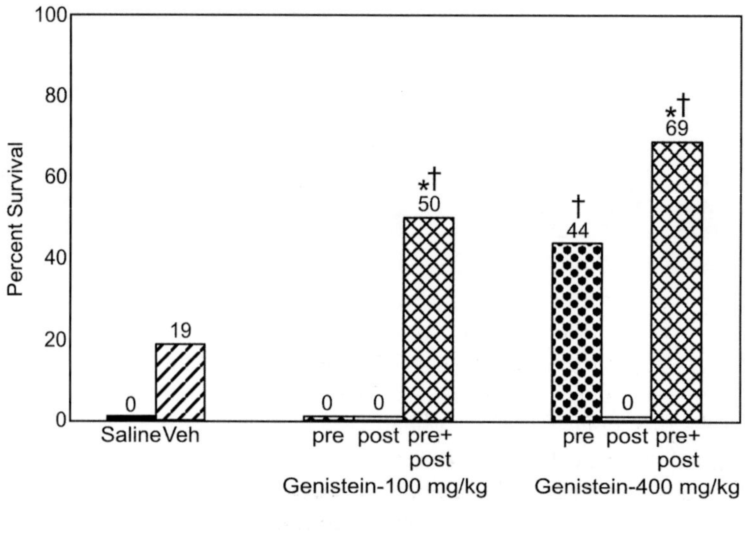

Fig. 12.3. Histogram illustrating 30-day survival of irradiated mice treated with multiple, daily oral gavage with saline, polyethylene glycol 400 (PEG) vehicle or genistein (100mg/kg or 400mg/kg). Mice were given genistein for 4 days before irradiation (pre), 4 days after irradiation (post), or 4 days before and 4 days after irradiation (pre + post). The radiation dose was 9.5Gy ^{60}Co. *$p < 0.05$ from PEG vehicle control group; †$p < 0.05$ from the saline group.

The post-irradiation dosing began 1hr after irradiation. Two control groups received either saline or PEG, both before and after irradiation. The survival rates for saline and PEG groups were only 0% and 19%, respectively. The survival rates for genistein 100mg/kg pre, post, and pre + post were 0%, 0%, and 50%, respectively, while survival rates for the genistein 400mg/kg pre, post, and pre + post groups were 44%, 0% and 69%, respectively. Oral genistein protected animals against a lethal dose of gamma irradiation, but it was only effective when given before or before and after irradiation (Landauer *et al.*, 2000) (Fig. 12.3).

Using a similar experimental paradigm, others (Zhou and Mi, 2005) have reported that when vehicle or genistein (160mg/kg) was administered by gavage for seven consecutive days before irradiation, survival rates were 17% and 53%, respectively, in male BALB/c mice.

This was similar to our results for four daily gavages for vehicle (PEG) and genistein (400mg/kg) of 19% and 44%, respectively, in male CD2F1 mice (Landauer *et al.*, 2000). Both oral studies yielded significantly lower survival rates compared to the 91% survival observed when genistein was administered in a single SC dose of 200mg/kg 24 hr before irradiation.

Discussion

Together, these data demonstrate that genistein is a non-toxic, naturally occurring compound that protects against radiation-induced lethality. While genistein is radioprotective when administered SC 24 hr before irradiation or by daily oral administration either before or before and after radiation, no protection was observed when genistein was administered only after irradiation. This absence of protection after irradiation has also been observed for the related flavonoid compounds, orientin and vicenin (Uma Devi *et al.*, 1999).

The doses used in the above studies are pharmacological doses. Pre-clinical research (Faqi *et al.*, 2004; Landauer *et al.*, 2003b) demonstrated that genistein has an excellent safety profile and has undergone a number of human clinical trials (Fischer *et al.*, 2004; Takimoto *et al.*, 2003). Based on differences in surface area and the more rapid metabolism of isoflavones in mice, a conversion factor of 12 can be used to estimate the human equivalent dose from the mouse dose (Freireich *et al.*, 1966). With this allometric scaling figure, doses of 100 and 200mg/kg genistein in the mouse would be equal to approximately 8 and 16mg/kg, respectively, in humans. Clinical trials have used oral doses as high as 16mg/kg body weight of genistein. These doses were reported to be well tolerated with no clinically related drug toxicity (Bergan *et al.*, 2001; Fischer *et al.*, 2004). Consequently, the genistein doses used in our mouse experiments appear to be clinically feasible.

While genistein was found to be an effective radiation protective agent when administered by SC or PO before irradiation to normal mice, this effect appears to be different for malignant cells. Genistein can act as a radiosensitizer under *in vitro* and *in vivo* conditions in a variety of cancer cell lines. For example, genistein potentiated the effects of irradiation in a human prostate carcinoma cell line (Hillman *et al.*, 2001) and also acted as a radiosensitizer in leukemic cells (Papazisis *et al.*, 2000) and cervical cancer cells (Yashar *et al.*, 2005; Zhang *et al.*, 2006) exposed to ionizing radiation. Using an orthotopic tumour model in mice, it was shown that genistein combined with irradiation of prostate tumours (Hillman *et al.*, 2004; Raffoul *et al.*, 2007; Wang *et al.*, 2006) or kidney tumours (Hillman *et al.*, 2007) resulted in greater inhibition of primary tumour growth than with irradiation alone.

The mechanism of genistein's radioprotective effect is probably multifactorial. Genistein-induced radioprotection involves stimulation of the haematopoietic system. In an experiment from our laboratory, we evaluated the recovery of bone marrow cells and peripheral blood haematology in mice that received a single SC injection of genistein

(200mg/kg) 24 hr prior to a lethal, total body irradiation dose (8.75Gy) of ^{60}Co gamma radiation (Davis *et al.,* 2007). In this study, 97% of genistein-treated mice survived after 30 days while 31% of vehicle-treated and 0% of untreated mice survived. The improvement in survival was related to accelerated neutrophil and platelet recovery, resulting from earlier and more pronounced multilineage, hematopoietic progenitor cell reconstitution in the femoral marrow compartment.

The antioxidative and free radical scavenging properties of isoflavones have been well documented (Jefremov *et al.,* 2007; Kruk *et al.,* 2005). However, the time delay of 24 hr for effective genistein radioprotection does not support a mechanism of direct antioxidant protection that would theoretically be most effective when the drug is administered in close proximity to the time of irradiation. However, genistein may act indirectly through the induction of antioxidant enzymes such as superoxide dismutase and glutathione peroxidase (Cai and Wei, 1996). An indirect mechanism such as cytokine release may also be responsible for the radioprotective effects of genistein (Neta, 1998; Singh and Yadav, 2005).

Like other protectants such as interleukin-1 (Neta, 1998) and glucan (Patchen, 1998), radioprotection by genistein may stimulate quiescent primitive haematopoietic stem-progenitor cells to arrest at specific points in the cell cycle. This arrest is hypothesized to allow time for DNA repair mechanisms before replication in rapidly growing progenitor cells. Recently, in an *ex-vivo* radiation model system, genistein has been shown to upregulate the gene expression levels of the DNA repair gene, Gadd45a (Grace *et al.,* in press). The late S-phase of the cell cycle is the most radioresistant (Denekamp, 1986), and genistein has been demonstrated to increase the percentage of cells in the S-phase in unirradiated mice (Morris *et al.,* 2003).

The research presented in this paper demonstrates that the prophylactic administration of non-toxic doses of the soy isoflavone, genistein, enhances survival and stimulates haematopoietic recovery in irradiated mice. Pre-treatment with genistein may be useful in mitigating radiation-induced injury to normal tissues for individuals undergoing radiotherapy or for emergency personnel entering radiation-contaminated areas.

Acknowledgements

The contributions and discussions with J. Harre, M. Whitnall, S. Parker, C. Lissner and T. Pellmar who read earlier versions of this manuscript are gratefully acknowledged.

References

Akiyama, T., Ishida, J., Nakagawa, S., Ogawara, H., Watanabe, S., Itoh, N. Shibuya, M. and Fukami, Y. (1987) Genistein, a specific inhibitor of tyrosine-specific protein kinases. *Journal of Biological Chemistry* 262, 5592–5595.

Arora, R., Gupta, D., Chawla, R., Sagar, R., Sharma, A., Kumar, R., Prasad, J., Singh, S., Samanta, N. and Sharma, R.K. (2005) Radioprotection by plant products: Present status and future prospects. *Phytotherapy Research* 19, 1–22.

Bergan, R.C., Waggle, D.H., Carter, S.K., Horak, I., Slichenmyer, W. and Meyers, M. (2001) Tyrosine kinase inhibitors and signal transduction modulators, rationale and current status as chemopreventive agents for prostate cancer. *Urology* 57, 77–80.

Bump, E.A. and Malaker, K. (1998) *Radioprotectors, Chemical, Biological, and Clinical Perspectives,* CRC Press, Washington, D.C., p. 431.

Buntzel, J., Glatzel, M., Mucke, R., Micke, O. and Bruns, F. (2007) Influence of amifostine on late radiation-toxicity in head and neck cancer-a follow-up study. *Anticancer Research* 27, 1953–1956.

Cai, Q. and Wei, H. (1996) Effect of dietary genistein on antioxidant enzyme activities in SENCAR mice. *Nutrition and Cancer* 25, 1–7.

Davis, T.A., Clarke, T.K., Mog, S.R and Landauer, M.R. (2007) Subcutaneous administration of genistein prior to lethal irradiation supports multilineage, hematopoietic progenitor cell recovery and survival. *International Journal of Radiation Biology* 83, 141–151.

Denekamp, J. (1986) Cell kinetics and radiation biology. *International Journal of Radiation Biology and Related Studies in Physics, Chemistry and Medicine* 49, 357–380.

Faqi, A.S, Johnson, W.D, Morrissey, R.L. and McCormick, D.L. (2004) Reproductive toxicity assessment of chronic dietary exposure to soy isoflavones in male rats. *Reproductive Toxicology* 18, 605–611.

Fischer, L., Mahoney, C., Jeffcoat, A.R., Koch, M.A., Thomas, B.F., Valentine, J.L., Stinchcombe, T., Boan, J., Crowell, J.A. and Zeisel, S.H. (2004) Clinical characteristics and pharmacokinetics of purified soy isoflavones, Multiple-dose administration to men with prostate neoplasia. *Nutrition and Cancer* 48, 160–170.

Flynn, D.F. and Goans, R.E. (2006) Nuclear terrorism, triage and medical management of radiation and combined-injury casualties. *Surgical Clinics of North America* 86, 601–636.

Food and Drug Administration (2007) Drug Preparedness and Response to Bioterrorism, Radiation Emergencies. http//www.fda.gov/cder/drugprepare/default.htm#Radiation; accessed July 29, 2007.

Freireich, E.J, Gehan, E.A., Rall, D.P., Schmidt, L.H. and Skipper, H.E. (1966) Quantitative comparison of toxicity of anticancer agents in mouse, rat, hamster, dog, monkey, and man. *Cancer Chemotherapy Reports* Part 1 50, 219–244.

Grace, M.B., Blakely, W.F. and Landauer, M.R. Genistein effects on gene expression biomarkers of radiation response. *Radiation Measurements* (in press).

Hillman, G.G., Forman, J.D., Kucuk, O., Yudelev, M., Maughan, R.L., Rubio, J., Layer, A., Tekyi-Mensah, S., Abrams, J. and Sarkar, F.H. (2001) Genistein potentiates the radiation effect on prostate carcinoma cells. *Clinical Cancer Research* 7, 382–390.

Hillman, G.G., Wang, Y., Che, M., Raffoul, J.J., Yudelev, M., Kucuk, O. and Sarkar, F.H. (2007) Progression of renal cell carcinoma is inhibited by genistein and radiation in an orthotopic model. *BMC Cancer* 7, 4.

Hillman, G.G., Wang, Y., Kucuk, O., Che, M., Doerge, D.R., Yudelev, M., Joiner, M.C., Marples, B., Forman, J.D. and Sarkar, F.H. (2004) Genistein potentiates inhibition of tumour growth by radiation in a prostate cancer orthotopic model. *Molecular Cancer Therapeutics* 3, 1271–1279.

Hong, H., Landauer, M.R., Foriska, M.A. and Ledney, G.D. (2006) Antibacterial activity of the soy isoflavone genistein. *Journal of Basic Microbiology* 46, 329–335.

Jefremov, V., Zilmer, M., Zilmer, K., Bogdanovic, N. and Karelson, E. (2007) Antioxidative effects of plant polyphenols, from protection of G protein signaling to prevention of age-related

pathologies. *Annals of the New York Academy of Sciences* 1095, 449–457.
Kruk, I, Aboul–Enein, H.Y., Michalska, T., Lichszteld, K. and Kladna, A. (2005) Scavenging of reactive oxygen species by the plant phenols genistein and oleuropein. *Luminescence* 20, 81–89.
Landauer, M.R., Davis, H.D., Dominitz, J.A. and Weiss, J.F. (1987) Dose and time relationships of the radioprotector WR–2721 on locomotor activity in mice. *Pharmacology, Biochemistry and Behavior* 27, 573–576.
Landauer, M.R. (2002) Radiation–induced performance decrement. *Military Medicine* 167, 128–130.
Landauer, M.R., Castro, C.A., Benson, K.A., Hogan, J.B. and Weiss, J.F. (2001) Radioprotective and locomotor responses of mice treated with nimodipine alone and in combination with WR-151327. *Journal of Applied Toxicology* 21, 25–31.
Landauer, M.R., Davis, H.D., Kumar, K.S. and Weiss, J.F. (1992) Behavioral toxicity of selected radioprotectors. *Advances in Space Research* 12, 273–283.
Landauer, M.R., Srinivasan, S.V., Shapiro, A., Takimoto, C. and Seed, T.M. (2000) Protection against lethal irradiation by genistein. *International Journal of Toxicology* 19(6), 37.
Landauer, M.R., Srinivasan, S.V., Shapiro, A., Takimoto, C.H., Wang, P.S. and Seed, T.M. (2002) Effect of dose, time, and route of administration of genistein on radiation protection. *The Toxicologist* 66, 88.
Landauer, M.R., Srinivasan, V. and Seed, T.M. (2003b) Genistein treatment protects mice from ionizing radiation injury. *Journal of Applied Toxicology* 23, 379–385.
Landauer, M.R., Srinivasan, V., Nguyen, A.M. and Seed, T.M. (2003a) Soy phytoestrogens protect against gamma irradiation. *The Toxicologist* 72, 162.
McCue, P. and Shetty, K. (2004) Health benefits of soy isoflavonoids and strategies for enhancement, a review. *Critical Reviews in Food Science and Nutrition* 44, 361–367.
McPherson, C.W. (1963) Reduction of *Pseudomonas aeruginosa* and coliform bacteria in mouse drinking water following treatment with hydrochloric acid and chlorine. *Laboratory Animal Care* 13, 737–744.
Montoro, P., Braca, A., Pizza, C. and De Tommasi, N. (2005) Structure-antioxidant activity relationship of flavonoids isolated from different plant species. *Food Chemistry* 92, 349–355.
Morris, S.M., Akerman, G.S., Warbritton, A.R., Patton, R.E., Doerge, D.R., Ding, X. and Chen, J.J. (2003) Effect of dietary genistein on cell replication indices in C57BL6 mice. *Cancer Letters* 195, 139–145.
Neta, R. (1998) Modulation of the radiation response by cytokines. In: Bump, E.A and Malaker, K. (eds) *Radioprotectors, Chemical, Biological, and Clinical Perspectives.* CRC Press, New York, USA, pp. 237–252.
Newman, D.J. and Cragg, G.M. (2007) Natural products as sources of new drugs over the last 25 years. *Journal of Natural Products* 70, 461–477.
Papazisis, K.T., Zambouli, D., Kimoundri, O.T., Papadakis, E.S., Vala, V., Geromichalos, G.D., Voyatzi, S., Markala, D., Destouni, E., Boutis, L. and Kortsaris, A.H. (2000) Protein tyrosine kinase inhibitor, genistein, enhances apoptosis and cell cycle arrest in K562 cells treated with gamma-irradiation. *Cancer Letters* 160, 107–113.
Patchen, M. (1998) Immunomodulators and cytokines, their use in the mitigation of radiation-induced hemopoietic injury. In: Bump, E.A. and Malaker, K. (eds) *Radioprotectors, Chemical, Biological, and Clinical Perspectives.* CRC Press, New York, pp. 213–236.
Pellmar, T.C. (2006) Radiological/nuclear preparedness. Military Medical Technology (Online Edition) http//www.military-medical-technology.com/article.cfm?DocID=1430 10.
Pietta, P.G. (2000) Flavonoids as antioxidants. *Journal of Natural Products* 63, 1035–1042.
Rades, D., Fehlauer, F., Bajrovic, A., Mahlmann, B., Richter, E. and Alberti, W. (2004) Serious adverse effects of amifostine during radiotherapy in head and neck cancer patients. *Radiotherapy and Oncology* 70, 261–264.
Raffoul, J.J., Banerjee, S., Che, M., Knoll, Z.E., Doerge, D.R., Abrams, J., Kucuk, O., Sarkar, F.H. and Hillman, G.G. (2007) Soy isoflavones enhance radiotherapy in a metastatic prostate cancer model. *International Journal of Cancer* 120, 2491–2498.

Shimizu, S. (2004) Routes of administration. In: Hedrich, H.J. and Bullock, G. (eds) *The Laboratory Mouse*, Elsevier Academic Press, San Diego, CA, USA, pp. 527–542.

Singh, V.K. and Yadav, V.S. (2005) Role of cytokines and growth factors in radioprotection. *Experimental and Molecular Pathology* 78, 156–169.

Stone, H.B., Moulder, J.E., Coleman, C.N., Ang, K.K., Anscher, M.S., Barcellos-Hoff, M.H., Dynan, W.S., Fike, J.R., Grdina, D.J., Greenberger, J.S., Hauer–Jensen, M., Hill, R.P., Kolesnick, R.N., Macvittie, T.J., Marks, C., McBride, W.H., Metting, N., Pellmar, T., Purucker, M., Robbins, M.E., Schiestl, R.H., Seed, T.M., Tomaszewski, J.E., Travis, E.L., Wallner, P.E., Wolpert, M. and Zaharevitz, D. (2004) Models for evaluating agents intended for the prophylaxis, mitigation and treatment of radiation injuries. Report of an NCI Workshop, December 3–4, 2003. *Radiation Research* 162, 711–728.

Storer, J., Fry, R. and Ullrich, R. (1982) Somatic effects of radiation. In: Foster, H., Small, J. and Fox, J. (eds) *The Mouse in Biomedical Research*, Volume IV Experimental Biology and Oncology, Academic Press, New York, USA, pp. 133–146.

Sugano, M. (2006) *Soy in Health and Disease Prevention*, Taylor and Francis, New York, USA.

Takimoto, C.H., Glover, K., Huang, X., Hayes, S.A., Gallot, L., Quinn, M., Jovanovic, B.D., Shapiro, A., Hernandez, L., Goetz, A., Llorens, V., Lieberman, R., Crowell, J.A., Poisson, B.A. and Bergan, R.C. (2003) Phase I pharmacokinetic and pharmacodynamic analysis of unconjugated soy isoflavones administered to individuals with cancer. *Cancer Epidemiology, Biomarkers and Prevention* 12, 1213–1221.

Uma Devi, P., Ganasoundari, A., Rao, B.S. and Srinivasan, K.K. (1999) *In vivo* radioprotection by Ocimum flavonoids, survival of mice. *Radiation Research* 151, 74–78.

Valachovicova, T., Slivova, V. and Sliva, D. (2004) Cellular and physiological effects of soy flavonoids. *Mini Reviews in Medicinal Chemistry* 4, 881–887.

Verdrengh, M., Jonsson, I.M., Holmdahl, R. and Tarkowski, A. (2003) Genistein as an anti-inflammatory agent. *Inflammation Research* 52, 341–346.

Wang, Y., Raffoul, J.J., Che, M., Doerge, D.R., Joiner, M.C., Kucuk, O., Sarkar, F.H. and Hillman, G.G. (2006) Prostate cancer treatment is enhanced by genistein *in vitro* and *in vivo* in a syngeneic orthotopic tumour model. *Radiation Research* 166, 73–80.

Waselenko, J.K., MacVittie, T.J., Blakely, W.F., Pesik, N., Wiley, A.L., Dickerson, W.E., Tsu, H., Confer, D.L., Coleman, C.N., Seed, T., Lowry, P., Armitage, J.O. and Dainiak, N. (2004) Medical management of the acute radiation syndrome, recommendations of the Strategic National Stockpile Radiation Working Group. *Annals of Internal Medicine* 140, 1037–1051.

Winkel-Shirley, B. (2002) Biosynthesis of flavonoids and effects of stress. *Current Opinion in Plant Biology* 5, 218–223.

Yashar, C.M., Spanos, W.J., Taylor, D.D. and Gercel-Taylor, C. (2005) Potentiation of the radiation effect with genistein in cervical cancer cells. *Gynecologic Oncology* 99, 199–205.

Zhang, B., Liu, J.Y., Pan, J.S., Han, S.P., Yin, X.X., Wang, B. and Hu, G. (2006) Combined treatment of ionizing radiation with genistein on cervical cancer HeLa cells. *Journal of Pharmacological Sciences* 102, 129–135.

Zhou, Y. and Mi, M.T. (2005) Genistein stimulates hematopoiesis and increases survival in irradiated mice. *Journal of Radiation Research* 46, 425–433.

Chapter 13

Propolis and Related Flavonoids as Radioprotective Agents

Nada Oršolić, Vesna Benković, Anica Horvat-Knežević and Ivan Bašić

Introduction

The potential use of flavonoid compounds as radioprotectors is of increasing interest. This review focuses on the radioprotective efficacy of naturally occurring antioxidants, specifically propolis and related flavonoids, and their influence on various endpoints of radiation damage. Results from animal experiments indicate that plant products, such as propolis and related flavonoid compounds, protect against lethality and other types of radiation effects. Naturally occurring, antioxidant nutrients and phytochemicals have the advantage of low toxicity as compared to synthetic protectors and are generally protective when administered at pharmacological doses (Arora, 2006; Arora, 2007; Arora *et al.,* 2005, 2006a,b, 2008; Oršolić and Bašić, 2005a,b). Besides such compounds may also provide an extended window of protection against low-dose, low-dose-rate irradiation, including their therapeutic potential when administered after irradiation. A number of phytochemicals, including caffeic acid, quercetin, naringin, chrysin and propolis, have multiple physiological effects, as well as antioxidant activity, which result in radioprotection *in vitro* and *in vivo*. The suggested protective effects of flavonoids, together with their potent antioxidative and free radical scavenging activities observed in *in vitro* and *in vivo* studies have increased public interest in the use of flavonoids for their potential health benefits.

Propolis and related flavonoids as radioprotectors

Propolis (bee glue) is the generic name for the resinous substance collected by honeybees from various plant sources and used by bees to seal holes in their honeycombs, smooth out the internal walls, and protect the entrance against intruders. It is rich in biochemical constituents, including mostly a mixture of polyphenols, flavonoid aglycones, phenolic acid and their esters, and phenolic aldehydes and ketones, terpenes, sterols, vitamins, amino acids etc. (Walker and Crane 1987; Kosalec *et al.,* 2004; Oršolić *et al.,* 2006a,b). Healing properties of propolis are known in folk medicine from antiquity. Recently, the interest for propolis as a "harmless medicine" is increasing.

Our data suggest that propolis preparations (water or ethanolic extract of propolis; WSDP or EEP) and propolis polyphenolic compounds (caffeic acid, naringin, chrysin or quercetin) possess promising radioprotective effects, which is comparable to the well-established radioprotector aminoethyl isothiourea (AET). AET at a dose of 1mM/kg was used as positive control. Propolis and its flavonoids given to mice before or after whole-body γ-irradiation (WBI) (9Gy) protect mice at molecular (Tables 13.1–13.3), and organism level (Fig. 13.1).

Propolis and its polyphenolic compounds cause a decrease in the tail length, tail intensity and tail moment showing radioprotective effects on DNA. Results showed that the test components possess radioprotective properties similar or in some instances even better than AET. WSDP and EEP were most effective in reducing DNA damage to lymphocytes, indicating synergistic antioxidative effects of different polyphenolic flavonoid components present in EEP or WSDP. Pre-treatment of mice with either WSDP or EEP or flavonoids produced a reduction in oxidative DNA damage of lymphocytes, as compared with control. They were ranked in decreasing order of potency as follows: naringin (2.98%); chrysin (16.84%); quercetin (33.67%); AET (48.52%); caffeic acid (49.51%); EEP (53.47%); and WSDP (54.46%), respectively. Treatment with propolis or its polyphenolic/flavonoids compounds after irradiation also resulted in a significant reduction of DNA damage as follows: caffeic acid (32.31%); AET (75.39%); naringin (78.46%); EEP (80%); chrysin (83.08%); quercetin (84.62%); and WSDP (89.24%), respectively. Based on these results, it could be concluded that the natural compounds WSDP and EEP as well as their flavonoids are highly effective in cytoprotection. Despite treatment with test compounds, on the third day after lethal whole-body irradiation (WBI) severe disturbances in the leukocyte counts in mice were present.

Haematopoietic syndrome occurs at doses ranging between 2.5–8Gy and is manifested by haematopoietic stem cell depletion, and ultimately by depletion of mature haemopoietic and immune cells (Hall, 1994; Orsolic and Basic, 2005a). Therefore, the majority of white blood cells measured by the alkaline comet assay in our study could be considered as newly produced since highly damaged cells after irradiation died via apoptosis/necrosis and were obviously destroyed. The test components did not induce significant genotoxicity to the white blood cells of non-irradiated mice, but offered a measurable protection against ionizing radiation-induced DNA damage (Table 13.3).

Moreover, our studies indicated statistically significant differences in the survival times of WBI mice pre-treated with test compounds, as compared to control (solvent: H_2O or ethanol). The most effective compound was quercetin (QU), showing protection similar to that achieved by the AET. Such a huge protective effect of QU could result from its chemical structure, which consists of one of the most suitable structural forms for scavenging free radicals (Russo *et al.,* 2000; Molina *et al.,* 2003). Treatment with test components [including propolis preparations (EEP and WSDP)] after irradiation was ineffective. All other polyphenolic compounds used, including propolis, were effective in rendering protection against radiation-induced damage. The similarity in radiation protection between

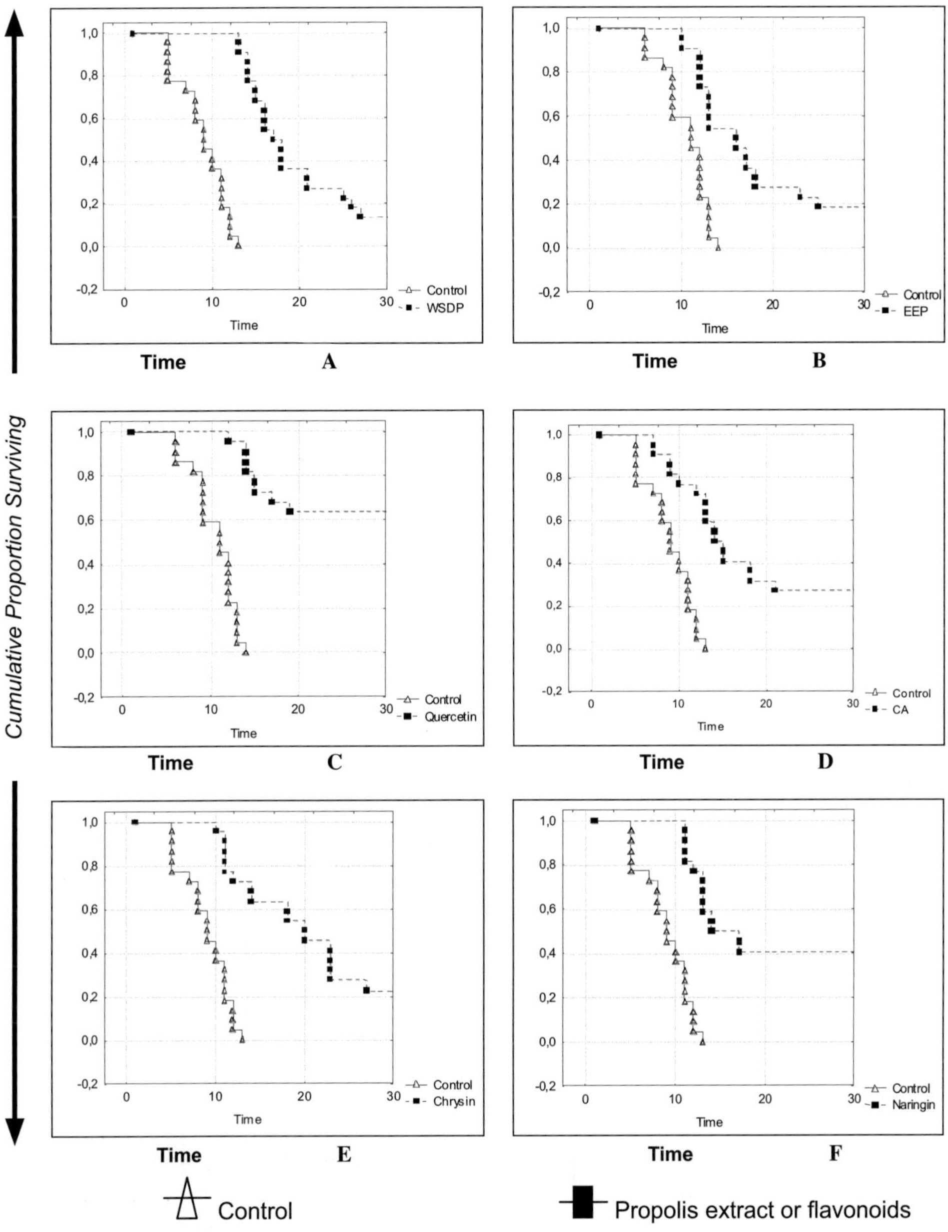

Fig. 13.1. Kaplan-Meier survival curves for mice (n = 22) treated with WSDP (A), EEP (B), quercetin (C), caffeic acid (D), chrysin (E) and naringin (F) before irradiation. The test compounds were given to mice i.p. daily for 3 consecutive days, and the daily dose was 100mg/kg body weight; one hour after the last treatment, mice were exposed to 9Gy. The results of log rank test showed that the test compounds significantly reduced the total body injury at the organism level ($p < 0.001$) and increased the life span of the mice after irradiation. [Reproduced with permission from Oršolić et al., Biological and Pharmaceutical Bulletin 30(5), 946–951. (2007) Pharmaceutical Society of Japan.]

Table 13.1. The total leukocyte number and the results of alkaline comet assay on peripheral blood leukocytes of mice treated preventively and irradiated with a dose of 9Gy[a].

Group	Leukocytes ($\times 10^6$) X ± S.E.	COMET PARAMETERS							
		TAIL LENGTH (μm)				TAIL INTENSITY		TAIL MOMENT	
		Mean ± S.E.	Min.	Max.	% LTN	Mean ±S.E.	Max.	Mean ±S.E.	Max.
WSDP	7.01±1.10	17.99±0.27*	10.9	32.69	45.6■	3.25±0.23*	26.86	0.46±0.03*	3.27
EEP	6.69±0.74	19.88±0.29*	10.90	38.46	90.8	3.02±0.22*	22.35	0.47±0.03*	3.29
Quercetin	8.95±0.68	19.75±0.27*	12.82	41.02	96.8	4.66±0.27*	32.38	0.67±0.04*	3.94
Caffeic acid	7.25±1.54	21.43±0.31*	12.18	39.74	78.8	3.12±0.23*	20.33	0.51±0.04*	3.52
Chrysin	6.95±0.86	22.56±0.35	14.10	46.15	84.4	5.36±0.27	23.29	0.84±0.04	3.52
Naringin	7.15±0.66	22.75±0.39	14.10	46.15	81.2	6.44±0.35	31.27	0.98±0.05	4.81
Positive control[b]	7.45±1.32	20.11±0.26*	13.46	38.46	72.0	3.32±0.24*	23.77	0.52±0.04*	3.78
Negative control	7.86±0.45	25.42±0.63	12.18	67.95	78.0	5.96±0.39	36.17	1.01±0.07	6.84

[a]The test compounds were given to mice i.p. daily for 3 consecutive days, and the daily dose contained 100mg/kg body weight. One hour after the last treatment, mice were exposed to an acute whole-body gamma radiation dose of 9Gy.
[b]AET (2-aminoetilizotiourea dihydrobromide- a chemical protector) was given to mice i.p. at a dose of 1mM/kg one hour after injection, mice were exposed to an acute whole-body gamma radiation dose of 9Gy.
LTN: comets with a long-tailed nucleus, i.e., the length over the 95th percentile of the distribution of the tail lengths among controls.* $p < 0.05$ compared with negative control (irradiated mice) (ANOVA).
■$p < 0.001$ compared with negative control (irradiated mice) (χ^2 test); 250 comets per group were evaluated; minimum tail intensity and minimum tail movement were 0 in all groups.

Table 13.2. The total leukocyte number and the results of alkaline comet assay on peripheral blood leukocytes of mice treated therapeutically after irradiation with a dose of 9Gy[a].

Group	Leukocyte ($x10^6$/L) X ± S.E.	COMET PARAMETERS							
		TAIL LENGTH (μm)				TAIL INTENSITY		TAIL MOMENT	
		Mean ± S.E.	Min.	Max.	% LTN	Mean ± S.E.	Max.	Mean ± S.E.	Max.
WSDP	0.22 ±0.05	14.54 ±0.13*	9.62	21.15	12.0	0.54±0.09*	15.42	0.07±0.01*	1.38
EEP	0.20±0.04	14.70±0.12*	10.26	19.23	15.2▪	1.01±0.14*	18.70	0.13±0.02*	2.16
Quercetin	0.41±0.11	15.64±0.21*	10.26	28.85	23.2▪	0.74±0.01*	23.81	0.10±0.02*	3.05
Caffeic acid	0.31±0.07	17.83±0.34*	11.54	36.54	40.8	2.90±0.25*	21.82	0.44±0.04*	3.64
Chrysin	0.22±0.05	14.75±0.13*	10.26	19.87	16.4	0.83±0.09*	11.03	0.11±0.01*	1.34
Naringin	0.36±0.10	14.87±0.14*	10.26	21.79	15.6	1.02±0.11*	15.86	0.14±0.01*	1.93
Positive control[b]	0.33±0.07	14.73±0.17*	10.26	30.13	12.8	1.20±0.13*	17.62	0.16±0.02*	1.92
Negative control	0.63±0.62	20.36±0.71	10.90	87.82	49.6	4.23±0.44	46.90	0.65±0.07	7.45

[a]Mice were exposed to an acute whole-body gamma radiation dose of 9Gy; thirty minutes after irradiation the test compounds were given to mice i.p. daily for 3 consecutive days, and the daily dose contained 100mg/kg body weight.
[b]AET (2-aminoetilizotiourea dihydrobromide- a chemical protector) was given to mice i.p. at a dose of 1mM/kg. One hour after injection mice were exposed to an acute whole-body gamma radiation dose of 9Gy.
LTN: comets with a long-tailed nucleus, i.e., the length over the 95th percentile of the distribution of the tail lengths among controls.
* $p < 0.05$ compared with control (irradiated mice) (ANOVA); ▪$p < 0.001$ compared with negative control (irradiated mice) (χ^2 test); 250 comets per group were evaluated; minimum tail intensity and minimum tail movement were 0 in all groups.

Table 13.3. The total leukocyte number and the results of alkaline comet assay on peripheral blood leukocytes of mice treated with tested compounds[a].

Group	Leukocytes (x 10^6/l) X ± S.E.	COMET PARAMETERS							
		TAIL LENGTH (µm)				TAIL INTENSITY			TAIL MOMENT
		Mean ± S.E.	Min.	Max.	% LTN	Mean ± S.E.	Max.	Mean ± S.E.	Max.
WSDP	8.03 ± 0.66	13.65 ± 0.18	10.26	42.31	16.9	1.55 ± 0.14	14.37	0.19 ± 0.02	1.66
EEP	7.30 ± 0.68	15.21 ± 0.33	10.26	41.67	8.4	3.09 ± 0.31	28.92	0.41 ± 0.04	3.71
Quercetin	7.70 ± 1.18	14.49 ± 0.32	10.26	49.36	3.6	2.82 ± 0.35	37.76	0.37 ± 0.05	5.62
Caffeic acid	8.13 ± 1.00	14.39 ± 0.19	10.26	32.69	13.0	2.90 ± 0.31	31.47	0.36 ± 0.04	3.91
Chrysin	6.53 ± 1.00	15.06 ± 0.22	10.90	44.23	17.7	2.55 ± 0.27	34.43	0.33 ± 0.03	4.41
Naringin	7.20 ± 0.77	14.60 ± 0.27	9.62	39.74	15.9	2.30 ± 0.30	42.47	0.30 ± 0.04	4.90
Positive control[b]	9.63 ± 1.28	15.38 ± 0.21	10.90	31.41	12.6	3.00 ± 0.31	32.08	0.390.04	3.70
Negative control	9.84 ± 0.98	14.11 ± 0.18	10.26	28.85	11.5	2.19 ± 0.23	30.25	0.27 ± 0.03	4.07

[a]The test compounds were given to mice i.p. daily for 3 consecutive days, and the daily dose contained 100mg/kg body weight; the comet assay was performed on the third day after finishing the treatment.
[b]AET (2-aminoetilizotiourea dihydrobromide-a chemical protector; 1mM/kg) was given to mice i.p.; the comet assay was performed after finishing the treatment.
LTN: comets with a long-tailed nucleus, i.e., the length over the 95th percentile of the distribution of the tail lengths among controls. 250 comets per group were evaluated; minimum tail intensity and minimum tail movement were 0 in all groups.

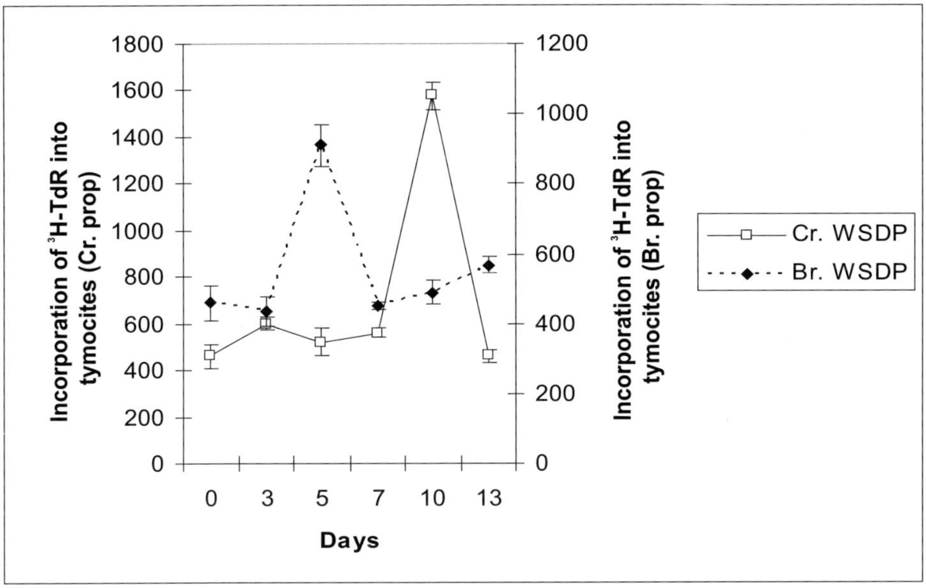

Fig. 13.2. *In vitro* IL-1 activity of peritoneal macrophages after treatment of CBA mice with WSDP. Mice (n = 6) were treated i.p. with WSDP (50mg/kg, once on days, for 3 consecutive days) and macrophages were collected at different times. Macrophages (2 x 10^6/ml) were plated into 35-mm diameter wells and the level of IL-1 in the 24 h macrophage supernatants was determined by augmentation of ^{3}H-thymidine (^{3}H-TdR) incorporation in mouse thymocytes. From Oršolić and Bašić (2005a).

WSDP and EEP could be explained on the basis of the higher content of polyphenols present in both WSDP and EEP preparations (Oršolić *et al.*, 2007). Kaplan-Meier analysis and log-rank test revealed the following surviving time of mice: quercetin, WSDP ($p < 0.0001$), naringin ($p = 0.0001$), caffeic acid, chrysin ($p = 0.0003$) and EEP ($p = 0.0081$).

A single whole-body exposure to ionizing radiation results in a complex set of symptoms the onset, nature and severity of which is a function of both radiation dose and radiation quality. In this study, the pathological cellular emptiness of bone marrow and spleen in mice, which died on day 5 after lethal WBI, could be attributed to the haemopoietic syndrome. Administration of several immunomodulators including propolis and propolis-derived compounds has been shown to stimulate haemopoietic recovery and enhance the survival of irradiated animals (Oršolić and Bašić, 2003a,b, 2005a, 2007; Orsi *et al.,* 2005; Russo *et al.,* 2004).

Dimov *et al.* (1991, 1992) observed that oral and parenteral administration of WSDP enhanced the survival rate and the mean survival time in experimental animals and reduced bacterial and fungal infections in normal and immunodepressed mice. The authors suggested that the broad therapeutic spectrum of propolis includes a pronounced immunomodulatory activity, directed mainly towards augmenting of non-specific anti-infectious resistance via macrophage activation. Our recent findings imply that the antitumour activity of WSDP and polyphenolic compounds of propolis enhance host resistance in the Ehrlich ascites tumour model by increasing the activities of macrophages, cytotoxic T cells, B cells and NK cells as well (Oršolić *et al.,* 2005e). Moreover, WSDP stimulated peritoneal macrophages to produce IL-1 (Fig. 13.2), which serves as a differentiation and maturation-inducing agent for a variety of cells. There are also indications that IL-1 could serve as a signal that initiates radioprotective events *in vivo* (Neta *et al.,* 1986a,b). Our recent observations (Oršolić and Bašić, 2005a,b) and those by others (Sadzuka *et al.,* 1997; Suzuki *et al.,* 2002; Lahouel *et al.,* 2004) proved the protective effect of propolis on bone marrow and lymphoid tissues of mice to cytotoxic drugs and radiation. Augmented immunological activity as seen in increased activity of macrophages, cytotoxic T cells, B cells and NK cells by propolis and related compounds (Oršolić *et al.,* 2005a,b,c,d) seems to play a central role in preventing secondary infections associated with irradiation, contributing to further acceleration of haemopoietic regeneration and increased survival following radiation-induced lympho- and myelo-suppression.

The exact mechanism of action in protecting mice from the lethal effects of acute whole-body irradiation by propolis and related flavonoids is not fully known. In addition to modulation of immunohaematopoiesis, scavenging of radiation-induced free radicals could be the important mechanisms of radiation protection. Chen *et al.* (2004) have also shown that propolis and related flavonoids exercise their activity through the scavenging of hydroxyl, superoxide free radicals and lipid peroxides. The antioxidant activities of propolis and its polyphenolic/flavonoid components are related to their ability to chelate metal ions and to scavenge singlet oxygen, superoxide anions, peroxyl radicals, hydroxyl radicals and peroxynitrite (Bors *et al.,* 1997, 2004). Jeon *et al.* (2002) showed that flavonoids from propolis elevate catalase, superoxide dismutase and glutathione peroxidase mRNA synthesis; the elevation of these enzymes by flavonoids was considered to be responsible for the observed protection against radiation-induced damage. By increasing the activities of antioxidant enzymes, flavonoids from propolis reduce the number of free radicals and ROS and increase the production of molecules capable of protecting against oxidative stress. It is possible that propolis and its polyphenolic/flavonoid components may influence the survival of the WBI mice via increased activities of SOD, CAT, GPx, GR and GSH (Molina *et al.,* 2003; Oršolić *et al.,* 2005a) through the free radical scavenging ability, tissue regeneration properties and immunostimulatory effects. It is likely that several different mutually co-operative and synergistic mechanisms of propolis and its polyphenolic compounds are involved in the protection of the whole organism against radiation.

Other natural radioprotectors are known to elicit their action by various mechanisms, such as:

i) suppression of formation of reactive species
ii) detoxification of radiation induced species
iii) target stabilization and
iv) enhancing the repair and recovery processes (Arora *et al.,* 2005, 2006a,b; Chawla *et al.,* 2006, 2007).

Several amino thiol radioprotectors, such as cysteamine, guanidoethyldisulfide and glutathione disulfide, bind to DNA and their DNA binding paralleled the radioprotective potency of propolis and its flavonoids. Moreover, it was shown that thiols, such as GSH, might be involved in the repair of DNA single-strand breaks. Cells genetically deficient in GSH synthesis or cells in which GSH deficiency is produced by dl-Buthionine-sulfoxime or by hypoxia or misonidazole show a lack of DNA single-strand break repair (Edgreen, 1983; Revesz and Edgreen, 1984; Revesz, 1985). Our results indicate that preventive treatment of irradiated mice with WSDP and flavonoids significantly decreased the levels of primary DNA damage in their white blood cells, as compared to untreated animals. Other studies have confirmed the role of flavonoids in the deactivation of the free radicals (Heim *et al.,* 2002; Galati and O'Brien 2004). Flavonoids possess antioxidant activity based on their ability of direct free radical scavenging, or stabilizing the ROS by reacting with the reactive compound of the radical. Because of the high reactivity of the hydroxyl substituents of the flavonoids, the radicals are made inactive (Chen *et al.,* 2004). Flavonoids can also increase the function of the endogenous antioxidant enzyme systems (Jeon *et al.,* 2002; El-khawaga *et al.,* 2003). Furthermore, antioxidant effects may be a result of a combination of radical scavenging and an interaction with enzyme functions. Our results strongly support these observations. Therefore, stable doses of WSDP or flavonoids present in an organism prior to irradiation could possibly diminish the frequency of DNA-strand breakage or stimulate the repair processes with consequent decrease of the frequency of double-strand breaks (considered to be the primary lesion involved in cellular death). It is also possible that both WSDP and flavonoid components stimulate other intracellular enzyme systems capable of protecting cells against free radicals and DNA damage (Russo *et al.,* 2000; Manach *et al.,* 2004).

Similar radioprotective potential of other flavonoids on radiation-induced chromosomal damage in human lymphocytes [e.g., genistein (Jacquet, *et al.,* 1995), luteolin (Shimoi *et al.,* 1996), naringin (Jagetia *et al.,* 2003), quercetin (Oliveira *et al.,* 2000), apigenin (Rithidech *et al.,* 2005), curcumin (Jagetia and Rajanikant, 2005), orientin and vicenin (Nayak and Devi, 2005)] has been reported. However, the genotoxicity of these flavonoids has also been documented (da Silva *et al.,* 2002). Several studies have shown an increase in micronucleus (MN) frequency at high concentrations of flavonoids (Yen *et al.,* 2003; Rithidech *et al.,* 2005). A study conducted by Snyder and Gillies (2002) also showed clastogenic activity of apigenin at a high dose level (i.e., 100μM) in Chinese hamster V79 cells.

We conducted experiments to investigate potential genotoxic activity of propolis preparations and its polyphenolic compounds (caffeic acid, quercetin, chrysin, naringin) in human lymphocytes and compared the effect of test components with that of AET. In these experiments, the well-established *in vitro* cytokinesis-block micronucleus (CBMN) assay was used to determine whether test components are capable of inducing chromosomal damage (measured as the presence of micronucleus (MN) in binucleated cells in treated human lymphocytes). Results indicated that WSDP, quercetin and caffeic acid bring about a decrease in the frequency of MN as compared to that in the group with no test components. However, EEP, chrysin and naringin slightly increased the frequency of MN, as compared to control. AET was very toxic to human lymphocytes and induced the highest increase in the frequency of MN (see Table 13.4 and statistical analysis in Table 13.4a).

Flavonoid compounds can behave as both prooxidants and antioxidants, depending on concentration and free radical source. Flavonoids autooxidize in aqueous medium and may form highly reactive OH radicals in the presence of transition metals. Moreover, polyphenols may act as a substrate for peroxidases and other metalloenzymes, yielding quinone or quinomethide type prooxidant (Metodiewa *et al.,* 1999; Yamashita and Kawanishi, 2000).

Recent reports have linked polyphenols to ROS production, especially hydrogen peroxide and subsequent apoptosis. It is likely that γ-radiation (4Gy) may have induced the prooxidative effect of flavonoids in *in vitro* studies conducted on human lymphocytes.

Following treatment with EEP, caffeic acid, chrysin and naringin, a significant increase in MN frequency was found, while after treatment with WSDP and quercetin a decrease in MN frequency in irradiated samples *in vitro* was noticed (see Table 13.5 and statistical analysis in Table 13.5a).

Table 13.4. Number of human binucleated lymphocytes in peripheral blood with micronucleus[a] after pre-incubation of lymphocytes with propolis (100μg/ml) and its polyphenolic/flavonoid components (50μM) *in vitro* for 30 minutes.

	Number of cells		Number of cells			Total MN
	Without MN	With MN	1 MN	2 MN	3 MN	
Control (unirradiated)	995	5	4	1	-	6
Control (ethanol)	995	5	4	1	-	6
WSDP	999	1	1	-	-	1
EEP	986	14	14	-	-	14
Quercetin	999	1	1	-	-	1
Caffeic acid	998	2	2	-	-	2
Chrysin	992	8	7	1	-	9
Naringin	988	12	12	-	-	12
AET	963	37	31	10	1	44

[a] Micronucleus frequencies were assessed by scoring 1000 binucleate lymphocytes.

Table 13.4a. Statistical analysis for Table 13.4 (χ^2 test).

	Control	Control (EtOH)	WSDP	EEP	Quercetin	Caffeic acid	Chrysin	Naringin	AET
Control		NZ	NZ	$p < 0.05$	NZ	NZ	NZ	NZ	$p < 0.05$
Control (ethanol)			NZ	$p < 0.05$	NZ	$p < 0.05$	NZ	NZ	$p < 0.05$
WSDP				$p < 0.05$	NZ	NZ	$p < 0.05$	$p < 0.05$	$p < 0.05$
EEP					$p < 0.05$	$p < 0.05$	NZ	NZ	$p < 0.05$
Quercetin						NZ	$p < 0.05$	$p < 0.05$	$p < 0.05$
Caffeic acid							NZ	$p < 0.05$	$p < 0.05$
Chrysin								NZ	$p < 0.05$
Naringin									$p < 0.05$
AET									

Table 13.5. Number of human binucleated lymphocytes in peripheral blood with micronucleus[a]. Lymphocytes were pre-incubated with propolis (100μg/ml) and its polyphenolic/flavonoid components (50μM) for 30 minutes and then irradiated with 4Gy *in vitro*.

	Number of cells		Number of cells				Total MN
	Without MN	With MN	1 MN	2 MN	3 MN	4 MN	
Control (unirradiated)	995	5	4	1	-	-	6
Control (irradiated)	641	359	262	90	25	8	549
Control (ethanol)	598	402	326	66	10	-	488
WSDP	755	245	198	38	9	-	301
EEP	389	611	399	182	27	3	844
Quercetin	608	392	298	77	16	1	504
Caffeic acid	570	430	318	99	13	-	555
Chrysin	553	447	324	101	17	5	597
Naringin	530	470	321	101	39	10	680
AET	664	336	222	85	24	5	484

[a] Micronuclei frequencies were assessed scoring 1000 binucleate lymphocytes.

Table 13.5a. Statistical analysis for Table 13.5 (χ^2 test).

	Control unirradiated	Control irradiated	Control (ethanol)	WSDP	EEP	Quercetin	Caffeic acid	Chrysin	Naringin	AET
Control unirradiated		$p < 0.05$	$p < 0.05$	$p < 0.05$	$p < 0.05$	$p < 0.05$	$p < 0.05$	$p < 0.05$	$p < 0.05$	$p < 0.05$
Control irradiated			$p < 0.05$	$p < 0.05$	$p < 0.05$	$p < 0.05$	$p < 0.05$	$p < 0.05$	$p < 0.05$	NZ
Control (ethanol)				$p < 0.05$	$p < 0.05$	NZ	NZ	$p < 0.05$	$p < 0.05$	$p < 0.05$
WSDP					$p < 0.05$	$p < 0.05$	$p < 0.05$	$p < 0.05$	$p < 0.05$	$p < 0.05$
EEP						$p < 0.05$	$p < 0.05$	$p < 0.05$	$p < 0.05$	$p < 0.05$
Quercetin							NZ	$p < 0.05$	$p < 0.05$	$p < 0.05$
Caffeic acid								$p < 0.05$	NZ	$p < 0.05$
Chrysin									NZ	$p < 0.05$
Naringin										$p < 0.05$
AET										

Our results are in concordance with other studies in which described prooxidative mechanism of flavonoids and flavonoid-induced impairment of the antioxidant defence system comprised GSH and GST (Lodovici *et al.,* 2001; Yen *et al.,* 2003; Geetha *et al.,* 2005; Fujisawa and Kadoma, 2006). The hydroxyl radical generated by the autooxidation and redox cycling of the polyphenolic compounds may indicate peroxidation in the nuclear membrane; the level of lipid peroxides and intermediate radicals may then be amplified by a chain reaction.

The pro-oxidant activity of flavonoids may deplete the nuclear antioxidant defence and lead to oxidative DNA damage, which may be responsible for their mutagenicity.

The possible mechanism of radioprotective and haematostimulative action of Propolis preparation and related flavonoids

In general, immunotherapeutic agents, cytokines or growth factors provide a large window of protection, although protection against lethality is lower than that afforded by phosphorothioates. Administration of cytokines and growth factors after radiation exposure, combined with pre-irradiation administration of phosphorothioates appears to synergistically reduce radiation damage (Weiss *et al.,* 1990; Patchen *et al.,* 1992). Propolis and its flavonoids have a pronounced immunomodulatory and haemostimulative activity directed mainly towards augmenting of non-specific activity via macrophage activation. One possible mechanism of haematopoietic action of WSDP is its influence on macrophage activity and production of interleukin-1 (IL-1) by these cells (Fig. 13.2). It is known that IL-1 stimulates stem cells and haematopoiesis. Increased level of IL-1 activity, produced by the WSDP-activated macrophages, correlated directly with haematopoietic activity of treated mice as well as macrophage cytotoxicity to tumour cells (Oršolić *et al.,* 2005a). It was shown that IL-1 might protect mice from the lethal effects of irradiation (Neta *et al.,* 1996).

It has been suggested that radioprotective activity conferred by immunomodulators can be attributed to their capacity to enhance haematopoietic and immune functions (Neta *et al.,* 1996). Results described here suggest the protective role of the WSDP against radiation-induced destruction of the lymphoid and haematopoietic system (Table 13.6), as well as against the toxicity of the chemotherapeutic agents, which induce bone marrow aplasia and other haematological failures in experimental animals and humans.

Table 13.6. Effect of whole-body irradiation on the spleen weight and spleen and bone marrow cellularity in normal and WSDP-treated CBA mice.

	Spleen				Bone marrow	
Days after 6Gy WBI [b]	Weight (mg) [a]		Cellularity (×10^6) [a]		Cellularity (×10^6) [a]	
	Normal mice	WSDP-treated mice [c]	Normal mice	WSDP-treated mice	Normal mice	WSDP-treated mice [c]
1	35.1 ± 5.4	89.2 ± 3.4	19.8 ± 3.2	91.0 ± 5.4	2.8 ± 0.9	4.4 ± 1.7
4	28.4 ± 3.1	46.1 ± 3.1	0.9 ± 0.7	37.2 ± 2.7	1.4 ± 0.3	3.1 ± 1.6
9	23.2 ± 1.9	63.3 ± 3.6	8.7 ± 1.9	51.3 ± 0.4	1.5 ± 0.4	5.2 ± 1.9

[a]Mean ± SD; [b]Whole-body irradiation was applied 24 hours following last WSDP treatment; [c]50mg/kg WSDP was given p.o. for 20 consecutive days. Groups comprised 6–7 mice each. From Oršolić and Bašić (2005a).

Table 13.7. The effect of WSDP on the formation of types of spleen colonies in lethally irradiated mice.

Days	20					40				
Colony type	Erythroid	Myeloid	Mega karyocytic	Undiffere-ntiated	Mixed	Erythroid	Myeloid	Mega karyocytic	Undiffere-ntiated	Mixed
Control	3.8 ± 1.9 [a] (44.2%)	3.4 ± 0.5 (39.5%)	0.9 ± 0.3 (10.5%)	0.1 ± 0.08 (1.2%)	0.4 ± 0.02 (4.6%)	5.7 ± 1.8 (60.6%)	2.6 ± 0.4 (27.6%)	0.3 ± 0.18 (3.2%)	0.6 ± 0.2 (6.4%)	0.2 ± 0.1 (2.1%)
Bone marrow	5.8 ± 0.6 (32%)	9.3 ± 1.3 (51.4%)	0.4 ± 0.02 (2.2%)	2.0 ± 1.1 (11%)	0.6 ± 0.38 (3.3%)	6.2 ± 0.9 (26.3%)	10.2 ± 1.1 [b] (43.2%)	3.4 ± 0.6 [b] (14.6%)	1.2 ± 0.3 (5%)	2.6 ± 0.1 (11%)
Control	1.8 ± 1.2 (43.9%)	1.4 ± 0.3 (34.1%)	0.3 ± 0.19 (7.3%)	0.2 ± 0.04 (4.9%)	0.4 ± 0.2 (9.7%)	1.9 ± 1.0 (73%)	0.3 ± 0.19 (11.5%)	0.25 ± 0.18 (9.6%)	0.08 ± 0.07 (3%)	0.07 ± 0.06 (2.7%)
Spleen	5.7 ± 2.3 (49.1%)	3.8 ± 1.2 (32.7%)	0.8 ± 0.2 (6.9%)	1.0 ± 0.4 (8.6%)	0.3 ± 0.2 (2.6%)	6.0 ± 0.6 (49.2%)	3.5 ± 0.12 [b] (28.6%)	2.1 ± 0.24 [b] (17.2%)	0.26 ± 0.04 (2.1%)	0.34 ± 0.1 (2.8%)
Control	10.4 ± 2.9 (70.7%)	3.2 ± 0.2 (21.8%)	0.5 ± 0.42 (3.4%)	0.2 ± 0.18 (1.4%)	0.4 ± 0.2 (2.7%)	10.3 ± 0.4 (81.1%)	2.1 ± 0.6 (16.5%)	0.21 ± 0.14 (1.6%)	0.06 ± 0.04 (0.5%)	0.03 ± 0.02 (0.2%)
Whole blood	16.0 ± 1.4 (44.4%)	14.3 ± 1.1 (39.7%)	0.8 ± 0.11 (2.2%)	2.7 ± 0.29 (7.5%)	2.2 ± 0.3 (6.1%)	12.3 ± 2.8 (36.7%)	14.4 ± 1.2 [b] (42.9%)	4.3 ± 0.2 [b] (12.8%)	0.7 ± 0.24 (2%)	1.8 ± 0.06 (5.4%)

[a]Mean ± SD; [b]$p < 0.05$ (Mann–Whitney U test); From Oršolić and Bašić (2005a).

Table 13.8. The effect of WSDP given p.o. on peripheral WBC number/ mm^3 during six weeks of treatment [From Oršolić and Bašić (2005a)].

Weeks	0	1	2	3	4	5	6
Control	7207 ± 1771.1	6367 ± 737.1	5965 ± 1499.8	7575 ± 2104.3	6202 ± 7770.1	6286 ± 829.6	6640 ± 969.2
WSDP	7156 ± 1312.5 [a]	6967 ± 760.7	10375 ± 1946.0	11583 ± 2104.3	10799±3018.5	10390 ± 2309.1	11032 ± 1406.7
Mann–Whitney U test	1.000	0.130	<0.05	<0.05	<0.05	<0.05	<0.05

Numerous microbial components such as bacterial lipopolysaccharide (LPS), muramyl dipeptide, *Mycobacterium bovis* strain BCG and glucan have been shown to exhibit radioprotective effect when administered before irradiation (Behling, 1983). There is a similarity between our studies and those described in the literature (Behling, 1983; Maruyama *et al.,* 1977; Gordon *et al.,* 1977). WSDP given orally throughout the period of 20 consecutive days induced extensive proliferation of nucleated cells, mainly macrophages (Oršolić and Bašić, 2003a,b) as well as haematopoietic cells (Tables 3.6) in the spleen and bone marrow of treated mice. Stimulated haematopoietic activity in WSDP-treated animals, as evidenced by the increased number of cells capable of producing haematopoietic colonies in the spleen of lethally irradiated recipients (Table 13.7) has also been evidenced in animals treated with other biological response modifiers e.g., *Corynebacterium parvum* (Maruyama *et al.,* 1977; Gordon *et al.,* 1977; Bašić and Milas 1979).

WSDP caused a significant elevation of leukocytes in peripheral blood of treated mice (Table 13.8). WBI of mice with either 5Gy or 7Gy decreased the number of leukocytes in the same proportion in normal and WSDP-treated mice respectively. However, the recovery of leucopenia was three times faster in WSDP-treated mice, as compared to control, suggesting that proliferation of leukocyte precursors from pluripotent stem cells is increased in mice after treatment with WSDP. In these studies no changes in erythrocyte cell count in peripheral blood of WSDP-treated mice were noticed during the single or combined treatment with WSDP plus WBI.

Concerning the specific type of CFUs (Table 13.8), the only difference noticed between untreated and WSDP-treated mice (receiving treatment for 40 consecutive days) in exogenous spleen CFU assay was that their spleen cells injected to WBI recipients gave rise to more myeloid and megakaryocytic colonies. These findings prove faster recovery of leukocyte count in peripheral blood after WBI. Data on the effect of WSDP showing increased resistance of mice to lung metastasis (Oršolić and Bašić, 2003a) suggested that the antitumour effect of WSDP is mediated by the effects of WSDP on macrophages and production of IL-1 (Bezwoda *et al.,* 1986). In addition, anti-ILR antibody reduced the survival of untreated and irradiated mice, indicating that natural levels of IL-1 contribute to radioresistance to mice. Furthermore, IL-1 has been shown to have a role in stimulating bone marrow by overcoming the myelosuppressive effect of radiation (Constine *et al.,* 1991) and is able to protect bone marrow progenitor cells from radiation injury *in vitro* (Gallicchio *et al.,* 1989).

Pre-clinical and clinical studies have also demonstrated that a large number of cytokines accelerate bone marrow restoration after treatment with cytotoxic drugs or radiation, which ablates the bone marrow (Singh and Yadav, 2005). It is possible that cytokines such as IL-1, IL-4 or IFN-γ also have restorative effects on the bone marrow. Moreover, it has been shown that IL-1 plays an important role in the regulation of normal haematopoiesis directly, by stimulating the most primitive stem cells, and indirectly, by raising the production of other haematopoietic factors such as G-CSF, M-CSF, GM-CSF, and IL-6 (Maisin *et al.,* 1998). IL-1 is the most extensively investigated cytokine for radioprotection in animal models (Neta *et al.,* 1986a,b, 1996) and even a single dose of IL-1 given to mice prior to irradiation protected several strains of mice (DBA/1, C3H/HeN, B6D2F1, CDF1 and Balb/c) from radiation damage (Dinarello and Neta, 1989; Neta, 1990, 1997; Neta and Oppenheim, 1988; Neta *et al.,* 1986a,b, 1988; Wu *et al.,* 1989; Wu and Miyamoto, 1990).

Studies on radioprotection by immunomodulators have indicated that endotoxin provides protection when administered 20 to 24 hr before irradiation and/or shortly before or after exposure (Ainsworth, 1988). The possible mechanisms of protection by cytokines have been reviewed (Neta *et al.,* 1988, Neta, 1997). The potential utility of cytokines and growth factors as therapeutic agents and/or protective agents is great, and future studies should lead to a specific agent to treat specific tissue damage (MacVittie *et al.,* 1996).

It has been shown that tumour cell glutathione cycle is the rate-limiting factor of their survival after treatment with different agents possessing antitumour cytotoxic property (Meister, 1994). Inhibition of the pathways of glutathione synthesis makes tumour cells more susceptible to the action of different antitumour agents; glutathione deficiency sensitizes cells to the disadvantageous effect of radiation (Dethmers and Meister, 1981; Meister, 1991).

It has been demonstrated that the tumour cell resistance to various antitumour agents is partially associated with an overproduction of glutathione synthesis in those cells, and that its production can be reversed by treatment with selective inhibitors of glutathione synthesis (Hamilton *et al.,* 1990). Our findings (Oršolić *et al.,* 2004) showing that the glutathione content of tumourigenic cell line is higher than in the non-transformed cell line are in accordance with the above-mentioned findings. The increase in glutathione in bone marrow cells after treatment with WSDP suggests that the elevation of glutathione level may spare normal haematopoietic cells from the deleterious effects of radiation or chemotherapy. Our results (Oršolić and Bašić, 2005a) show that WSDP protected more of these cells, possibly due to elevation of glutathione synthesis in their bone marrow which was not compromised by either 3Gy or 6Gy WBI. It is likely that treatment with WSDP increases the ability of haematopoietic tissue to synthesize glutathione in bone marrow compartment, making the treated mice more resistant to radiation damage.

These data are in agreement with other workers (Kavanagh *et al.*, 1990; Terradez *et al.,* 1993; Burdon, 1995), suggesting that intracellular glutathione level may be implicated in the control of cell proliferation. More CFUs in haematopoietic tissues of mice, as shown by exogenous spleen CFU assay, indicate that WSDP used p.o. for a long period of time exercised stimulative effect on haematopoiesis. The stimulative effect of WSDP on haematopoiesis may be due to IL-1 production and its action to stem cells (Neta *et al.,* 1986) or through its influence on glutathione level. Moreover, some data suggest that IL-1α and IL-β may increase mRNA of MnSOD (Wong and Goeddel, 1988). It is likely that the antioxidative effect of flavonoids in biological systems is related to a great deal of events including:

i) their ability to scavenge ROS including $^1O_2$, $OH^\bullet$, H_2O_2, $O_2^{\bullet-}$, $HO_2^\bullet$, lipid radical ($LO^\bullet$) and lipid peroxy radical ($LOO^\bullet$)
ii) ability to scavenge nitric reactive radical (HOONO, NO, NO_3)
iii) inhibition of oxidative enzymes
iv) metal ion chelation (Cu^{2+}, Fe^{2+}, Zn^{2+}, Mg^{2+})
v) increase the activity of antioxidant enzymes and their protection (Galati and O'Brien, 2004; Manach *et al.,* 2004; Jagetia and Reddy, 2005).

Our studies demonstrate that propolis and flavonoids present in propolis have radioprotective, stimulative and regenerative properties on haematopoiesis and suggest a clinically potential use of propolis and related flavonoids in the treatment of various cytopenias induced by radiation and/or chemotherapy. In addition, these studies have provided the first evidence that propolis and its flavonoids act directly on haematopoietic bone marrow and spleen cells and enhance their growth and differentiation into colony forming cells.

References

Ainsworth, E.J. (1988) From endotoxins to newer immunomodulators: survival-promoting effects of microbial polysaccharide complexes in irradiated animals. *Pharmacology and Therapeutics* 39, 223–241.

Arora, R. (2007) Botanicals as potential radioprotective and radiorecovery agents: Current status and emerging directions for clinical applications. *Proceedings of the World Ayurveda Conference*, Pune, India.

Arora, R., Chawla, R., Singh, S., Sagar, R.K., Kumar, R., Sharma, A., Prasad, J., Singh, S., Gurudatta, G.U. and Sharma, R.K. (2006a) Bioprospection for Radioprotective Molecules from Indigenous Flora. In: Govil, J.N., Singh, V.K. and Bhardwaj, R. *Recent Progress in Medicinal Plants*. Studium Press, LLC, USA, Vol. 16, pp. 179–219.

Arora, R., Chawla, R., Singh, S., Sagar, R.K., Kumar, R., Sharma, A.K., Singh, S., Prasad, J., Sharma, R.K. and Tripathi, R.P. (2006b) Radioprotection by Himalayan High-Altitude Region Plants. In: Sharma, R.K. and Arora, R. (eds) *Herbal Drugs: A Twenty First Century Perspective*. Jaypee Brothers Medical Publishers Private Ltd., Delhi, India, pp. 301–325.

Arora, R., Gupta, D., Chawla, R., Sagar, R., Sharma, A., Kumar, R., Prasad, J., Singh, S., Samanta, N. and Sharma, R.K. (2005) Radioprotection by plant products: present status and future prospects. *Phytotherapy Research* 19, 1–22.

Arora, R., Singh, S., Puri S.C. and Sharma, R.K. (2008) Himalayan Mayapple (*Podophyllum hexandrum* Royle): Traditional Uses, Clinical Indications and Future Prospects. In: Watson, R. and Preedy, V.R. *The Encyclopedia of Herbal Medicine in Clinical Practice*, CABI Publishing, Wallingford, Oxon, United Kingdom, p. 26.

Bašić, I. and Milas, L. (1979) Effect of whole-body irradiation on mice treated with *Corynebacterium parvum*. *European Journal of Cancer* 13, 901–908.

Behling, U.H. (1983) The radioprotective effect of bacterial endotoxin. In: Nowotny, A. (ed) *Beneficial Effects of Endotoxin*, Plenum Press, New York, USA, pp. 127–131.

Bezwoda, W.R., Lewis, D. and Livini, N. (1986) Bone marrow involvement in anaplastic small cell lung cancer. Diagnosis, hematologic features, and prognostic implications. *Cancer* 58, 1762–1767.

Bors, W., Michel, C. and Stettmaier, K. (1997) Antioxidant effects of flavonoids. *Biofactors* 6(4), 399–402.

Bors, W., Michel, C., Stettmaier, K., Lu, Y. and Fo, Y. (2004) Antioxidant mechanisms of polyphenolic caffeic acid oligomers, constituents of *Salvia officinalis*. *Biological Research* 37, 301–311.

Burdon, R.H. (1995) Superoxide and hydrogen peroxide in relation to mammalian cell proliferation. *Free Radical Biology and Medicine* 18, 775–794.

Chawla, R., Arora, R., Singh, S., Sagar, R., Sharma, R.K., Puri, S.C., Khan, H.A., Shakir, M., Shawl, A.S., Kumar, R., Sharma, A., Singh, S., Prasad, J., Tripathi, R.P. and Qazi, G.N. (2006) Variation in aryltetralin lignan content of *Podophyllum hexandrum* influences radioprotective efficacy. *Evidence Based Complementary and Alternative Medicine* 3(4), 503–511.

Chawla, R., Arora, R., Singh, S., Sagar, R.K., Sharma, R.K., Kumar, R., Sharma, A., Gupta, M.L., Singh, S., Prasad, J., Khan, H.A., Swaroop, A., Sinha, A.K., Gupta, A.K., Tripathi, R.P. and Ahuja, P.S. (2007) Radioprotective and antioxidant activity of fractionated extracts of berries of *Hippophae rhamnoides*. *Journal of Medicinal Food* 10(1), 101–109.

Chen, C.N., Weng M.S., Wu C.-L. and Lin J.K. (2004) Comparison of radical scavenging activity, cytotoxic effects and apoptosis induction in human melanoma cells by Taiwanese propolis from

different sources. *Evidence Based Complementary and Alternative Medicine* 1, 175–185.
Constine, L.S., Harwell, S., Keng, P., Lee, F., Rubin, P. and Siemann, D. (1991) Interleukin-1a stimulates haemopoiesis but not tumour cell proliferation and protects mice from lethal total body irradiation. *International Journal of Radiation Oncology, Biology and Physics* 20, 447–456.
da Silva, J., Herrmann, S.M., Heuser, V., Peres, W., Possa Marroni, N., Gonzalez-Gallego, J. and Erdtmann, B. (2002) Evaluation of the genotoxic effect of rutin and quercetin by comet assay and micronucleus test. *Food and Chemical Toxicology* 40, 941–947.
Dethmers, J.K. and Meister, A (1981). Glutathione exports by human lymphoid cells: depletion of glutathione by inhibition of its synthesis decreases export and increases sensitivity to irradiation. *Proceedings of the National Academy of Sciences, USA* 78, 7492–7497.
Dimov, V., Ivanovska, N., Bankova, V. and Popov, S. (1992) Immunomodulatory action of propolis. IV. Prophylactic activity against gram-negative infections and adjuvant effect of the water-soluble derivate. *Vaccine* 10, 817–823.
Dimov, V., Ivanovska, N., Manolova, N., Bankova, V., Nikolov, N. and Popov, S. (1991) Immunodulatory action of propolis. Influence on anti-infectious protection and macrophage function. *Apidologie* 22, 155–162.
Dinarello, C.A. and Neta, R. (1989) An overview on interleukin-1 as a therapeutic agent. *Biotherapy* 1, 245–254.
Edgreen, M. (1983) Depletion of cellular thiols by misonidazole pretreatment prevents some post irradiation repair processes. *British Journal of Radiology* 56, 211–212.
El-khawaga, O.A., Salem, T.A. and Elshal, M.F. (2003) Protective role of Egyptian propolis against tumour in mice. *Clinica Chimica Acta* 338, 11–16.
Fujisawa, S. and Kadoma, Y. (2006) Anti- and pro-oxidant effects of oxidized quercetin, curcumin or curcumin-related compounds with thiols or ascorbate as measured by the induction period method. *In vivo* 20(1), 39–44.
Galati, G. and O'Brien, P.J. (2004) Potential toxicity of flavonoids and other dietary phenolics: Significance for their chemopreventive and anticancer properties. Serial Review: Flavonoids and Isoflavones (Phytoestrogens): Absorption, Metabolism, and Bioactivity. *Free Radical Biology and Medicine* 37, 287–303.
Gallicchio, V.S., Hulette, B.C., Messino, M.J., Gass, C., Bieschke, M.W. and Doukas, M.A. (1989) Effect of various interleukins (IL-1, IL-2, and IL-3) on the *in vitro* radioprotection of bone marrow progenitors (CFU-GM and CFU-MEG). *Journal of Biological Response Modifiers* 8, 479–487.
Geetha, S., Singh, V., Ram, M.S., Ilavazhagan, G., Banerjee, P.K. and Sawhney, R.C. (2005) Immunomodulatory effects of seabuckthorn (*Hippophae rhamnoides* L.) against chromium (VI) induced immunosuppression. *Molecular and Cellular Biochemistry* 278(1-2), 101-109.
Gordon, M.Y., Aguado, M. and Blackett, N.M. (1977) Effect of BCG and *Corynebacterium parvum* on the hematopoietic precursor cell in continuously irradiated mice: possible mechanisms of action in immunotherapy. *European Journal of Cancer* 13, 229–233.
Hall, E.J. (1994) Molecular biology in radiation therapy: the potential impact of recombinant technology on clinical practice. *International Journal of Radiation Oncology, Biology and Physics* 30(5), 1019–1028.
Hamilton, T., O'Dwyer, P. and Yong, R. (1990) Phase I trial of buthionine sulfoximine (BSO) plus melphalan (L-PAM) in patients with advanced cancer. *Proceedings of a Meeting of the American Society of Clinical Oncology* 9, A281.
Heim, K.E., Tagliaferro, A.R. and Bobilya, D.J. (2002) Flavonoid antioxidants: chemistry, metabolism and structure activity relationships. *Journal of Nutritional Biochemistry* 13, 572–584.
Jacquet, P. de., Saint-Georges, L., Barrio, S. and Baugnet-Mahieu, L. (1995) Morphological effects of caffeine, okadaic acid and genistein in one–cell mouse embryos blocked in G2 by X-irradiation. *International Journal of Radiation Biology* 67, 347–358.
Jagetia, G.C. and Rajanikant, G.K. (2005) Curcumin treatment enhances the repair and regeneration of wounds in mice exposed to hemibody gamma-irradiation. *Plastic and Reconstructive Surgery* 115(2), 515–28.
Jagetia, G.C. and Reddy, K.T. (2005). Modulation of radiation-induced alteration in the antioxidant

status of mice by naringin. *Life Science* 77, 780–794.
Jagetia, G.C., Venkatesha, V.A. and Reddy, T.K. (2003) Naringin, a citrus flavonone, protects against radiation–induced chromosome damage in mouse bone marrow. *Mutagenesis* 18, 337–343.
Jeon, S.M., Bok, S.H., Jang, M.K., Kim, Y.H., Nam, K.T., Jeong, T.S., Park, Y.B. and Choi, M.S. (2002) Comparison of antioxidant effects of naringin and probucol in cholesterol-fed rabbits. *Clinica Chimica Acta* 317(1–2), 181–90.
Kavanagh, T.J., Grossmann, A., Jaecks, E.P., Jinneman, J.C., Eaton, D.L., Martin, G.M. and Rabinovitch, P.S. (1990) Proliferative capacity of human peripheral blood lymphocytes sorted on the basis of glutathione content. *Journal of Cell Physiology* 145, 472–480.
Kosalec, I., Bakmaz, M., Pepeljnjak, S. and Vladimir-Knežević, S. (2004) Quantitative analysis of the flavonoids in raw propolis from northern Croatia. *Acta Pharmaceutica* 54, 65–72.
Lahouel, M., Boulkour, S., Segueni, N. and Fillastre. J.P. (2004) Protective effect of flavonoides against the toxicity of vinblastine, cyclophoshamide and paracetamol by inhibition of lipid-peroxydation and increase of liver glutathione. *Hematology 7*, 59–67.
Lodovici, M., Guglielmi, F., Meoni, M. and Dolara, P. (2001) Effect of natural phenolic acids on DNA oxidation *in vitro. Food and Chemical Toxicology* 39(12), 1205–1210.
MacVittie, T.J, Weiss, J.F. and Browne, D. (1996) *Advances in the Treatment of Radiation Injuries*, Pergamon, Oxford, UK.
Maisin, J.R. (1998) Bacq and Alexander Award Lecture–Chemical radioprotection: past present and future. *International Journal of Radiation Biology* 73, 443–450.
Manach, C., Schalbert, A., Morand, C., Rémésy, C. and Jiménez, L. (2004) Polyphenols: food sources and bioavailability. *American Journal of Clinical Nutrition* 79(5), 727–747.
Maruyama, V., Magura, C. and Feoba, J. (1977) *Corynebacterium parvum*-induced, biosensitivity and cyclic changes of haematopoietic spleen colony-forming units. *Journal of the National Cancer Institute* 59, 173–178.
Meister, A. (1991) Glutathione deficiency produced by inhibition of its synthesis and its reversal; application in research and therapy. *Pharmacology and Therapeutics* 51, 155–159.
Meister, A. (1994) Glutathione, ascorbate, and cellular protection. *Cancer Research* 54, 1969–1975.
Metodiewa, D., Jaiswal, A.K., Cenas, N., Dickancaite, E. and Aguilar, J.S. (1999) Quercetin may act as a cytotoxic prooxidant after its metabolic activation to semiquinone and quinoidal product. *Free Radical Biology and Medicine* 26, 107–116.
Molina, M.F., Sanchez-Reus, I., Iglesias, I. And Benedi, J. (2003) Quercetin, a flavonoid antioxidant, prevents and protects against ethanol-induced oxidative stress in mouse liver. *Biological & Pharmaceutical Bulletin* 26(10), 1398–1402.
Nayak, V. and Devi, P.U. (2005) Protection of mouse bone marrow against radiation-induced chromosome damage and stem cell death by the *Ocimum* flavonoids orientin and vicenin. *Radiation Research* 163(2), 165–171.
Neta, R. (1990) Radioprotection and therapy of radiation injury with cytokines. *Progress in Clinical and Biological Research* 352, 471–478.
Neta, R. (1997) Modulation of radiation damage by cytokines. *Stem Cells* 15(S2), 87–94.
Neta, R. and Oppenheim, J.J. (1988) Cytokines in therapy of radiation injury. *Blood* 72, 1093–1095.
Neta, R., Douches, S. and Oppenheim, J.J. (1986a) Interleukin 1 is a radioprotector. *Journal of Immunology* 136, 2483–2485.
Neta, R., Keller, J.R., Ali, N., Blanchette. F. and Dubois, C.M. (1996) Contrasting mechanisms of the myeloprotective effects of interleukin-1 against ionizing radiation and cytotoxic 5-fluorouracil. *Radiation Research* 145, 624–631.
Neta, R., Oppenheim, J.J. and Douches, S.D. (1988) Interdependence of the radioprotective effects of human recombinant interleukin-1a, tumour necrosis factor-α, granulocyte colony-stimulating factor, and murine recombinant granulocyte–macrophage colony-stimulating factor. *Journal of Immunology* 140, 108–111.
Neta, R., Vogel, S.N., Oppenheim, J.J. and Douches, S.D. (1986b) Cytokines in radioprotection. Comparison of the radioprotective effects of IL-1 to IL2, GM-CSF and IFN-γ. *Lymphokine Research* 5(S1), S105–S110.

Oliveira, N.G., Neves, M., Rodrigues, A.S., Monteiro, Gil O., Chaveca, T. and Rueff J. (2000) Assessment of the adaptive response induced by quercetin using the MNCB peripheral blood human lymphocytes assay. *Mutagenesis* 15, 77–83.

Orsi, R.O., Sforcin, J.M., Funari, S.R. and Bankova, V. (2005) Effects of Brazilian and Bulgarian propolis on bactericidal activity of macrophages against *Salmonella typhimurium*. *International Immunopharmacology* 5(2), 359–68.

Oršolić, N. and Bašić, I. (2003a) Immunomodulation by water-soluble derivative of propolis (WSDP) a factor of antitumour reactivity. *Journal of Ethnopharmacology* 84, 265–273.

Oršolić, N. and Bašić, I. (2003b) Immunomodulatory action of water-soluble derivative of propolis, caffeic acid and caffeic acid phenethyl ester: nitric oxide and macrophage function. *Mellifera* 3, 56–64.

Oršolić, N. and Bašić, I. (2005a) Antitumour, hematostimulative and radioprotective action of water-soluble derivative of propolis (WSDP). *Biomedicine and Pharmacotherapy* 59(10), 561–570.

Oršolić, N. and Bašić, I. (2005b) Water-soluble derivative of propolis and its polyphenolic compounds enhance tumouricidal activity of macrophages. *Journal of Ethnopharmacology* 102, 37–45.

Oršolić, N. and Bašić, I. (2007) Cancer chemoprevention by propolis and its polyphenolic compounds in experimental animals. *Recent Progress in Medicinal Plants* 17, 55–113.

Oršolić, N., Bendelja, K., Brbot-Šaranović, A. and Bašić, I. (2004) Effects of caffeic acid phenethyl ester and caffeic acid, antioxidants from propolis, on inducing apoptosis in HeLa human cervical carcinoma and Chinese hamster lung V79 fibroblast cells. *Periodicum Biologorum* 106, 367–372.

Oršolić, N., Benkovic, V., Horvat- Knežević, A., Kopjar, N., Kosalec, I., Bakmaz, M., Mihaljevic, Z., Bendelja, K. and Bašić, I. (2007) Assessment by survival analysis of the radioprotective properties of Propolis and its polyphenolic compounds. *Biological and Pharmaceutical Bulletin* 30(5), 946–951.

Oršolić, N., Benković, V., Horvat-Knežević, A., Kopjar, N. and Bašić, I. (2006a) Effect of propolis and its polyphenolic/flavonoids compounds on DNA damage induced by radiation to mouse lymphocytes. *Pharmacology Online* 3, 690–697.

Oršolić, N., Kosalec, I. and Bašić, I. (2005c) Synergystic antitumour effect of polyphenolic components of water-soluble derivative of propolis against Ehrlich ascites tumour. *Biological & Pharmaceutical Bulletin* 28, 694–700.

Oršolić, N., Štajcar, D. and Bašić, I. (2006b) Cytotoxicity of propolis and its polyphenolic compounds on primary culture of human urinary bladder transitional cell carcinoma. *Pharmacology Online* 3, 408–415.

Oršolić, N., Šver, L., Terzić, S. and Bašić, I. (2005d) Peroral aplication of water-soluble derivative of propolis (WSDP) and its related polyphenolic compounds and their influence on immunological and antitumour activity. *Veterinary Research Communications* 29, 575–593.

Oršolić, N., Terzić, S., Šver, L. and Bašić, I. (2005e) Polyphenolic compounds from propolis modulate immune responses and increase host resistance to tumour cells. *Food and Agricultural Immunology* 16, 165–179.

Patchen, M.L., MacVittie, T.J. and Souza, L.M. (1992) Post–irradiation treatment with granulocyte colony-stimulating factor and pre-irradiation WR-2721 administration synergize to enhance haemopoietic reconstitution and increase survival. *International Journal of Radiation Oncology, Biology and Physics* 22, 773–779.

Revesz, L. (1985) The role of endogenous thiols in intrinsic radioprotection. *International Journal of Radiation Biology* 47, 361–368.

Revesz, L. and Edgreen, M. (1984) Glutathione dependent yield and repair of single strand DNA breaks in irradiated cells. *British Journal of Cancer* 49(6), 55–60.

Rithidech, K.N., Tungjai, M. and Whorton, E.B. (2005) Protective effect of apigenin on radiation-induced chromosomal damage in human lymphocytes. *Mutation Research* 585(1–2), 96–104.

Russo, A., Acquaviva, R., Campisi, A., Sorrenti, V., Di Giacomo, C., Virgata, G., Barcellona, M.L. and Vanella, A. (2000) Bioflavonoids as antiradicals, antioxidants and DNA cleavage protectors. *Cell Biology and Toxicology* 16(2), 91–98.

Russo, A., Cardile, V., Sanchez, F., Troncoso, N., Vanella. A. and Garbarino, J.A. (2004) Chilean propolis: antioxidant activity and antiproliferative action in human tumour cell lines. *Life Science* 76(5), 545–558.

Sadzuka, Y., Sugiyama, T., Shimoi, K., Kinae, N. and Hirota, S. (1997) Protective effect of flavonoids on doxorubicin-induced cardiotoxicity. *Toxicology Letters* 92, 1–7.

Shimoi, K., Masuda, S., Shen, B., Furugori, M. and Kinae, N. (1996) Radioprotective effects of antioxidative plant flavonoids in mice. *Mutation Research* 350, 153–161.

Singh, V.K. and Yadav, S. (2005) Role of cytokines and growth factor in radioprotection. *Experimental and Molecular Pathology* 8, 156–169.

Snyder, R.D. and Gillies, P.J. (2002) Evaluation of the clastogenic, DNA intercalative, and topoisomerase II–interactive properties of bioflavonoids in Chinese hamster V79 cells. *Environmental and Molecular Mutagenesis* 40, 266–276.

Suzuki, I., Hayashi, I., Takaki, T, Groveman, D.S. and Fujimiya, Y. (2002) Antitumour and anticytopenic effects of aqueous extracts in combination with chemotherapeutic agents. *Cancer Biotherapy and Radiopharmaceuticals* 17, 553–562.

Terradez, P., Asensi, M., Lasso de la Vega, M.C., Puertes, I.R., Vina, J. and Estrela, J.M. (1993) Depletion of tumour glutathione *in vivo* by buthionine sulphoximine: modulation by the rate of cellular proliferation and inhibition of cancer growth. *Biochemical Journal* 292, 477–83.

Walker, P. and Crane, E. (1987) Constituents of propolis. *Apidologie* 18, 327–334.

Weiss, J.F., Kumar, K.S., Walden, T.L., Neta, R., Landauer, M.R. and Clark, E.P. (1990) Advances in radioprotection through the use of combined agent regimens. *International Journal of Radiation Biology* 57, 709–722.

Wong, G.H. and Goeddel, D.V. (1988) Induction of manganous superoxide dismutase by tumour necrosis factor: possible protective mechanism. *Science* 242, 941–944.

Wu, S.G. and Miyamoto, T. (1990) Radioprotection of the intestinal crypts of mice by recombinant human interleukin-1a. *Radiation Research* 123, 112–115.

Wu, S.G., Tuboi, A. and Miyamoto, T. (1989) Radioprotection of C3H mice by recombinant human interleukin-1a. *International Journal of Radiation Biology* 56, 485–492.

Yamashita, N. and Kawanishi, S. (2000) Distinct mechanisms of DNA damage in apoptosis induced by quercetin and luteolin. *Free Radical Research* 33, 623–633.

Yen, G.C., Duh, P.D., Tsai, H.L. and Huang, S.L. (2003) Pro-oxidative properties of flavonoids in human lymphocytes. *Bioscience Biotechnology and Biochemistry* 67(6), 1215–1222.

SECTION IV

ELUCIDATION OF MECHANISM OF ACTION OF HERBAL RADIOPROTECTORS (A STEPPING STONE FOR DESIGNING OF NOVEL RADIOMODULATORS)

Chapter 14

Radioprotective Effects of *Ginkgo biloba* via its Antioxidant Action

Göksel Şener, Abdullah Sakarcan and Berrak Ç. Yeğen

Introduction

Ginkgo biloba is a tall, sturdy, extremely long-lived tree that is extensively cultivated. The tree represents the genus Ginkgo and, with its leaves that are divided into two distinct lobes in young specimens, it takes the species name of "biloba". Although the seeds are most commonly used in Traditional Chinese Medicine, in recent years, standardized extracts of the leaves have been widely sold as a natural medicine in Europe and as a dietary supplement in the United States. In relation to modern therapeutics, the extract termed "EGb 761", present in many other recently commercialized products, is a well-defined preparation of the leaves of *Ginkgo biloba* that is generally used to treat brain disorders including dementias, neurosensory problems and peripheral circulatory disturbances (DeFeudis, 2003; 2005). The extract of the leaves of the *Ginkgo biloba*, a mixture mainly composed of flavonoid glycosides and terpenoids (ginkgolides and bilobalide), has been shown to exhibit a variety of pharmacological actions (VanBeek, 2005).

In order to discuss the radioprotective effects of EGb 761, it seems necessary to provide an introduction about free radicals and their role in the radiation-induced oxidative damage.

Reactive oxygen and nitrogen species (ROS/RNS)

A free radical is an atom or a molecule that contains an unpaired electron. Usually electrons associated with atoms or molecules are paired where these molecules are relatively stable and non-reactive. Conversely, the loss or addition of an electron leaves the atom or molecule unstable, making it relatively more reactive than its non-radical counterpart (Reiter *et al.*, 1995). Oxygen free radicals or, more generally, reactive oxygen species (ROS), as well as reactive nitrogen species (RNS) are products of normal cellular metabolism. ROS and RNS are well recognized for playing a dual role, as they can be either harmful or beneficial to living systems (Valko *et al.*, 2007). Beneficial effects of ROS occur at low/moderate concentrations and involve physiological roles in cellular response to noxious stimuli, for example in defence against infectious agents as a function of a number of cellular signalling systems. One further beneficial example of ROS at low/moderate concentrations is the induction of mitogenic response (Valko *et al.*, 2006). The harmful effect of free radicals causing potential biological damage is termed as oxidative and nitrosative stress (Retier *et al.*, 2001). This occurs in biological systems when there is an overproduction of ROS/

RNS on one side and a deficiency of enzymatic antioxidants on the other. The excess ROS can damage cellular lipids, proteins or DNA inhibiting their normal function. Recently, oxidative stress and related tissue injury have been implicated in a number of human diseases, as well as in the aging process. Furthermore, oxidative stress plays an important role in the toxicities, including those induced by chemical agents (e.g. drugs, heavy metals) and by physical factors, as induced by irradiation.

Ionizing radiation has many uses in medicine, aiding both in diagnosis and treatment. X-rays, CT scanners and fluoroscopes are used to form images of the inside of the body. Similarly, nuclear medicine also uses radioactive isotopes in both diagnosis and treatment, in which radioactive elements get localized to specific tissues of the body and give off tiny amounts of radiation. On the other hand, the use of nuclear bombs during the Second World War and accidents at nuclear reactors, which led to multiple secondary effects such as myelosuppression, cancer, hypothyroidism and cancer, have demonstrated the vulnerability of living cells to radiation, emphasizing the importance of exposure time and radiation dose in radiation-induced changes in the tissues (Greenstock, 1993; Riley, 1994).

The importance of radiotherapy

With ongoing improvements in cancer therapy and health care, the population of long-term cancer survivors continues to grow, making cancer a chronic disease. In the treatment of this chronic disease, radiation is an important modality and in some instances it may be the single best agent for treatment. However, a major problem associated with cancer radiotherapy is the immediate or delayed severe side effects that result from normal tissue damage, limiting the effectiveness of the therapy. Tissue injury due to ionizing radiation is believed to be a consequence of a cascade of cytokine activity, which ultimately begins with oxidative stress from radiolytic hydrolysis and formation of ROS (Rubin *et al.*, 1995; Stark, 2005). Radiation has been reported to cause tissue injury in various ways. At the cellular level, ionizing radiation can induce damage in biologically important macromolecules, such as DNA, proteins, lipids and carbohydrates in various organs. While some damage may be expressed early, the other may be expressed over a period of time depending upon cell kinetics and tolerance of the tissues to radiation. The differentiation level and anti-oxidative defence capabilities of various tissues play an important role in the extent of radiation-induced injury (Ikushima *et al.*, 1996; Parasassi *et al.*, 1994). Radiation can damage every tissue in the body; however, the fastest growing tissues are the most vulnerable, because radiation triples its effects during the growth phase. Since the proliferating cells are highly sensitive to irradiation, the highly proliferating germinal epithelium, gastrointestinal epithelium and the bone marrow progenitor cells are affected the most by irradiation. The germinal cells in the testes and ovaries can be rendered useless with very small doses of radiation. More resistant are the lining cells of the body-skin and intestines, while the brain cells are the most resistant cells (Greenstock, 1993; Riley, 1994). Thus, an integrative approach for managing a patient with cancer should target the multiple biochemical and physiological pathways that inhibit tumour development while minimizing normal tissue toxicity. Accordingly, agents, which can protect normal tissues against radiation-induced damage, can increase the patient's tolerance to radiotherapy and ameliorate radiation sickness (Jagetia *et al.*, 2004).

Tissue injury from radiation

It is well known that free radicals are naturally generated in aerobic cells as a result of energy metabolism, while the healthy cells have mechanisms to scavenge them. These mechanisms are represented by a complex redox buffering system. Ionizing radiation generates high levels of ROS that overwhelms the oxidative countering systems (Agrawal *et al.*, 2001) and these hit the major building blocks of the cell, such as lipids and proteins (Verma *et al.*, 1991). The half-life of free radicals is extremely short, on the order of 10^{-6}–10^{-8} seconds; however, these radicals immediately react with any biomolecule in the vicinity and produce a highly site-specific oxidative damage. It is estimated that 60–70% of tissue damage induced by ionizing radiation is caused by $^{\bullet}OH$. The major by-products of such interactions are hydroperoxides, which damage hydrophobic links between phospholipids. Especially, radiation injury to the bilipid layer of the cellular membrane generates products that reduce the fluidity of the membrane, creating a dysfunctional medium. This would make the cellular membrane permeable to unintended amounts of water (Parasassi *et al.*, 1994; Edwards *et al.*, 1984).

Another highly injury-prone element of the cellular membrane is unsaturated acyl chains (Samuni *et al.*, 1997). Due to such damage, the structure of cellular membrane could collapse and lead to cell death. The interaction between polyunsaturated fats and free radicals due to irradiation generates thiobarbituric acid reactive substances (TBARS), malondialdehyde (MDA) and 4-hydroxyalkelnals (4-HAD) (Manda *et al.*, 2007). The increase in TBARS is specifically noted in later stages of ionizing radiation injury, but not immediately after the oxidative stress (Ward, 1988). Free radicals generated by exposure to ionizing radiation also induce detectable changes in the structure and function of intrinsic proteins (Jhun *et al.*, 1991).

The by-products of protein and lipid damagc could also indirectly damage the DNA, which is the most vulnerable and critical part of cellular structure in case of ionizing radiation, since it is present in the nucleus, centre of cellular operations, and mitochondria, energy hub of cellular operations (Greenstock, 1993; Riley, 1994). Under such circumstances, the genes that control apoptosis could be stimulated (Pervan *et al.*, 2005).

Irradiation could also cause oxidation of guanine bases in different tissues (Tchou and Grollman, 1993). The most vulnerable nucleotide is guanine, which is affected by OH through the radiolysis of water (Ward, 1988). Absorption of ionizing radiation in water leads to the formation of the primary radicals $^{\bullet}OH$, $H^{\bullet}$ and e^{-}_{aq}. In the presence of oxygen, the hydrogen radical $H^{\bullet}$ and the hydrated electron e^{-}_{aq} react with O_2 and form the hydroperoxyl radical $HO_2^{\bullet}$ and the superoxide radical $O_2^{\bullet -}$ (Stark, 2005).

Single or double bond breaks could either arise from direct effect of ionizing radiation or free radical effect. They could lead to genetic changes linked to mutagenesis and carcinogenesis (Elia *et al.*, 1991). Thesc changes in germ cell lines could have lasting effects on subsequent generations in the form of chromosomal aberrations and micronuclei (Kawata *et al.*, 2004; Coates *et al.*, 2004).

Radioprotectors

Since radiation-induced cellular damage is attributed primarily to the harmful effects of free radicals, agents that have direct free radical scavenging activity are particularly promising as radioprotectors. The first report dealing with the ability of certain substances as radioprotectors was published in 1949 (Dale *et al.,* 1949). Following this study, sulfhydryl compounds such as cysteine and cysteamine were introduced as radioprotective agents (Patt *et al.,* 1949; Bacq *et al.,* 1951). However, these chemicals have caused serious side effects, such as nausea and vomiting, which are considered to be toxic at the doses required for radioprotection. Since then, several compounds with varied chemical structures and pharmacological properties have been screened for their radioprotective ability in humans.

On the other hand, agents that activate or support the antioxidant systems are also nominated as putative radioprotectors. Treatment with antioxidants as an adjuvant therapy added to radiotherapy would protect the normal tissues against radiation-induced side effects and also enhance the efficiency of radiotherapy. Various cellular defence mechanisms (i.e., antioxidant vitamins, such as vitamin C and vitamin E; antioxidant enzyme systems, such as superoxide dismutase (SOD), glutathione peroxidase and catalase) are activated under the conditions that have been associated with excessive formation of reactive oxygen species, to prevent against the imposed "oxidative stress". However, in many conditions such defences are overburdened and supplementary antioxidant protection may be provided by "exogenous antioxidant" (DeFeudis, 2005). Several antioxidants, such as melatonin, lycopene, ferulic acid, cafeic ester etc. have been shown to be cytoprotective in the setting of ionizing radiation (Vijayalaxmi *et al.,* 1995; Srinivasan *et al.,* 2007; Mansour, 2006).

G. biloba extract has also been reported to be protective against radiation-induced oxidative damage through both its free radical scavenger and antioxidant effects. In this chapter, the antioxidant properties of *Ginkgo biloba* and its cytoprotective role in ionizing radiation will be outlined in the context of previous literature and our associated research findings.

Ginkgo biloba as an Antioxidant

There are many herbs that were traditionally used for anticancer treatment and are proved to possess anti-angiogenic properties through multiple interdependent processes, including effects on gene expression, signal processing and enzyme activities. In different animal and organ models of oxidant injury, *Ginkgo biloba* has been reported to be anti-angiogenic and cytoprotective through different mechanisms. The leaf extract acts as a scavenger of reactive oxygen species; Pincemail *et al.,* (1989) demonstrated that *Ginkgo* extract appears to have both an $O_2^{\bullet -}$ scavenging effect and also a superoxide dismutase activity. Similarly, Maitra *et al.* (1995) have examined the ability of EGb 761 to scavenge peroxyl radicals, involved mainly in the propagation step of lipid peroxidation. On the other hand, the effect of extract on nitric oxide-mediated activity is controversial. According to *in vitro* experiments, conducted by Delaflotte *et al.* (1984) over two decades ago, it was shown that the vasorelaxant effect of EGb 761 on rabbit isolated aortic ring strips is partly mediated by the release of EDRF/NO, and that a 30 minute pre-treatment of aortic strips with EGb 761 can inhibit the endothelium-dependent vasorelaxation elicited by EGb761 itself. Thus, various constituents of EGb 761 might either enhance or inhibit the EDRF/NO mechanism. These conclusions have recently been supported by the findings showing that EGb 761 has an NO scavenging activity. Furthermore, the extract and its ingredients exhibit an antagonistic effect on platelet-activating factor (Lamant *et al.,* 1987). They also exhibit protective effect on tissue abnormalities that include myocardial ischemia/reperfusion injury (Shen and Zhou, 1995), ischemic brain damage (Zhang *et al.,* 2000) and neuronal apoptosis (Ahlemeyer and Krieglstein, 2003). These effects are supposed to be beneficial in cardiovascular, cerebrovascular and neurological disorders (DeFeudis, 2003; Yoshikawa *et al.,* 1999).

In a rat model of cisplatinum-induced ototoxicity, the scanning electron microscopy observation indicated severe outer hair cell loss in the cochleae of cisplatin-treated rats, whereas in the rats treated with EGb 761 plus cisplatin, outer hair cells remained intact (Huang *et al.,* 2007). Zhou *et al.* (2006) demonstrated that in rats with colitis, *Ginkgo biloba* extract significantly elevated activity of SOD, reduced the contents of MDA and inhibited mRNA expression of TNF-alpha. Thus, it appears that *Ginkgo biloba* has modulated the inflammatory response not only by its antioxidant effect, but also through its inhibitory effect on pro-inflammatory cytokine production. In another study, Naik and Panda (2007) showed the hepatoprotective effects of *Ginkgo biloba* in CCl_4-induced hepatotoxicity through its antioxidant properties. Phase II enzymes play important roles in the antioxidant system by reducing electrophiles and reactive oxygen species. In quinone oxidoreductase 1 (NQO1)-antioxidant response element (ARE) reporter assay, *Ginkgo biloba* has been shown to induce Keap1-Nrf2-ARE signalling pathway, which is a very critical component of the antioxidant control system (Liu *et al.,* 2007).

In support of these findings, we have recently demonstrated the protective effects of EGb 761 through its antioxidant properties on different experimental models of hepatic injury, drug- and chemical-induced toxicity and thermal injury models. In rats with mercury-induced or bile duct ligation-induced hepatic injury, *Ginkgo biloba* significantly elevated the hepatic GSH levels and decreased the MDA levels (Sener *et al.,* 2005a, 2007; Fig. 14.1), while liver enzymes were restored back to control levels indicating the functional recovery (Table 14.1). Similarly, in acetaminophen-induced liver toxicity in mice, acetaminophen caused a significant decrease in GSH levels, while MDA levels were increased in liver tissues and these changes were reversed by EGb treatment (Sener *et al.,* 2006b).

The other study focusing on the systemic antioxidant effect of EGb extract was based on a model of naphthalene-induced multiple organ injury (Tozan *et al.,* 2007). The results revealed that EGb extract balanced the oxidant-antioxidant status and inhibited the generation of pro-inflammatory cytokines and neutrophil infiltration (Fig. 14.2). Furthermore, we demonstrated that EGb treatment in the rats with thermal trauma significantly increased the depleted GSH level and decreased the burn-induced elevation in the MDA level of both hepatic and renal tissues (Sakarcan *et al.,* 2005). Similarly, serum alanine aminotransferase, aspartate aminotransferase and blood urea nitrogen levels, as well as lactate dehydrogenase and tumour necrosis factor-alpha, were increased in the burn group as compared with the control group (Table 14.1). However, treatment with EGb reversed all these biochemical indices, as well as histopathological alterations that were induced by thermal trauma (Fig. 14.3). Our results showed that thermal trauma-induced oxidative damage in hepatic and renal tissues is ameliorated through the antioxidant effects of EGb.

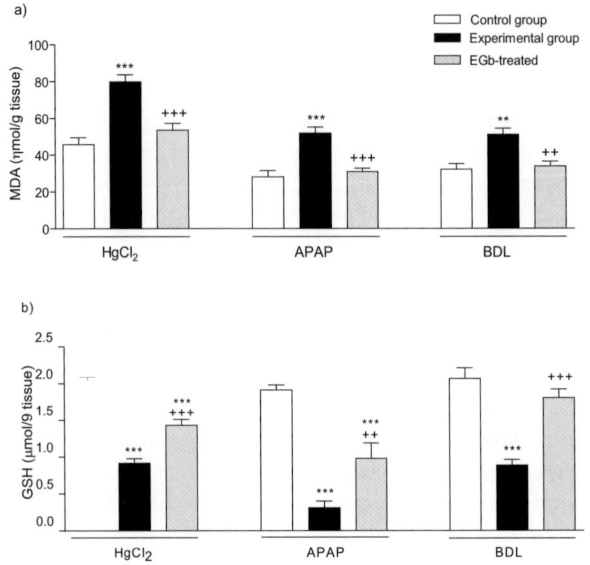

Fig. 14.1. Malondialdehyde (MDA) and glutathione (GSH) levels in the hepatic tissues of rats treated with mercury ($HgCl_2$) or acetaminophen (APAP) or rats with bile duct ligation (BDL), as compared to control rats. $^{**}p < 0.01$, $^{***}p < 0.001$: compared to control group; $^{++}p < 0.01$, $^{+++}p < 0.001$, compared to experimental group (Sener *et al.*, 2005a, 2006b, 2007; reproduced with permission).

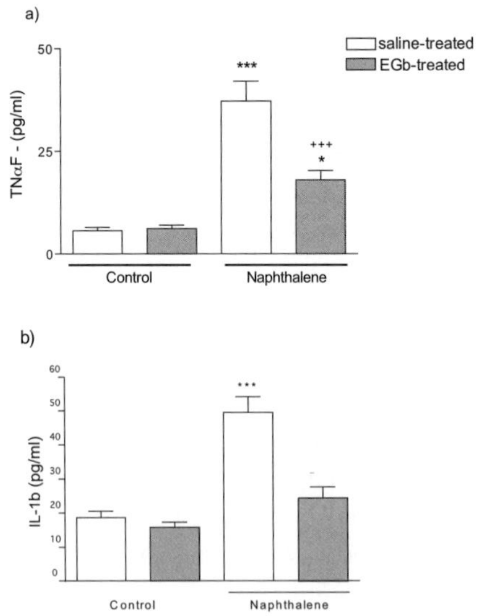

Fig. 14.2. Plasma **a)** Tumour necrosis factor-alpha (TNF-α), and **b)** Interleukin 1-beta (IL-1 β levels of Egb-treated rats with naphthalene toxicity, as compared to saline-treated or control groups. Each group consists of 12 animals. ***: $p < 0.001$, *: $p < 0.05$ versus saline-treated control group; +++: $p < 0.001$ versus saline-treated naphthalene group (Tozan *et al.*, 2007; reproduced with permission).

In a rat model of ischemia-reperfusion-induced renal injury, we have also shown that EGb exerts renoprotective effects (Table 14.1; Fig. 14.4) probably by the radical scavenging and antioxidant activities (Sener *et al.,* 2005b).

Ginkgo biloba against Radiation Injury

Ginkgo biloba has been postulated to possess a therapeutic effect on free radical-induced cancer and an enhancing effect on radiation therapy (Eli and Fasciano, 2006). Lamprouglu *et al.* (2000) have studied the effect of ionizing radiation on learning of rats with irradiation-induced acute encephalopathy. They observed that acute learning dysfunction caused by low-level total-body irradiation was reversed by EGb 761 started at 24 hours of irradiation. As an evidence for EGb's ability to modulate the antioxidant system, Hedayat *et al.* (2005) demonstrated that carrageenan-induced paw oedema was decreased when rats were exposed to irradiation, and the administration of EGb 761 further lessened the severity of this inflammatory response in irradiated rats.

These effects were suggested to be related, in part, to the inhibition MDA production, and partly to augmentation of GSH content in the inflamed paw. Similarly, Ertekin *et al.* (2004) showed that EGb supplementation significantly increased the activities of SOD and GSH-Px enzymes and decreased the MDA levels in the lenses of the rats exposed to a single dose of total cranium irradiation and prevented the generation of cataracts.

In parallel with the studies mentioned above, we have also shown in a rat model of whole-body irradiation (800cGy) that EGb acts as a scavenger of superoxide radicals and modulates cytokine generation and antioxidant system activity (Sener *et al.,* 2006a). Ionizing radiation-induced free oxygen radicals, due to their high reactivity and short lives, are very difficult to measure accurately. However, using luminol- and lucigenin-enhanced chemiluminescence (CL), a simple and reproducible tool that measures the amount of free radical generation, we have shown that the increased tissue CL levels were attenuated by EGb treatment. Moreover, EGb treatment continued for 2 weeks after the irradiation prevented elevations in tissue MDA in the lung, liver, kidney and ileum of irradiated rats. These results suggest that EGb ameliorates ionizing radiation-induced oxidative damage by scavenging the reactive oxygen radicals (Sener *et al.,* 2006a).

Oxidative stress leads to the generation of pro-inflammatory cytokines from activated tissue macrophages and monocytes (Anscher and Vujaskovic, 2005; Kettle and Winterbourn, 1997). In our study with irradiated rats, serum TNF-α levels were increased, indicating the role of this cytokine in radiotherapy-induced toxicity; while in the rats that received daily EGb treatment the TNF-α response was depressed (Sener *et al.,* 2006a) (Fig. 14.5a).

Thus, it seems likely that the amelioration of ionizing radiation-induced oxidative damage by EGb involves the suppression of pro-inflammatory mediators produced by leukocytes and macrophages. In accordance with that, EGb effectively prevented the neutrophil infiltration in the affected tissues, as assessed by elevated MPO activity (Fig. 14.5b,c,d). These results prove that cytoprotective effect of EGb against radiation-induced injury involves its inhibitory effect on tissue neutrophil infiltration and neutrophil-associated TNF-alpha response.

Table 14.1. Plasma TNF- α (pg/ml), and serum lactate dehydrogenase (LDH) (U/l), aspartate aminotransferase (AST), alanine aminotransferase (ALT), blood urea nitrogen (BUN) and creatinine levels in the rats exposed to mercury ($HgCl_2$) or acetaminophen (APAP) toxicity or bile duct ligation (BDL)-, burn- or renal ischemia/reperfusion (I/R)-induced oxidative injury. **NS:** not studied (Sener *et al.,* 2005a,b, 2006b, 2007; Sakarcan *et al.,* 2005; reproduced with permission).

	TNF- α	LDH	AST (U/L)	ALT (U/L)	BUN (U/l)	Creatinine (U/l)
Control	5.7 ± 0.8	2308 ± 70.1	185 ± 5.9	46.2 ± 3.5	23.8 ± 2.3	0.47 ± 0.05
$HgCl_2$	38.3 ± 5.2 ***	4054 ± 115 ***	274.8 ± 38 *	67.3 ± 4.3 **	47.7 ± 4.5 ***	1.20 ± 0.22 **
$HgCl_2$ + EGb	16.3 ± 1.5+++	2969 ± 84.3 +++	195.8 ± 6.5 +	51.2 ± 4.3 +	32.2 ± 2.7 ++	0.65 ± 0.08 +
Control	7.0 ± 0.9		224 ± 13.2	98 ± 5.4		
APAP	31.2 ± 2.2 ***	NS	523 ± 51.9 ***	324 ± 24.5 ***	NS	NS
APAP + EGb	20.5 ± 1.5 +++		361 ± 22.5 ++	178 ± 18.5 +++		
Control	5.0 ± 1.1	1672 ± 240	148 ± 16.6	68 ± 7.3		
BDL	64. 6 ± 6.6 ***	5082 ± 225 ***	300 ± 18 ***	135 ± 9.7 ***	NS	NS
BDL + EGb	21.2 ± 1.6 +++	1763 ± 226 +++	155 ± 12 +++	67 ± 5.7 +++		
Control	6.0 ± 0.5	2149 ± 90	250 ± 12.5	66 ± 3.0	33.4 ± 3.1	0.50 ± 0.03
Burn	32.8 ± 2.4 ***	3603 ± 205 ***	550 ± 72 ***	166 ± 21.7 ***	73.0 ± 4.5 ***	0.65 ± 0.03 *
Burn + EGb	14.4 ± 1.4 +++	2292 ± 145 +++	312 ± 12 ++	101 ± 5.3 ++	45.2 ± 1.7 +++	0.50 ± 0.03 +
Control	3.3 ± 0.6	1426 ± 226			33.3 ± 1.5	0.52 ± 0.03
I/R	15.2 ± 2.1 ***	3520 ± 280 ***	NS	NS	134.3 ± 7.1 ***	1.55 ± 0.12 ***
I/R + EGb	4.8 ± 0.8 +++	1781 ± 185 +++			58.5 ± 3.9 **, +++	0.82 ± 0.07 +++

* $p < 0.05$, ** $p < 0.01$, *** $p < 0.001$: compared to respective control group; + $p < 0.05$, ++ $p < 0.01$, +++ $p < 0.001$: compared to respective oxidative injury group.

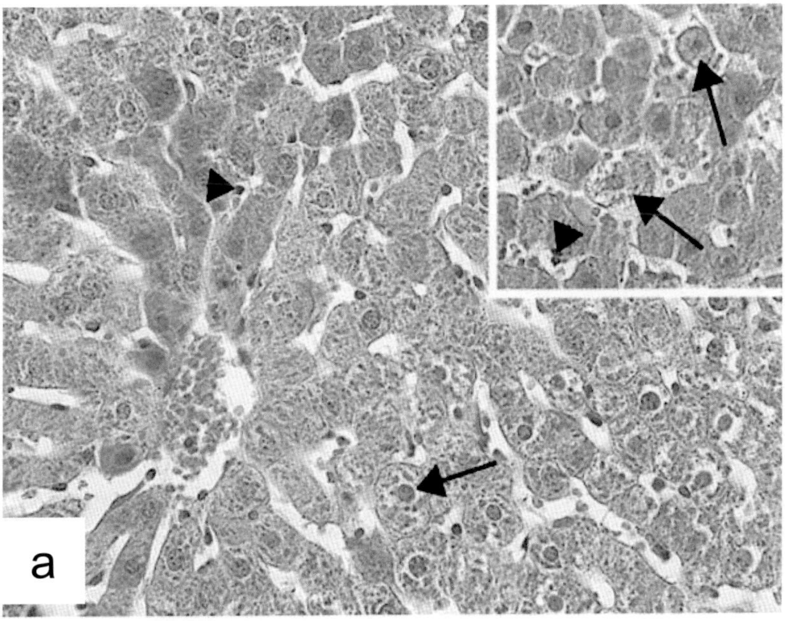
a

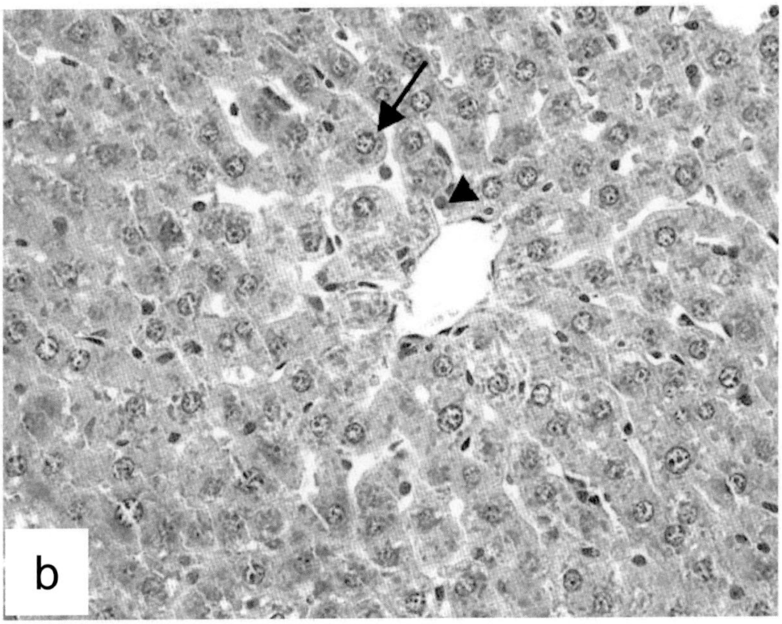
b

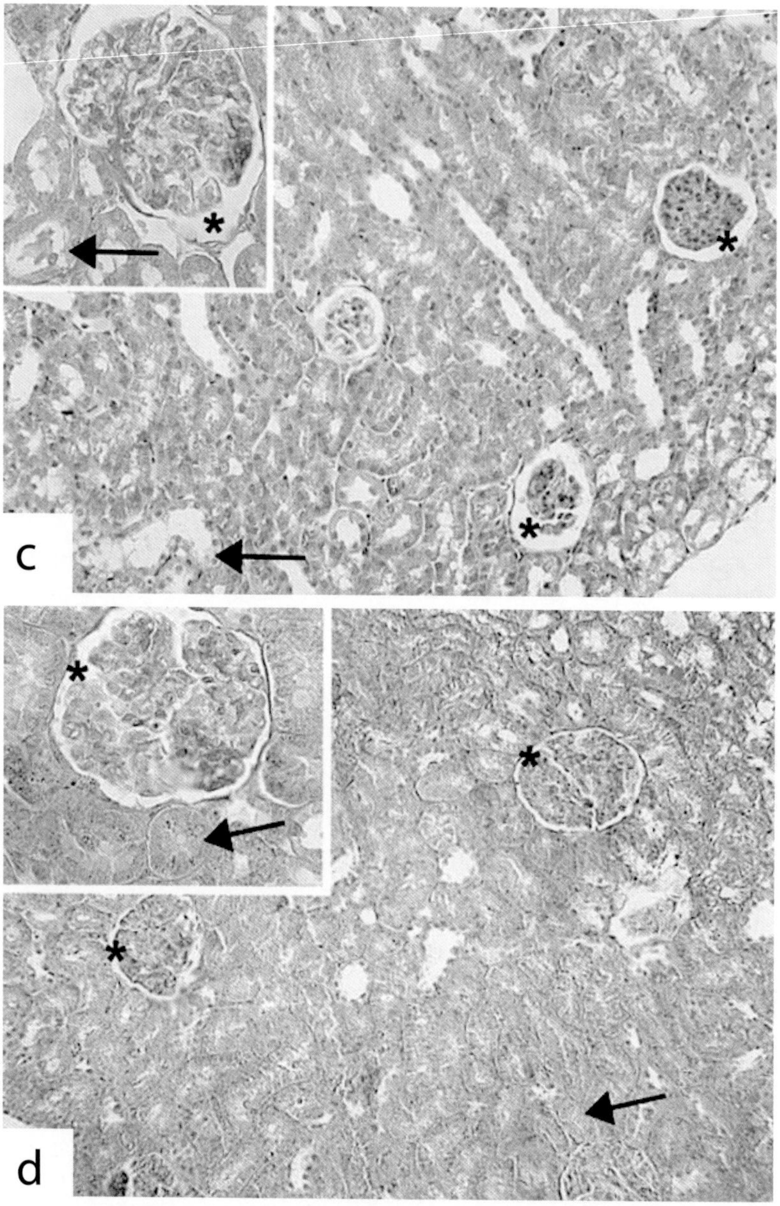

Fig. 14.3. Micrographs of the hepatic (**a**, **b**) and renal tissues (**c**, **d**) in **a**) Burn group; Disorganisation of hepatic cords, degeneration and increased eosinophilia in hepatocytes (→), inflammatory cell infiltration (▶), **b**) Burn + EGb group; Mild sinusoidal congestion, inflammatory cell infiltration (▶) and generally normal hepatocytes (➔). **c**) Burn group; Severe vascular congestion, degeneration in glomerular structures and dilatation in Bowman space (*) and degeneration in proximal tubules (➔), **d**) Burn + EGb group; Mild vascular congestion and normal morphology of glomerular structures and Bowman space (*) and mild degeneration of proximal tubuli (▶) in most regions. HE staining, original magnifications: X200, inserts: X400 (Sakarcan *et al.,* 2005).

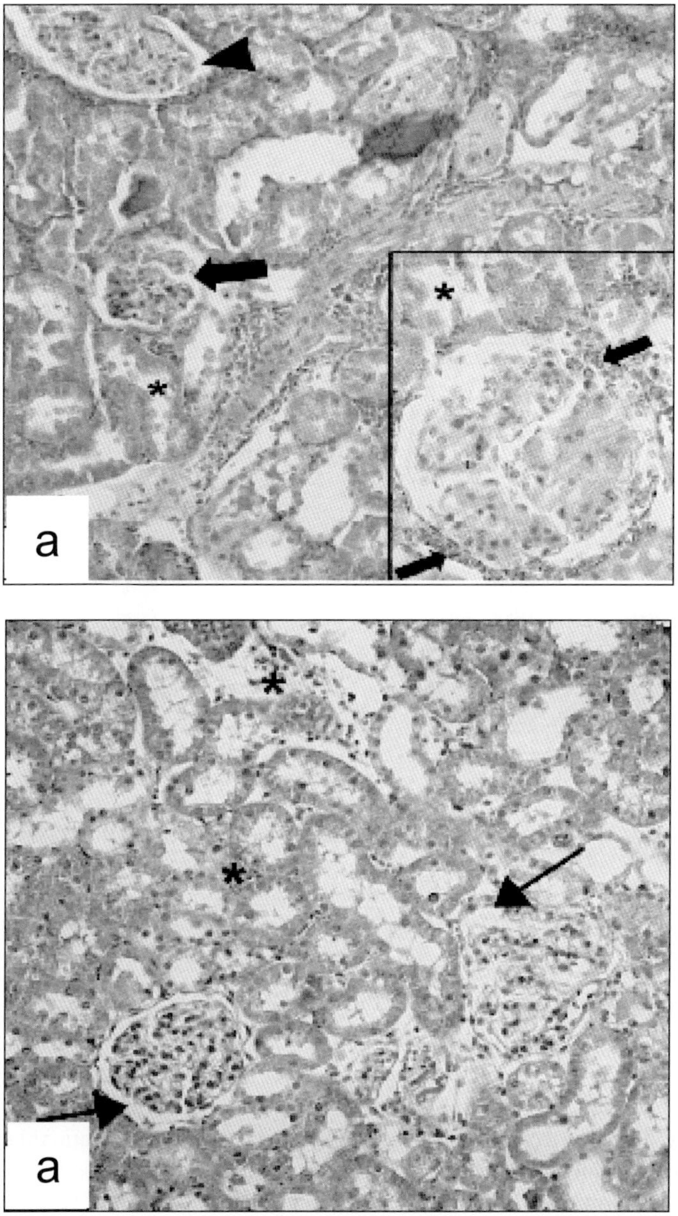

Fig. 14.4. Micrographs of the renal tissue: **a**) I/R group, severe vascular congestion in the interstitium and degenerated glomerulus (➔), urinary space is present but the configuration of glomerulus is changed (arrowhead), cellular debris in the tubulus (*), there is extensive haemorrhage in the urinary space (inset, ➔), **b**) I/R and *Ginkgo biloba* group; the parenchyme maintained its regular morphology but there is still mild haemorrhage in some interstitial spaces (*). Note the regenerating urinary spaces (➔). HE staining, original magnification, X 200, inset X 400 (Sener *et al.,* 2005b; reproduced with permission).

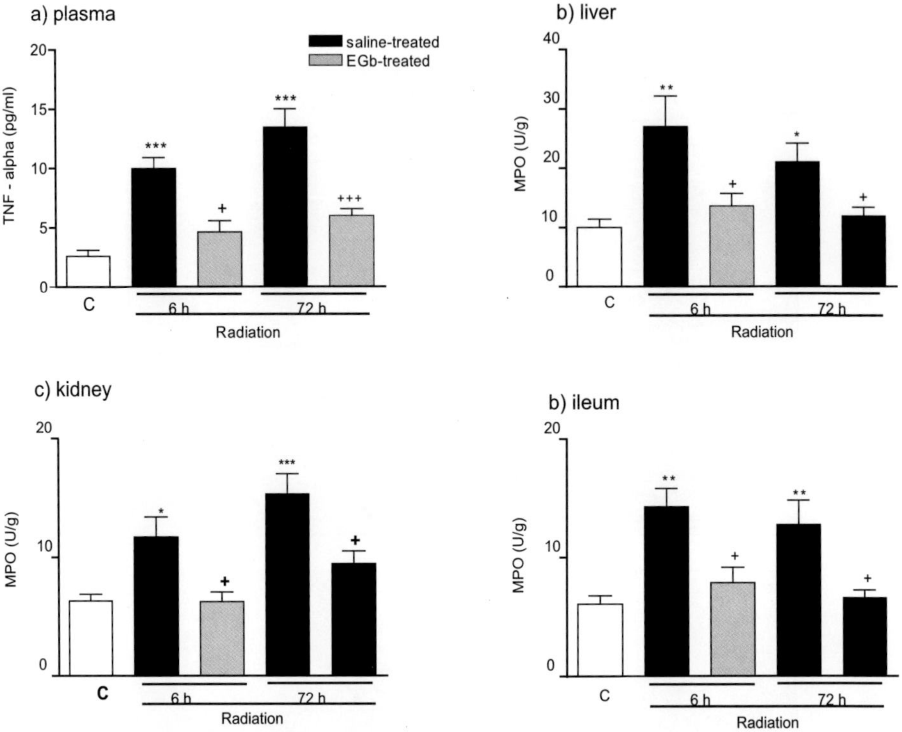

Fig. 14.5. Plasma TNF-α (a) and myeloperoxidase (MPO) activity in the (b) hepatic, (c) renal and (d) ileal tissues of control (C) and vehicle- or EGb-treated groups decapitated at 6 or 72h after irradiation. Each group consists of 6 rats. $^{*}p < 0.05$, $^{**}p < 0.01$, $^{***}p < 0.001$: compared to control group. $^{+}p < 0.05$, $^{+++}p < 0.001$ compared to saline-treated group (Sener *et al.,* 2006a; reproduced with permission).

It is well known that GSH provides major protection against oxidative damage by participating in the cellular defence. Thus, the tissue GSH levels and the activities of glutathione reductase and glutathione peroxidase, which are critical constituents of GSH-redox cycle, are significantly reduced by oxidative stress, permitting enhanced free radical-induced tissue damage (Reiter *et al.,* 2001). EGb, as a promoter of antioxidative enzymes (e.g., superoxide dismutase, glutathione peroxidase and glutathione reductase), appears to be involved in determining GSH homeostasis and the total amount of GSH within the cell (Huang 2005; Ilhan *et al.,* 2004). In our whole-body irradiated rats, the decreased GSH content in the tissues may be due to its consumption during irradiation-induced oxidative stress, while the protection accomplished by EGb treatment involves the maintenance of GSH stores (Sener *et al.,* 2006a).

Accordingly, Lin and Chang (1997) have also suggested that the antioxidant effects of EGb against UVB irradiation are through increased activities of antioxidant enzyme systems.

The risk of irradiation-induced fibrosis occurring in the normal tissues several months to years after radiotherapy remains a major debilitating factor. Fibrosis is a common form of normal tissue damage after exposure to irradiation (Martin *et al.*, 2000), caused by oxidative injury that triggers fibrogenic responses. Thus, fibrosis of the irradiated organs following radiotherapy is inevitable. In our study, collagen contents of the studied tissues were significantly increased at 72h following irradiation, indicating the presence of tissue fibrotic activity, while EGb treatment attenuated the fibrotic activity by its antioxidant properties (Sener *et al.*, 2006a). In parallel with our research, it was recently shown that the leaf extract could interfere with the overproduction of extracellular matrix proteins, by suppressing the production of fibronectin probably through its inhibitory effects on ROS generation (Akiba *et al.*, 2004).

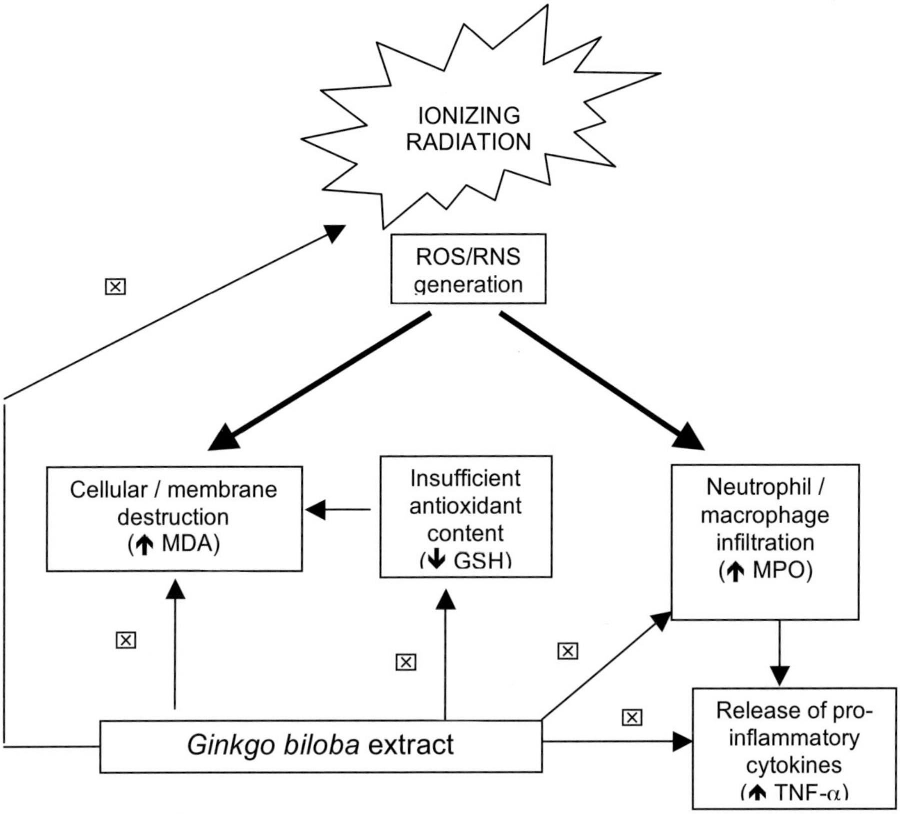

Fig.14.6. The inhibitory effects (☒) of EGB on ionizing radiation-induced and ROS/RNS-mediated oxidative stress.

Experimental data verify that ionizing radiation causes tissue injury by generating oxygen free radicals and by promoting inflammatory processes. On the other hand, EGb has the ability to counteract those effects by inhibiting pro-inflammatory cytokines, restoring cellular antioxidant defence system and, probably, by inducing genes that would switch these systems on (Fig 14.6). Additional experiments and clinical trials in human subjects are required to further elucidate the radioprotective effects of *Ginkgo biloba.* Its future use as an adjuvant in radiotherapy probably lies in its synergistic use in combination with conventional therapies that would potentially enhance the efficacy of the treatment.

References

Agrawal, A., Chandra, D. and Kale, R.K. (2001) Radiation–induced oxidative stress, II studies in liver as a distant organ of tumour bearing mice. *Molecular and Cellular Biochemistry* 224, 9–17.

Ahlemeyer, B. and Krieglstein, J. (2003) Pharmacological studies supporting the therapeutic use of *Ginkgo biloba* extract for Alzheimer's disease. *Pharmacopsychiatry* 36 Suppl 1, S8–14.

Akiba, S., Chiba, M., Mukaida, Y., Tamura, A. and Sato T. (2004) The leaf extract of *Ginkgo biloba* L. suppresses oxidized LDL-stimulated fibronectin production through an antioxidant action in rat mesangial cells. *British Journal of Pharmacology* 142, 419–24.

Anscher, M.S. and Vujaskovic, Z. (2005) Mechanisms and potential targets for prevention and treatment of normal tissue injury after radiation therapy. *Seminars in Oncology* 32(2 Suppl 3), S86–91.

Bacq, Z.M., Herve, A., Lecomte, J., Fischer, P. and Blavier, J. (1951) Protection against x–rays by beta–mercaptoethylamine. *Archives Internationales de Physiologie* 59, 442–447.

Coates, P.J, Lorimore, S.A. and Wright, E.G. (2004) Damaging and protective cell signalling in the untargeted effects of ionizing radiation. *Mutation Research* 568, 5–20.

Dale, W.M., Gray, L.H. and Meredith, W.J. (1949) The inactivation of an enzyme (Carboxypeptidase) by X- and α-radiation. *Philosophical Transactions of the Royal Society of London. A Math Physics, Sciences* 242, 33–62.

DeFeudis, F.V. (2003) A brief history of EGb 761 and its therapeutic uses. *Pharmacopsychiatry* 36 Suppl 1, S2–7.

DeFeudis, F.V. (2005) Free radical scavenging, antioxidant, and related mechanisms of action of EGb761 and its constituents. In: *Ginkgo biloba Extract (EGb761). From Chemistry to the Clinic.* Ullstein Medical Verlagsgesellschaft mbH & Co., Wiesbaden, Germany, pp. 119–133.

DeFeudis, F.V. (2005) Biosynthetic mechanisms and pharmacokinetic studies of EGb761 and its constituents. In: *Ginkgo biloba extract (EGb761). From Chemistry to the Clinic*, Ullstein Medical Verlagsgesellschaft mbH & Co., Wiesbaden, Germany, pp. 17–23.

Delaflotte, S., Auguet, M., DeFeudis, F.V., Baranes, J., Clostre, F., Drieu, K. and Braquet, P. (1984) Endothelium-dependent relaxations of rabbit isolated aorta produced by carbachol and by *Ginkgo biloba* extract. *Biomedica Biochimica Acta* 43, S212–6.

Edwards, J.C., Chapman, D., Cramp, W.A. and Yatvin, M.B. (1984) The effects of ionizing radiation on biomembrane structure and function. *Progress in Biophysics and Molecular Biology* 43, 71–93.

Eli, R. and Fasciano, J.A. (2006) An adjunctive preventive treatment for cancer, ultraviolet light and *Ginkgo biloba*, together with other antioxidants, are a safe and powerful, but largely ignored, treatment option for the prevention of cancer. *Medical Hypotheses* 66, 1152–1156.

Elia, M.C., DeLuca, J.G. and Bradley, M.O. (1991) Significance and measurement of DNA double strand breaks in mammalian cells. *Pharmacology and Therapeutics* 51, 291–327.

Ertekin, M.V., Kocer, I., Karslioglu, I., Taysi, S., Gepdiremen, A., Sezen, O., Balci, E. and Bakan, N. (2004) Effects of oral *Ginkgo biloba* supplementation on cataract formation and oxidative stress occurring in lenses of rats exposed to total cranium radiotherapy. *Japanese Journal of Ophthalmology* 48, 499–450.

Greenstock, C.L. (1993) Radiation and aging, free radical damage, biological response and possible antioxidant intervention. *Medical Hypotheses* 41, 473–482.

Hedayat, I., Salam, O.M. and Baiuomy, A.R. (2005) Effect of *Ginkgo biloba* extract on carrageenan-induced acute local inflammation in gamma irradiated rats. *Pharmazie* 60, 614–619.

Huang, S.Z., Luo, Y.J., Wang, L. and Cai, K.Y. (2005) Effect of *Ginkgo biloba* extract on livers in aged rats. *World Journal of Gastroenterology* 11, 132–135.

Huang, X., Whitworth, C.A. and Rybak, L.P. (2007) *Ginkgo biloba* extract (Egb 761) protects against cisplatin-induced ototoxicity in rats. *Otology & Neurotology* Apr 19 [In press].

Ikushima, T., Aritomi, H. and Morisita, J. (1996) Radioadaptive response, efficient repair of radiation-induced DNA damage in adapted cells. *Mutation Research* 358, 193–198.

Ilhan, A., Gurel, A., Armutcu, F., Kamisli, S., Iraz, M. and Akyol, O. (2004) *Ginkgo biloba* prevents mobile phone-induced oxidative stress in rat brain. *Clinica Chimica Acta* 340, 153–162.

Jagetia, G.C., Malagi, K.J., Baliga, M.S., Venkatesh, P. and Veruva, R.R. (2004) Triphala, an ayurvedic rasayana drug, protects mice against radiation-induced lethality by free-radical scavenging. *Journal of Alternative and Complementary Medicine* 10, 971–978.

Jhun, E., Jhun, B.H., Jones, L.R. and Jung, C.Y. (1991) Direct effects of ionizing radiation on integral membrane proteins. Noncovalent energy transfer requires specific interpeptide interactions. *Journal of Biological Chemistry* 266, 9403–9407.

Kawata, T., Ito, H., George, K., Wu, H. and Cucinotta, F.A. (2004) Chromosome aberrations induced by high-LET radiations. *Biological Science Space* 18, 216–223.

Kettle, A.J. and Winterbourn, C.C. (1997) Myeloperoxidase, a key regulator of neutrophil oxidant production. *Redox Reports* 3, 3–15.

Lamant, V., Mauco, G., Braquet, P., Chap, H. and Douste–Blazy, L. (1987) Inhibition of the metabolism of platelet activating factor (PAF–acether) by three specific antagonists from *Ginkgo biloba*. *Biochemical Pharmacology* 36, 2749–2758.

Lamproglou, I., Boisserie, G., Mazeron, J.J., Bok, B., Baillet, F. and Drieu, K. (2000) Effect of *Ginkgo biloba* extract (EGb 761) on rats in an experimental model of acute encephalopathy after total body irradiation. *Cancer Radiotherapie* 4(3), 202–206.

Lin, S.Y. and Chang, H.P. (1997) Induction of superoxide dismutase and catalase activity in different rat tissues and protection from UVB irradiation after topical application of *Ginkgo biloba* extracts. *Methods and Findings in Experimental and Clinical Pharmacology* 19, 367–371.

Liu, X.P., Goldring, C.E., Copple, I.M., Wang, H.Y., Wei, W., Kitteringham, N.R. and Park, B.K. (2007) Extract of *Ginkgo biloba* induces phase 2 genes through Keap1-Nrf2-ARE signalling pathway. *Life Sciences* 80, 1586–1591.

Maitra, I., Marcocci, L., Droy-Lefaix, M.T. and Packer, L. (1995) Peroxyl radical scavenging activity of *Ginkgo biloba* extract EGb 761. *Biochemical Pharmacology* 49, 1649–1655.

Manda, K., Ueno, M. and Anzai, K. (2007) AFMK, a melatonin metabolite, attenuates X-ray-induced oxidative damage to DNA, proteins and lipids in mice. *Journal of Pineal Research* 42, 386–393.

Mansour, H.H. (2006) Protective role of carnitine ester against radiation–induced oxidative stress in rats. *Pharmacological Research* 54, 165–171.

Martin, M., Lefaix, J.-L. and Delanian, S. (2000) TGF-alpha 1 and radiation fibrosis, a master switch and a specific therapeutic target? *International Journal of Radiation Oncology, Biology and Physics* 47, 277–290.

Naik, S.R. and Panda, V.S. (2007) Antioxidant and hepatoprotective effects of *Ginkgo biloba* phytosomes in carbon tetrachloride-induced liver injury in rodents. *Liver International* 27, 393–399.

Parasassi, T., Giusti, A.M., Gratton, E., Monaco, E., Raimondi, M., Ravagnan, G. and Sapora, O. (1994) Evidence for an increase in water concentration in bilayers after oxidative damage of phospholipids induced by ionizing radiation. *International Journal of Radiation Biology* 65, 329–334.

Patt, H.M., Tyree, E.B., Straube, R.L. and Smith, D.E. (1949) Cysteine protection against x-irradiation. *Science* 110, 213–214.

Pervan, M., Iwamoto, K.S. and McBride, W.H. (2005) Proteasome structures affected by ionizing radiation. *Molecular Cancer Research* 3, 381–390.

Pincemail, J., Dupuis, M., Nasr, C., Hans, P., Haag-Berrurier, M., Anton, R. and Deby, C. (1989) Superoxide anion scavenging effect and superoxide dismutase activity of *Ginkgo biloba* extract. *Experientia* 45, 708–712.

Reiter, R.J., Tan, D.X., Manchester, L.C. and Qi, W. (2001) Biochemical reactivity of melatonin with reactive oxygen and nitrogen species. A review of the evidence. *Cell Biochemistry and Biophysics* 34, 237–256.

Reiter, R.J., Melchiorri, D., Sewerynek, E., Poeggeler, B., Barlow-Walden, L., Chuang, J., Ortiz, G.G. and Acuna-Castroviejo, D. (1995) A review of the evidence supporting melatonin's role as an antioxidant. *Journal of Pineal Research* 18, 1–11.

Riley, P.A. (1994) Free radicals in biology: oxidative stress and the effects of ionizing radiation. *International Journal of Radiation Biology* 65, 27–33.
Rubin, P., Johnston, C.J., Williams, J.P., McDonald, S. and Finkelstein, J.N. (1995) A perpetual cascade of cytokines post irradiation leads to pulmonary fibrosis. *International Journal of Radiation Oncology, Biology and Physics* 33, 99–109.
Sakarcan, A., Sehirli, O., Velioglu-Ovunc, A., Ercan, F., Erkanl, G., Gedik, N. and Sener, G. (2005) *Ginkgo biloba* extract improves oxidative organ damage in a rat model of thermal trauma. *Journal of Burn Care and Rehabilitation* 26, 515–524.
Samuni, A.M. and Barenholz, Y. (1997) Stable nitroxide radicals protect lipid acyl chains from radiation damage. *Free Radical Biology and Medicine* 22, 1165–1174.
Sener, G., Kabasakal, L., Atasoy, B.M., Erzik, C., Velioglu-Ogunc, A., Cetinel, S., Gedik, N. and Yegen, B.C. (2006a) *Ginkgo biloba* extract protects against ionizing radiation-induced oxidative organ damage in rats. *Pharmacological Research* 53, 241–252.
Sener, G., Kabasakal, L., Yuksel, M., Gedik, N. and Alican, I. (2005a) Hepatic fibrosis in biliary-obstructed rats is prevented by *Ginkgo biloba* treatment. *World Journal of Gastroenterology* 11, 5444–5449.
Sener, G., Omurtag, G.Z., Sehirli, O., Tozan, A., Yuksel, M., Ercan, F. and Gedik, N. (2006b) Protective effects of *Ginkgo biloba* against acetaminophen-induced toxicity in mice. *Molecular and Cellular Biochemistry* 283, 39–45.
Sener, G., Sehirli, O., Tozan, A., Velioglu-Ovunc, A., Gedik, N. and Omurtag, G.Z. (2007) *Ginkgo biloba* extract protects against mercury (II)-induced oxidative tissue damage in rats. *Food and Chemical Toxicology* 45, 543–550.
Sener, G., Sener, E., Sehirli, O., Oğünç, A.V., Cetinel, S., Gedik, N. and Sakarcan, A. (2005b) *Ginkgo biloba* extract ameliorates ischemia reperfusion-induced renal injury in rats. *Pharmacological Research* 52, 216–222.
Shen, J.G. and Zhou, D.Y. (1995) Efficiency of *Ginkgo biloba* extract (EGb 761) in antioxidant protection against myocardial ischemia and reperfusion injury. *Biochemistry and Molecular Biology International* 35, 125–134.
Srinivasan, M., Sudheer, A.R., Pillai, K.R., Kumar, P.R., Sudhakaran, P.R. and Menon, V.P. (2007) Lycopene as a natural protector against gamma radiation-induced DNA damage, lipid peroxidation and antioxidant status in primary culture of isolated rat hepatocytes *in vitro*. *Biochimica et Biophysica Acta* 1770, 659–665.
Srinivasan, M., Sudheer, A.R., Pillai, K.R., Kumar, P.R., Sudhakaran, P.R. and Menon, V.P. (2007) Influence of ferulic acid on gamma-radiation-induced DNA damage, lipid peroxidation and antioxidant status in primary culture of isolated rat hepatocytes. *Toxicology* 228, 249–258.
Stark, G. (2005) Functional consequences of oxidative membrane damage. *Journal of Membrane Biology* 205, 1–16.
Tchou, J. and Grollman, A.P. (1993) Repair of DNA containing the oxidatively-damaged base, 8-oxoguanine. *Mutation Research* 299, 277–287.
Tozan, A., Sehirli, O., Omurtag, G.Z., Cetinel, S., Gedik, N. and Sener, G. (2007) *Ginkgo biloba* extract reduces naphthalene-induced oxidative damage in mice. *Phytotherapy Research* 21, 72–77.
Valko, M., Leibfritz, D., Moncol, J., Cronin, M.T., Mazur, M. and Telser, J. (2007) Free radicals and antioxidants in normal physiological functions and human disease. *International Journal of Biochemistry and Cell Biology* 39, 44–84.
Valko, M., Rhodes, C.J., Moncol, J., Izakovic, M. and Mazur, M. (2006) Free radicals, metals and antioxidants in oxidative stress-induced cancer. *Chemico-Biological Interactions* 160, 1–40.
VanBeek, T.A. (2005) Ginkgolides and bilobalide, their physical, chromatographic and spectroscopic properties. *Bioorganic and Medicinal Chemistry* 13, 5001–5012.
Verma, S.P. and Sonwalkar, N. (1991) Structural changes in plasma membranes prepared from irradiated Chinese hamster V79 cells as revealed by Raman spectroscopy. *Radiation Research* 126, 27–35.
Vijayalaxmi, Reiter, R.J., Sewerynek, E., Poeggeler, B., Leal, B.Z. and Meltz, M.L. (1995) Marked

reduction of radiation-induced micronuclei in human blood lymphocytes pre-treated with melatonin. *Radiation Research* 143, 102–106.

Ward, J.F. (1988) DNA damage produced by ionizing radiation in mammalian cells, identities, mechanisms of formation, and reparability. *Progress in Nucleic Acid Research and Molecular Biology* 35, 95–125.

Yoshikawa, T., Naito, Y. and Kondo, M. (1999) *Ginkgo biloba* leaf extract, review of biological actions and clinical applications. *Antioxidants and Redox Signalling* 1, 469–480.

Zhang, W.R., Hayashi, T., Kitagawa, H., Sasaki, C., Sakai, K., Warita, H., Wang, J.M., Shiro, Y., Uchida, M. and Abe, K. (2000) Protective effect of ginkgo extract on rat brain with transient middle cerebral artery occlusion. *Neurological Research* 22, 517–521.

Zhou, Y.H., Yu, J.P., Liu, Y.F., Teng, X.J., Ming, M., Lv, P., An, P., Liu, S.Q. and Yu, H.G. (2006) Effects of *Ginkgo biloba* extract on inflammatory mediators (SOD, MDA, TNF-alpha, NF-kappa β, p65, IL-6) in TNBS-induced colitis in rats. *Mediators of Inflammation* (5), 92642 .

Chapter 15

Novel Strategies for Protecting Mitochondria (the Cellular Powerhouse) Against Low-LET Radiation: A Review

Damodar Gupta, Rajesh Arora, Rajiv R. Mohan, Alexandru Almasan, Andrei V. Gudkov and Roger M. Macklis

Introduction

The use of ionizing radiations and radioisotopes in medicine, radiation therapy, agriculture and industry has substantially enhanced the probability of accidental exposure of human populations. Space exploration, reactor accidents, terror incidents and other strategic nuclear eventualities are no longer hypothetical problems/activities but are very much inevitable realities. In view of this, development of suitable agents for radioprotection of human beings has become an inevitable requirement.

Radiation: Physical and Chemical Aspects

Radiation consists of energetic particles or electromagnetic waves. Low linear energy transfer (LET) radiations, such as X- and γ-rays are high-energy photons that have no mass or charge and thus are highly penetrating in nature as compared to high LET radiation (α, β, proton, neutron, electrons). High LET radiation causes damage to bio-molecules directly, whereas low LET radiation causes damage by both direct and indirect effects (Conklin and Walker, 1987; Hall and Giaccia, 2006; Prasad, 1995). Its indirect effect is mainly by inducing generation of reactive oxygen species (ROS), such as superoxide radicals ($^{-}O_2$), hydrogen peroxide (H_2O_2), hydroxyl radical ($^{\bullet}OH$), nitric oxide ($^{\bullet}NO$) and peroxynitrite ($NOO^{\bullet}$) (Lyng *et al.*, 2001; Schmidt-Ulrich *et al.*, 2000) (Table 15.1). Radiation brings about changes in bio-molecules, such as DNA, lipids, proteins and carbohydrates, which eventually result in manifestation of deleterious effects, such as mutations leading to cancer, aging, neurodegenerative disorders, cardiac dysfunction etc. (Sun, 1990; Finkel and Holbrook, 2000). Exposure of mammals to different doses of radiation results in manifestation of different syndromes, such as haematopoietic, gastrointestinal and central nervous system syndromes (Conklin and Walker, 1987; Prasad, 1995; Bump and Malaker, 1998).

Table 15.1. Properties of free radicals and their cellular localization.

Species	Symbol	Half life at 37°C (secs)	Cellular localization	Properties and comments
Singlet oxygen	$^{\cdot}O_2$	1x 10^{-6}	Chloroplast, thylakoid, mitochondria, peroxisome	Powerful oxidizing agent, membrane impermeable and cell signalling
Super oxide anion	$^{\cdot}O_2$	1x 10^{-6}	Chloroplast, thylakoid, mitochondria, apoplast, peroxisome, membrane bound oxidases	Good reductant and oxidant, membrane impermeable, interacts with NO·
Hydrogen peroxide	H_2O_2	Stable	Chloroplast, thylakoid, mitochondria, apoplast, peroxisome	Lipophilic, strong oxidant, signalling molecule
Hydroxyl	·OH	1x 10^{-9}	Chloroplast, thylakoid, mitochondria, apoplast, peroxisome	Very strong oxidant, product of radiolysis/ metal catalyzing reaction, very low diffusion distance
Alkoxyl	RO·	1x 10^{-6}	Chloroplast, thylakoid, mitochondria, apoplast, peroxisome	Intermediate in their reactivity with lipid between ·OH and ROO
Peroxyl	ROO·	1x 10^{-6}	Chloroplast, thylakoid, mitochondria, apoplast, peroxisome	Low oxidizing capacity as compared to ·OH but great diffusion potential
Nitric oxide	NO·	1x 10^{-6}	Mitochondria, cytosol, peroxisome, apoplast	Important messenger and powerful oxidizing agent, can react with $^{\cdot}O_2$ SH-groups of proteins
Peroxynitril	ONOO··	1x 10^{-9}	Peroxisome, cytosol, mitochodria, apoplast	Very strong oxidant, very low diffusion distances
Hypochlorous acid	HOCl	1x 10^{-6}	Immune system cells (peroxisomes)	Very strong oxidant, helps in killing micro-organisms

Radiation Effects on Bio-molecules

At the cellular level, the radiation response is manifested in the form of irreversible changes, such as mutations leading to malignant transformation, development of abnormal cell forms, and even cell death, or as minor reversible structural changes and functional alterations in various biological processes (Ramakrishnan *et al.,* 1993; Somosy, 2000). The presence of high water content and oxygen tension in a typical cell enhances the frequency of radiation-mediated damage to bio-molecules by the generation of ROS (Zhang *et al.,* 1995, 1998a). Oxygen, an important electronegative molecule, which is utilized by the cell during oxidative phosphorylation, plays an important role in radiation-induced oxidative alterations in bio-molecules (Table 15.2) (Conklin and Walker, 1987; Bump and Malaker, 1998; Prasad, 1995; Hall and Giaccia, 2006). An important metabolic effect of the DNA damage is the induction of poly-adenosine diphosphate ribose synthesis (ADP-ribosylation) in nuclei, resulting in extensive depletion of cellular NADH pools, which directly affects energy production in mitochondria (Singh, 2006). ADP-ribosylation has been associated with repair of damaged DNA.

The lipid peroxidation is a self-propagating chain reaction in which the initial lesion on a lipid molecule is significantly amplified in the presence of molecular oxygen, and redox-active transition metals (Fe^{+++}, Fe^{++}, Cu^{+}, Cu^{++} etc). The oxidative modification of poly-unsaturated fatty acids results in impairment of fluidity and elasticity of membranes (Sandstrom *et al.,* 1995). The product(s) of lipid peroxidation is a known mutagen and also reacts with proteins and DNA (Bump and Malaker, 1998; Conklin and Walker, 1987; Hall and Giaccia, 2006; Prasad, 1995).

Biomolecule	Product	Comments
DNA Damage to bases Damage to sugar	8-Oxoguanine, 8-hydroxyadenine, 2-hydroxyadenine, 5-hydroxycytosine, 5-hydroxy uracil, uracil glycol, 5,6-dihydrouracil, 5,6-dihydrothymine, 5-formyluracil 1-malondialdehyde, sugar peroxyl radical, 5-methylene-2-furanone, nucleotide aldehyde	Oxidative modification of bases may lead to mutation (transition or transversion)
Lipids	Lipid hydroperoxides (LOOH), malondialdehyde (MDA), HNE, acrolein, crotonaldehyde, HNE via 2,3-epoxy-4-hydroxynonanal, glyoxal	Alters homeostasis of cells and known to induce mutation, cancer, cell death by cross-linking with DNA. Chain reaction of free radical generation
Proteins	Protein carbonyls, protein hydroperoxides	Inter and intra molecule cross linking, peptide fragments, presence of metal are major source of ROS generation, generates signalling molecule
Carbohydrates	Hyaluronic acid etc., oxidation of sugar (DNA, RNA etc.)	Chain reaction and generates signalling molecules

Table 15.2. Products of oxidative modification of biomolecules.

Effect of Radiation on Cell Organelles

The radiation-induced alterations in the supra-molecular organization and function of cells or cell organelles play a significant role in the development of acute radiation injury (Somosy, 2000; Szumiel, 1994). Both irreversible and reversible damage to cells are manifested at the sub-cellular level as structural and functional changes in all organelles including plasma membrane, cytoskeletal system, endoplasmic reticulum (ER), Golgi complex (Kubasova *et al.,* 1976), lysosome (Brandes *et al.,* 1967; Snyder and Eklund, 1978) and mitochondria (Christozova *et al.,* 1977; Snyder and Eklund, 1978; Somosy *et al.,* 1985, 1987) at cellular and inter-cellular level *viz.*, gap junction (Ishii and Watanabe, 1996; Trosko, 1996).

Membrane integrity is essential for maintenance of cell viability. Higher doses of radiation (30–50Gy) are known to produce substantial changes in cellular membranes (Koteles and Somosy, 2001; Ojeda *et al.,* 1994). The radiation sensitivity of cellular membranes depends on the amount of unsaturated lipids, metalloproteins and membrane bound antioxidants. The radiation-induced (1–3Gy) morphological and functional changes on the gap junction lead to altered cell communication (Ishii and Watanabe, 1996). Biological membranes are closely associated with the cytoskeletal element and exposure to 0.5–2.0Gy radiation disrupts the cytoskeleton and Ca^{++} ion homeostasis in the mitochondria and endoplasmic reticulum (ER) and/or alters phosphorylation/dephosphorylation status of proteins (Kantak *et al.,* 1993). The ionizing radiation causes dilation, vascularization, fragmentation of cisternae and degranulation of ER.

Mitochondria: The Powerhouse of the Cell is Adversely Affected by Ionizing Radiation

ROS are continuously produced in cells and mitochondria (Table 15.3) during normal physiological conditions, however, their concentration is greatly increased following irradiation leading to severe oxidative damage (Somosy, 2000). As a consequence, alterations in the structure (e.g., elongation, branching, reversible increase in size, development of giant form, vacuolization) and function of mitochondria are common manifestations during and after irradiation (Kergonou *et al.,* 1981b; Matsko *et al.,* 2001). The structural changes in molecular organization of inner and outer membranes of mitochondria result in opening of membrane permeability transition pores (MPTP), thereby uncoupling of the oxidative phosphorylation leading to enhanced generation of ROS by the electron transport chain (ETC) components (Gong *et al.,* 1998; Rafique *et al.,* 2001). Under conditions of extreme stress mitochondria undergo autocatalytic collapse, associated with the loss of the normal membrane potential, uncoupling of ETC and/or release of cytochrome c and thereby resulting in enhanced generation of ROS, the depletion of redox equivalent (e.g., NADPH, GSH) and a complete failure of ATP production. The decline in cellular energy production impairs normal physiological function and compromises the cellular ability to adapt to various physiological stresses (Gong *et al.,* 1998; Skulachev, 1999). The opening of MPTP is irreversible and is normally a prelude to cell death (Hancock *et al.,* 2001). Apoptosis is tightly regulated at numerous control points along the cascade. Key effectors include: (i) the tumour suppressor protein p53, which induces cells to undergo apoptosis when irreparable DNA damage is detected, and (ii) the proto-oncogene Bcl-2 family proteins, some of which prevent the permeability transition, suppress apoptosis and potentially allow the survival of damaged or cancerous cells, while other family members (Bax, Bcl-xl, PUMA, NOXA) have a pro-apoptotic role. Bcl-2 is concentrated in the mitochondrial outer membrane, where it is closely involved in regulating the MPTP. In brief, oxidative damage to mitochondrial DNA, lipids or proteins is the main cause of loss of electrochemical gradient, matrix components and enhanced generation of ROS and bio-energetic catastrophe.

Table 15.3. Enzymatic and non-enzymatic sources of ROS in mitochondria.

Name of enzyme(s)	Location	Catalysis	Free radical	Reference
Cytochrome b5 reductase	Mitochondria (outer membrane)	Oxidation of cytoplasmic NADPH by reduction of cytochrome b5	$^{\cdot}O_2$	Lee *et al.*, 2001; Nishino and Ito, 1986
Monoamine oxidase	Mitochondria (outer membrane)	Amine groups	H_2O_2	Hauptmann *et al.*, 1996; Simonson *et al.*, 1993
Dihydroorotate dehydrogenase	Mitochondria (cytosolic site of inner-mitochondrial membrane)	Dihydroorotate to pyrimidine base and orotatate	$^{\cdot}O_2$, H_2O_2	Forman and Kennedy, 1975; Loffler *et al.*, 1996
Glycerol-3-phosphate dehydrogenase	Mitochondria (cytosolic site of inner-mitochondrial membrane)	Oxidation of glycerol 3-phosphate to dihydroxy acetone phosphate using coenzyme Q	H_2O_2	Kwong and Sohal, 1998
Succinate dehydrogenase complex	Mitochondria (matrix site of inner-mitochondrial membrane)	Oxidation of succinate to fumarate using coenzyme Q	$^{\cdot}O_2$	McLennan and Degli Esposti, 2000; Zhang *et al.*, 1998b
m-aconitase	Mitochondria (matrix)	Citrate to isocitrate	$^{\cdot}O_2$, H_2O_2	Vasquez-Vivar, *et al.*, 2000
2-Oxoglutarate dehydrogenase	Mitochondria (matrix)	Ketoglutarate to succinyl CoA using NAD+	$^{\cdot}O_2$, H_2O_2	Maas and Bisswanger, 1990; Starkov *et al.*, 2004
NADH-Ubiquinone oxidoreductase complex	Mitochondria (inner membrane)	Oxidation of NADH and reduction of Coenzyme Q	$^{\cdot}O_2$	Kwong and Sohal, 2000; McLennan and Degli Esposti, 2000
Cytochrome c reductase	Mitochondria (inner membrane)	Oxidation of Coenzyme Q and reduction of cytochrome c	$^{\cdot}O_2$	Genova *et al.*, 2004; Zhang *et al.*, 1998b
Cytochrome c oxidase	Mitochondria (inner membrane)	Oxidation of cytochrome c and reduction of molecular oxygen	$^{\cdot}O_2$, H_2O_2	Kowaltowski *et al.*, 2001; Kwong and Sohal, 2000
Q-cycle	Mitochondria (inner membrane	Oxidation of QH2 and reduction of cytochrome c	$^{\cdot}O_2$	Genova *et al.*, 2003
Nitric oxide synthease (NOS1, 2 & 3)	Wide distribution	Catalyzes five electron oxidation of L-arginine to form L-Citrulline and nitric oxide	NO	Garcia-Mediavilla *et al.*, 2007; Kim, *et al.*, 2001
Fenton and Haber Weiss reaction	Wide distribution of transitional metals in cells and proteins	Iron or metal catalyzed reaction	$^{\cdot}OH$	M'Bemba-Meka *et al.*, 2007; Zhang, *et al.*, 1990
Indirect reactions	Wide distribution	Superoxide + NO	NOO-	

Protecting Mitochondria for Achieving Radioprotection

The investigation of acute and long-term effects of ionizing radiation on cells and tissues has an exceptional importance in diagnosis, radiotherapy, agriculture, industrial research, energy production etc., as well as in the management of accidental radiation exposures. Therefore, the activities of numerous investigators world-wide have been directed towards finding effective radioprotective means. The mitigation of radiation damage to a cell depends on several physical and biochemical factors, including radiation dose, dose rate, temperature, atmospheric pressure, pH, intracellular oxygen tension, water content, DNA content, metabolic status of cell, cell cycle phase and differentiation (Bump and Malaker, 1998; Conklin and Walker, 1987; Nair *et al.,* 2001; Prasad, 1995).

Research into development of radioprotectors began in 1948 with the discovery that cysteine protects mice against X-rays. Among the large number of compounds, the most thoroughly examined were cysteamine derivatives, i.e., compounds known as aminothiol radioprotectors. Some compounds that have been evaluated as radioprotectors are aminothiols, aminodisulfides, thiourea derivatives, thiosulfur and thiophosphoric acid, dithiocarbamates, thiazole, some biogen amines and their derivatives. Subsequently a major initiative for development of radioprotectors was undertaken at the Walter and Reed Army Research Institute (USA), where nearly 5000 compounds were screened and WR-2721 (amifostine) was found to be the most promising agent, which forms the active metabolite WR-1065 by dephosphorylation *in vivo*. WR 1065 can activate the redox-sensitive nuclear transcription factor ϰB (NFϰB) and elevate the expression of the antioxidant gene manganese superoxide dismutase (MnSOD) in human microvascular endothelial cells, which is strictly localized in mitochondria and thereby protects cells from lethal doses of radiation. The inhibition of NFϰB-mediated elevation of MnSOD by helenalin lowers the radioprotective potential of WR-1065 (Murley *et al.,* 2006). Antioxidant gene therapy using MnSOD plasmid liposomes has been shown to provide organ-specific radiation protection associated with delay or prevention of acute and late toxicity and also proven by the MnSOD transgene expression in cells of the organ microenvironment, which contributes significantly to the mechanism of protection (Epperly *et al.,* 2007; Greenberger *et al.,* 2003).

Phytoextracts protect mitochondria against ionizing radiation

The intrinsic mitochondrial components, such as quinols, β-carotene (Burton *et al.,* 1982), cytochrome c, vitamin E (Burton *et al.,* 1982; Konings and Drijver, 1979) and GSH (Korotchkina *et al.,* 2001) under normal conditions reduce oxidative damage but fail to cope with the additional oxidative burden induced by irradiation (Sandstrom *et al.,* 1995; Singh, 2006). Isolated natural compounds, such as chlorophyllin (Boloor *et al.,* 2000), caffeine (Kamat *et al.,* 2000a) and crude extracts of plants, such as *Asparagus racemosus* (Kamat *et al.,* 2000b), *Podophyllum hexandrum* (Gupta *et al.,* 2003, 2004), *Hippophae rhamnoides* (Goel *et al.,* 2005) have been reported to enhance levels of antioxidants and inhibit radiation damage to mitochondria. Chlorophyllin, nicotinamide and caffeine have been examined for their ability to inhibit radiation-induced damage to mitochondrial lipids and proteins and scavenging of hydroxyl radicals (Boloor *et al.,* 2000; Kamat *et al.,* 2000b; Kamat and Devasagayam, 1999). In contrast, crude herbal preparations from several medicinal plants have been shown to exhibit radiomodifying properties via multiple mechanisms *viz.*, induction of cell cycle arrest, enhancement of DNA repair and antioxidant profile and inhibition of oxidative modifications of bio-molecules. Crude extracts of *Asparagus*, *Podophyllum* (RP-1) and *Hippophae* (RH-3) have also been shown to protect cells by protection or inhibition of mitochondrial activity. Antioxidant potential of *Asparagus* extract lowers formation of lipid hydroperoxide (LOOH), thiobarbituric acid reactive substances (TBARS) and protein carbonyls (PC), which are markers of oxidative modification of lipids and proteins, respectively. Gupta *et al.* (2003a,b) reported that *Podophyllum* and *Hippophae* play a vital role in affording radiation protection both *in vitro* and *in vivo* via inhibition of radiation damage to mitochondria.

Mitochondria, being the powerhouse of the cell, are one of the most important targets of radiation because: (i) they are the site of oxidative phosphorylation and generation of ROS, (ii) they contain a high amount of unsaturated lipids and the Fenton catalysts, (iii) they possess naked DNA (DNA without histone/non-histone proteins) and inefficient DNA repair mechanism, (iv) they possess limited antioxidants, and (v) the ETC in mitochondria since it is imperfect, electron leakage and superoxide anion generation occurs, which further causes damage to the bio-molecules resulting in a vicious circle. Once the radiation damage is inflicted, repair and restoration of normal metabolic activities acquires great significance. Gupta and co-workers showed protection of HepG2 cells by RP-1 to 2Gy γ-radiation (Gupta *et al.*, 2003a, 2004). Further studies on cellular and mitochondrial physiology confirmed the temporary alterations in mitochondrial respiration by inhibition of oxidation of NAD and FAD-linked substrate, which is potentially important in affording radiation protection by RP-1 components *viz.*, podophyllotoxin, lignans, polyphenolic compounds etc. Radiation-induced lipid peroxidation and damage to ETC components leads to leakage of electrons, which increases the generation of ROS and enhances mitochondrial membrane potential (Gong *et al.*, 1998; Lyng *et al.*, 2001), leading to bio-energetic catastrophe. The transfer of electrons through ETC components, coupled with shuttling of protons across the inner-mitochondrial membranes, generates potential gradient across the membrane. Lowering of complex I and complex II activity (oxidation of NAD and FAD-linked substrate) results in decreased membrane potential and ROS generation. Pre-irradiation administration of RP-1 and RH-3 to mice decreased the levels of superoxide anions, which demonstrated its antioxidant potential, and its ability to increase levels of antioxidants and decrease oxidative damage to mitochondrial lipids and proteins. These properties of the phytoextracts further decreased leakage of electrons from ETC components (complex I to III and complex II to III, complex III to IV), thereby lowering the radiation-mediated alterations in MMP *in vivo* (Gupta *et al.*, 2004; Goel *et al.*, 2005).

Mitochondrial proteins and lipids are susceptible to oxidative damage because of the presence of high amounts of metal catalysts (Kergonou *et al.*, 1981a, Shigenaga *et al.*, 1994). Oxidative damage to proteins and lipids alters flow of electrons through ETC via change in conformation of proteins, fragmentation, aggregation, formation of cross linkages in the polypeptide chain, and decreases fluidity of membranes, which further amplify the generation of superoxide anions. The decrease of ROS generation and elevation of antioxidant and ATP levels by RP-1 and RH-3 helped cells to recover from radiation-mediated damage to mitochondria and thereby these may be important mechanisms to protect cells.

Results obtained with RP-1 and RH-3 suggested that these herbal preparations act in a multifaceted manner, affording radioprotection to mitochondria *in vivo*, and can also be used in the prevention and treatment of several free-radical ailments associated with oxidative stress.

Recently, a flavonoid-rich fraction of *Hippophae rhamnoides* (REC-1001), with potent antioxidant properties, has been shown to protect mtDNA from ionizing radiation (Shukla *et al.,* 2006; Chawla *et al.,* 2007; Chawla, 2007).

REC-2001, a fractionated extract of *Podophyllum hexandrum* has been shown to selectively eliminate cells containing deletions in mtDNA (Shukla, 2007). The sensitivity of cells with mtDNA deletion towards apoptosis by *P. hexandrum* can be explained by considering the fact that all mitochondrial DNA mutations causing generalized oxidative phosphorylation disturbance by inhibition of mitochondrial protein biosynthesis sensitize proliferating cells against apoptotic stimuli. This apparently leads to enhanced superoxide generation by complex I and III, which further leads to consumption/exhaustion of mitochondrial antioxidants e.g., glutathione (GSH) causing oxidative stress via release of metal ions from mitochondria.

Concluding Remarks and Future Directions

Several herbals render radioprotection by protecting the powerhouse of the cell. The diverse mechanisms by which herbal drugs protect mitochondria against the damaging effects of γ-radiation include inhibition of radiation-mediated alteration in mitochondrial membranes, specifically ETC complexes and consequent stabilization of radiation-induced changes in MMP, up-regulation of the neo-synthesis of both molecular (glutathione) and enzymatic antioxidants, decreased generation of superoxide anions, and inhibition of oxidative modification of mitochondrial lipids and proteins. However, most herbal preparations that are known to have radioprotective or radiomodifying effects, elicit their response by scavenging free radicals and initiating the inherent antioxidant defense system. The advantage of scavenging free radicals is that it reduces the effects of ROS, e.g., disruption of membrane integrity, disturbances in energy production and Ca^{++} homeostasis. Attention needs to be focused in the following areas in future *viz.*, devising of less toxic and more effective radiomodulating agents that can protect mitochondria and other cellular organelles, enhancement of biological activity of these herbal agents and development of rapid low-cost bioassays for evaluation of radimodulating efficacy, evaluation of the effects of herbal drugs on gene and protein expression and other molecular pathways operating in the mitochondria and the cytoplasm, utilizing modern technologies. Very limited work has been done in this context, and considering the tremendous radioprotective potential of the phytodiversity available in various parts of the world, this area is worth investigating. Understanding the intricate mechanisms of action that operate during radioprotection could lead to development of safe herbal radioprotectors for human application, and these could be used for the management of planned radiation exposures, especially during rescue missions and radiotherapy.

References

Boloor, K.K., Kamat, J.P. and Devasagayam, T.P.A. (2000) Chlorophyllin as a protector of mitochondrial membranes against gamma-radiation and photosensitization. *Toxicology* 155(1–3), 63–71.

Brandes, D., Sloan, K.W., Anton, E. and Bloedorn, F. (1967) The effect of X-irradiation on the lysosomes of mouse mammary gland carcinomas. *Cancer Research* 27(4), 731–746.

Bump, E.A. and Malaker, K. (1998) *Radioprotectors: Chemical, Biological and Clinical Perspectives*, CRC Press, Florida, USA, p. 431.

Burton, G.W., Joyce, A. and Ingold, K.U. (1982) First proof that vitamin E is major lipid–soluble, chain-breaking antioxidant in human blood plasma. *Lancet* 2(8293), 327.

Chawla, R. (2007) Radiation Response in Mammalian Model System and its Modulation by Certain Natural Plant Product(s). *PhD Thesis*. Institute of Nuclear Medicine and Allied Sciences and Jamia Hamdard, New Delhi, India, p. 242.

Chawla, R., Arora, R., Singh, S, Sagar, R.K., Sharma, R.K., Kumar, R., Sharma, A., Gupta, M.L., Singh, S., Prasad, J., Khan, H.A., Swaroop, A., Sinha, A.K., Gupta, A.K., Tripathi, R.P. and Ahuja, P.S. (2007) Radioprotective and antioxidant activity of fractionated extracts of berries of *Hippophae rhamnoides*. *Journal of Medicinal Food* 10(1), 101–109.

Christozova, A., Tileva, M.D. and Dragiev, M. (1977) Morphometric electron microscopy study of changes in endometrial epithelial cells after irradiation with a dose of 1,000 rads. *Folia Morphologica* 25(3), 289–294.

Conklin, J. and Walker, R.I. (1987) Military Radiobiology: A Perspective. In: Conklin, J.J. and Walker, R.I . (eds.) *Military Radiobiology*, Academic Press, Orlando, USA.

Epperly, M.W., Wegner, R., Kanai, A.J., Kagan, V., Greenberger, E.E., Nie, S. and Greenberger, J.S. (2007) Effects of MnSOD-plasmid liposome gene therapy on antioxidant levels in irradiated murine oral cavity orthotopic tumours. *Radiation Research* 167(3), 289–297.

Finkel, T. and Holbrook, N.J. (2000) Oxidants, oxidative stress and the biology of ageing. *Nature* 408(6809), 239–247.

Forman, J.H. and Kennedy, J. (1975) Superoxide production and electron transport in mitochondrial oxidation of dihydroorotic acid. *Journal of Biological Chemistry* 250(11), 4322–4326.

Garcia-Mediavilla, V., Crespo, I., Collado, P.S., Esteller, A., Sanchez-Campos, S., Tunon, M.J. and Gonzalez-Gallego, J. (2007) The anti-inflammatory flavones quercetin and kaempferol cause inhibition of inducible nitric oxide synthase, cyclooxygenase-2 and reactive C-protein, and down–regulation of the nuclear factor kappa β pathway in Chang Liver cells. *European Journal of Pharmacology* 557(2–3), 221–229.

Genova, M.L., Pich, M.M., Bernacchia, A., Bianchi, C., Biondi, A., Bovina, C., Falasca, A.I., Formiggini, G., Castelli, G.P. and Lenaz, G. (2004) The mitochondrial production of reactive oxygen species in relation to aging and pathology. *Annals of the New York Academy of Sciences* 101186–100.

Genova, M.L., Pich, M.M., Biondi, A., Bernacchia, A., Falasca, A., Bovina, C., Formiggini, G., Parenti Castelli, G. and Lenaz, G. (2003) Mitochondrial production of oxygen radical species and the role of Coenzyme Q as an antioxidant. *Experimental Biology and Medicine* 228(5), 506–513.

Goel, H.C., Gupta, D., Gupta, S., Garg, A.P. and Mala, B. (2005) Protection of mitochondrial system by *Hippophae rhamnoides* L. against radiation-induced oxidative damage in mice. *Journal of Pharmacy and Pharmacology* 57(1), 135–143.

Gong, B., Chen, Q. and Almasan, A. (1998) Ionizing radiation stimulates mitochondrial gene expression and activity. *Radiation Research* 150(5), 505–512.

Greenberger, J.S., Epperly, M.W., Gretton, J., Jefferson, M., Nie, S., Bernarding, M., Kagan, V. and Guo, H. L. (2003) Radioprotective gene therapy. *Current Gene Therapy* 3(3), 183–195.

Gupta, D., Arora, R., Garg, A.P. and Goel, H.C. (2003) Radiation protection of HepG2 cells by *Podophyllum hexandrum* Royale. *Molecular and Cellular Biochemistry* 250(1–2), 27–40.

Gupta, D., Arora, R., Garg, A.P., Mala, B. and Goel, H.C. (2004) Modification of radiation damage to mitochondrial system *in vivo* by *Podophyllum hexandrum*: mechanistic aspects. *Molecular and Cellular Biochemistry* 266(1–2), 65–77.

Gupta, D., Arora, R., Gupta, S., Gupta, S., Garg, A.P. and Goel, H.C. (2003b) Radiation-induced oxidative damage to mitochondria and its modification by endogenous antioxidants *in vivo*: role of some herbal preparations. *Indian Journal of Nuclear Medicine* 18, 14.

Hall, E.J. and Giaccia, A.J. (2006) *Radiobiology for the Radiologist.* Lippincott Williams and Wilkins, Philadelphia, USA.

Hancock, J.T., Desikan, R. and Neill, S. J. (2001) Role of reactive oxygen species in cell signalling pathways. *Biochemical Society Transactions* 29(Pt 2), 345–350.

Hauptmann, N., Grimsby, J., Shih, J.C. and Cadenas, E. (1996) The metabolism of tyramine by monoamine oxidase A/B causes oxidative damage to mitochondrial DNA. *Archives of Biochemistry and Biophysics* 335(2), 295–304.

Ishii, K. and Watanabe, M. (1996) Participation of gap-junctional cell communication on the adaptive response in human cells induced by low dose of X-rays. *International Journal of Radiation Biology* (3), 291–299.

Kamat, J.P. and Devasagayam, T.P. (1999) Nicotinamide (vitamin B3) as an effective antioxidant against oxidative damage in rat brain mitochondria. *Redox Reports* 4(4), 179–184.

Kamat, J.P., Boloor, K.K., Devasagayam, T.P.A. and Venkatachalam, S.R. (2000b) Antioxidant properties of *Asparagus racemosus* against damage induced by gamma-radiation in rat liver mitochondria. *Journal of Ethnopharmacology* 71(3), 425–435.

Kamat, J.P., Boloor, K.K., Devasagayam, T.P.A., Jayashree, B. and Kesavan, P.C. (2000a) Differential modification by caffeine of oxygen-dependent and independent effects of gamma-irradiation on rat liver mitochondria. *International Journal of Radiation Biology* 76(9), 1281–1288.

Kantak, S.S., Diglio, C.A. and Onoda, J.M. (1993) Low dose radiation-induced endothelial cell retraction. *International Journal of Radiation Biology* 64(3), 319–328.

Kergonou, J.F., Bernard, P., Braquet, M. and Rocquet, G. (1981a) Effect of whole-body gamma irradiation on lipid peroxidation in rat tissues. *Biochimie* 63(6), 555–559.

Kergonou, J.F., Braquet, M. and Rocquet, G. (1981b) Influence of whole-body gamma irradiation upon rat liver mitochondrial fractions. *Radiation Research* 88(2), 377–384.

Kim, K.M., Chun, S.B., Koo, M.S., Choi, W.J., Kim, T.W., Kwon, Y.G., Chung, H.T., Billiar, T.R. and Kim, Y.M. (2001) Differential regulation of NO availability from macrophages and endothelial cells by the garlic component S-allyl cysteine. *Free Radical Biology and Medicine* 30(7), 747–756.

Konings, A.W. and Drijver, E.B. (1979) Radiation effects on membranes. I. Vitamin E deficiency and lipid peroxidation. *Radiation Research* 80(3), 494–501.

Korotchkina, L.G., Yang, H., Tirosh, O., Packer, L. and Patel, M.S. (2001) Protection by thiols of the mitochondrial complexes from 4-hydroxy-2-nonenal. *Free Radical Biology and Medicine* 30(9), 992–999.

Koteles, G.J. and Somosy, Z. (2001) Radiation responses in plasma membrane. Review of the present state and future trends. *Cellular and Molecular Biology* 47(3), 473–484.

Kowaltowski, A.J., Castilho, R.F. and Vercesi, A.E. (2001) Mitochondrial permeability transition and oxidative stress. *FEBS Letters* 495(1–2), 12–15.

Kubasova, T., Varga, L. and Koteles, G.J. (1976) Studies on the radiosensitivity of the function of the Golgi complex. *International Journal of Radiation Biology and Related Studies in Physics, Chemistry, and Medicine* 29(6), 533–540.

Kwong, L.K. and Sohal, R.S. (1998) Substrate and site specificity of hydrogen peroxide generation in mouse mitochondria. *Archives of Biochemistry and Biophysics* 350(1), 118–126.

Kwong, L.K. and Sohal, R.S. (2000) Age-related changes in activities of mitochondrial electron transport complexes in various tissues of the mouse. *Archives of Biochemistry and Biophysics* 373(1), 16–22.

Lee, J.S., Huh, W.K., Lee, B.H., Baek, Y.U., Hwang, C.S., Kim, S.T., Kim, Y.R. and Kang, S.O. (2001) Mitochondrial NADH-cytochrome b(5) reductase plays a crucial role in the reduction of D-erythroascorbyl free radical in *Saccharomyces cerevisiae. Biochimica et Biophysica Acta* 1527(1–2), 31–38.

Loffler, M., Becker, C., Wegerle, E. and Schuster, G. (1996) Catalytic enzyme histochemistry and biochemical analysis of dihydroorotate dehydrogenase/oxidase and succinate dehydrogenase in mammalian tissues, cells and mitochondria. *Histochemistry and Cell Biology* 105(2), 119–128.

Lyng, F.M., Seymour, C.B. and Mothersill, C. (2001) Oxidative stress in cells exposed to low levels of ionizing radiation. *Biochemical Society Transactions* 29(Pt 2), 350–353.

Maas, E. and Bisswanger, H. (1990) Localization of the alpha-oxoacid dehydrogenase multienzyme complexes within the mitochondrion. *FEBS letters* 277(1–2), 189–190.

Matsko, C.M., Hunter, O.C., Rabinowich, H., Lotze, M.T. and Amoscato, A.A. (2001) Mitochondrial lipid alterations during Fas- and radiation-induced apoptosis. *Biochemical and Biophysical Research Communications* 287(5), 1112–1120.

M'Bemba-Meka, P., Lemieux, N. and Chakrabarti, S.K. (2007) Role of oxidative stress and intracellular calcium in nickel carbonate hydroxide–induced sister-chromatid exchange, and alterations in replication index and mitotic index in cultured human peripheral blood lymphocytes. *Archives of Toxicology* 81(2), 89–99.

McLennan, H.R. and Degli Esposti, M. (2000) The contribution of mitochondrial respiratory complexes to the production of reactive oxygen species. *Journal of Bioenergetics and Biomembranes* 32(2), 153–162.

Murley, J.S., Kataoka, Y., Weydert, C.J., Oberley, L.W. and Grdina, D.J. (2006) Delayed radioprotection by nuclear transcription factor kappa β-mediated induction of manganese superoxide dismutase in human microvascular endothelial cells after exposure to the free radical scavenger WR1065. *Free Radical Biology and Medicine* 40(6), 1004–1016.

Nair, C.K.K., Parida, D.K. and Nomura, T. (2001) Radioprotectors in radiotherapy. *Journal of Radiation Research* 42(1), 21–37.

Nishino, H. and Ito, A. (1986) Subcellular distribution of OM cytochrome b-mediated NADH-semidehydroascorbate reductase activity in rat liver. *Journal of Biochemistry* 100(6), 1523–1531.

Ojeda, F., Diehl, H.A. and Folch, H. (1994) Radiation induced membrane changes and programmed cell death: possible interrelationships. *Scanning Microscopy* (3), 645–651.

Prasad, K.N. (1995) *Handbook of Radiobiology*, CRC Press, Florida.

Rafique, R., Schapira, A.H. and Cooper, J.M. (2001) Sensitivity of respiratory chain activities to lipid peroxidation: effect of vitamin E deficiency. *The Biochemical Journal* 357(Pt 3), 887–892.

Ramakrishnan, N., McClain, D.E. and Catravas, G.N. (1993) Membranes as sensitive targets in thymocyte apoptosis. *International Journal of Radiation Biology* 63(6), 693–701.

Sandstrom, P.A., Pardi, D., Tebbey, P.W., Dudek, R.W., Terrian, D.M., Folks, T.M. and Buttke, T.M. (1995) Lipid hydroperoxide-induced apoptosis: lack of inhibition by Bcl-2 over-expression. *FEBS Letters* 365(1), 66–70.

Schmidt-Ulrich, R.K., Dent, P., Grant, S., Mikkelsen, R.B. and Valerie, K. (2000) Signal transduction and cellular radiation responses. *Radiation Research* 153(3), 245–257.

Shigenaga, M.K., Hagen, T.M. and Ames, B.N. (1994) Oxidative damage and mitochondrial decay in aging. *Proceedings of the National Academy of Sciences of the United States of America* 91(23), 10771–10778.

Shukla, S.K. (2007) Radiation-induced apoptosis and its modulation by some herbal preparations. *PhD Thesis*. Jamia Hamdard, New Delhi, India, p. 208.

Shukla, S.K., Chaudhary, P., Kumar, I.P., Samanta, N., Afrin, F., Gupta, M.L., Sharma, U.K., Sinha, A.K., Sharma, Y.K. and Sharma, R.K. (2006) Protection from radiation-induced mitochondrial and genomic DNA damage by an extract of *Hippophae rhamnoides*. *Environmental and Molecular Mutagenesis* 9, 647–656.

Simonson, S.G., Zhang, J., Canada, A.T., Jr., Su, Y.F., Benveniste, H. and Piantadosi, C.A. (1993) Hydrogen peroxide production by monoamine oxidase during ischemia-reperfusion in the rat brain. *Journal of Cerebral Blood Flow & Metabolism* 13(1), 125–134.

Singh, K.K. (2006) *Oxidative Stress, Disease and Cancer*, Imperial College Press, New York.

Skulachev, V.P. (1999) Mitochondrial physiology and pathology; concepts of programmed death of organelles, cells and organisms. *Molecular Aspects of Medicine* 20(3), 139–184.

Snyder, S.L. and Eklund, S.K. (1978) Radiation–induced alterations in the distribution of lysosomal

hydrolases in rat spleen homogenates. *Radiation Research* 75(1), 91–97.
Somosy, Z. (2000) Radiation response of cell organelles. *Micron* 31(2), 165–181.
Somosy, Z., Kubasova, T. and Koteles, G.J. (1987) The effects of low doses of ionizing radiation upon the micromorphology and functional state of cell surface. *Scanning Microscopy* 1(3), 1267–1278.
Somosy, Z., Pikli, A., Kubasova, T., Koteles, G.J. and Bodo, M. (1985) Alteration of nuclear diameter of murine lymphocytes upon the effect of X-irradiation *in vivo. Radiobiologia, Radiotherapia* 26(3), 305–310.
Starkov, A.A., Fiskum, G., Chinopoulos, C., Lorenzo, B.J., Browne, S.E., Patel, M.S. and Beal, M.F. (2004) Mitochondrial alpha-ketoglutarate dehydrogenase complex generates reactive oxygen species. *The Journal of Neuroscience* 24(36), 7779–7788.
Sun, Y. (1990) Free radicals, antioxidant enzymes, and carcinogenesis. *Free Radical Biology and Medicine* 8(6), 583–599.
Szumiel, I. (1994) Ionizing radiation-induced cell death. *International Journal of Radiation Biology* 66(4), 329–341.
Trosko, J.E. (1996) Role of low-level ionizing radiation in multi-step carcinogenic process. *Health Physics* 70(6), 812–822.
Vasquez-Vivar, J., Kalyanaraman, B. and Kennedy, M.C. (2000) Mitochondrial aconitase is a source of hydroxyl radical. An electron spin resonance investigation. *Journal of Biological Chemistry* 275(19), 14064–14069.
Zhang, H., Koch, C.J., Wallen, C.A. and Wheeler, K.T. (1995) Radiation-induced DNA damage in tumours and normal tissues. III. Oxygen dependence of the formation of strand breaks and DNA-protein crosslinks. *Radiation Research* 142(2), 163–168.
Zhang, H., Zheng, R.L., Wei, Z.Q., Li, W.J., Gao, Q.X., Chen, W.Q., Wang, Z.H., He, J., Liang, J. P., Han, G.W., Huang, T., Li, Q., Xie, H.M., Zhang, S.M. and Cai, X.C. (1998a) Effects of pre-exposure of mouse testis with low-dose (16)O^{8+} ions or ^{60}Co gamma-rays on sperm shape abnormalities, lipid peroxidation and superoxide dismutase (SOD) activity induced by subsequent high-dose irradiation. *International Journal of Radiation Biology* 73(2), 163–167.
Zhang, L., Yu, L. and Yu, C.A. (1998b) Generation of superoxide anion by succinate-cytochrome c reductase from bovine heart mitochondria. *Journal of Biological Chemistry* 273(51), 33972–33976.
Zhang, Y., Marcillat, O., Giulivi, C., Ernster, L. and Davies, K.J. (1990) The oxidative inactivation of mitochondrial electron transport chain components and ATPase. *Journal of Biological Chemistry* 265(27), 16330–16336.

Chapter 16

Andrographis paniculata: An Emerging Radioprotective Agent for Membrane Proteins

Rakshamani Tripathi and Jayashree P. Kamat

Introduction

The deleterious biological consequences of low linear energy transfer (LET) radiations on living cells and the generation of a wide range of harmful reactive oxygen species (ROS) like e^-_{aq}, ·OH, $O_2^{\cdot-}$, H_2O_2 etc., through the radiolytic cleavage of water, are well documented (Sonntag, 1987). Among the ROS, superoxide is not reactive enough to cause molecular damage, but is a precursor of ·OH-, the most aggressive and powerful oxidant formed during metal-catalyzed Fenton reaction, which oxidizes virtually all the cellular constituents (Sonntag, 1987; Halliwell and Gutteridge, 1997; Sies, 1997). Superoxide also reacts with nitric oxide and generates the highly damaging reactive species, peroxynitrite (Kamat, 2006). Though the cell possesses a network of multifunctional defensive factors, radiation-induced stressors still escape this surveillance, causing enormous alterations in the cellular genome and inducing serious pathological disorders (Sonntag, 1987). Among the various radioprotectors screened so far, WR-2721 has been found to be an effective radioprotector. However, it is synthetic, expensive and has its own inherent toxicity (Rades *et al.*, 2004). Consequently, exploration of herbs used in Ayurveda, which have less toxic effects, are cheap and possess diverse medicinal attributes, might be useful in effecting radioprotection. Several herbs are known to exhibit radioprotective effects (Uma Devi and Ganasoundari, 1995; Kamat *et al.*, 2000; Kamat and Venkatachalam, 2004; Hari Kumar and Kuttan, 2004; Baliga *et al.*, 2004; Bhattacharya *et al.*, 2005; Kamat and Mishra, 2006; Arora *et al.*, 2005a, 2006a,b, 2007).

The present study was primarily aimed to this end, wherein the radioprotective property of the popular Indian medicinal plant, *Andrographis paniculata*, commonly known as Kalmegh (Sanskrit) (Family: Acanthaceae), with wide acceptability to man and diverse medicinal properties (Rana and Avadhoot, 1991; Reyes *et al.*, 2006; Sheeja *et al.*, 2006) was examined. Andrographolide, the major active constituent, has been isolated from *Andrographis paniculata* and reported to exhibit a plethora of beneficial effects such as antihepatotoxic, immunomodulatory and anticancer effects. It has been suggested that anticancer effects are triggered by apoptosis via activation of caspase, with a series of sequential events (Zhou *et al.*, 2006; Rajagopal *et al.*, 2003). Its anti-inflammatory effect is expressed through inhibition of NF-kappa β, which reduces the expression of proinflammatory proteins, COX-2 (Hidalgo *et al.*, 2005). In recent years, this phytomedicine is gaining active interest in the development of chemical fingerprints, having several implications in quality control and the preparation of formulation-based drugs (Srivastava *et al.*, 2004).

Impact of oxidants on mitochondria, the major site of ROS, and the proteomic system of mitochondria is gaining momentum in recent years. Cellular proteins are shown to be the major initial targets for the attack of ·OH, generated during radiation (Gebicki and Du, 2004). Since living cells are under continuous siege in oxidative stress resulting in deleterious oxidative reactions in the cellular proteins, affecting their homeostasis maintenance (Stadman, 2006), it was of interest to examine the preventive effects of *A. paniculata* extract against mitochondrial proteomic system with the radiation exposure. The results obtained are promising and demonstrated significant radioprotective effects by *A. paniculata* against several biochemical parameters, linked to membrane protein damage. The study has also contributed to understanding the possible mechanisms involved in the observed radioprotection.

Materials and Methods

Animals

Wistar rats (male, 8–10 weeks) were bred in the Bhabha Atomic Research Centre (BARC) Laboratory Animal House Facility, Mumbai, India, and procured after obtaining clearance from the BARC Animal Ethics Committee. All the experiments were conducted with strict adherence to the ethical guidelines laid down by the Committee for the Purpose of Control and Supervision of Experiments on Animals (CPCSEA) constituted by the Animal Welfare Division of the Government of India on the use of animals in scientific research. Animals were maintained under controlled laboratory conditions (25±2°C; RH: 60±5%; 12h photo-period) and fed standard animal food and tap water *ad libitum*.

Plant extract

Aqueous extract of *Andrographis paniculata* (Ap) was a gift from Zhandu Research Laboratory, Mumbai, India.

Chemicals

Glutathione, guanidine hydrochloride, 5,5-dithio-bis (2-nitrobenzoic acid) (DTNB), vitamin C, GSH, metaphosphoric acid, hydrogen peroxide, diethylenetriamine pentaacetic acid (DTPA), nitroblue tetrazolium, superoxide dismutase (SOD), BHT, epinephrine, dinitrophenyl hydrazine (DNPH), NADH/NADPH, CDNB, FAD, cytochrome c, ABTS, phenazine methosulfate, mannitol, gallic acid, quercetin, deoxyribose and tripyridyl-s-triazin (TPTZ) were purchased from Sigma Chemical Company, USA. All other chemicals used in the study were of the highest purity commercially available.

Preparation of rat liver mitochondria

Rats were fasted overnight and sacrificed by cervical dislocation. Livers were removed and homogenized in 0.25M sucrose containing 1mM EDTA. The homogenate, thus obtained, was centrifuged at 3000 × g for 10 min and the resulting supernatant was centrifuged at 10,000 × g for 10 min to sediment mitochondria in a Sorvall RC5C centrifuge (Kamat and Mishra, 2006). The mitochondrial pellets thus obtained were washed thrice with 5mM potassium phosphate buffer pH 7.4 to remove sucrose and suspended in the same buffer at a concentration of 10mg protein/ml.

Irradiation studies

Liver mitochondria (2mg protein/ml) were irradiated with a dose of 450Gy at a dose rate of 15Gy/min using ^{60}Co source (Atomic Energy of Canada Limited), with and without Ap extract (5–100µg/ml) and immediately subjected to oxidative studies related to membrane protein damage. Attempts were made to understand the possible radioprotective action of the extract. Samples without irradiation served as control. In our earlier studies, the oxidative damage to rat liver mitochondria was examined in various doses of γ-radiation from 75–600Gy and optimum damage to membrane lipids was observed at 450Gy (Kamat *et al.*, 2000). Hence, protein damage was studied at this dose.

Measurement of oxidative damage in mitochondrial membrane proteins

Oxidative damage in membrane proteins was monitored using various biological markers. The activity of SOD was measured by autooxidation of epinephrine at 320nm (Kamat *et al.*, 2005). Catalase activity was monitored by measuring the rate of absorbance of decomposition of H_2O_2 at 240nm (Aebi, 1974). Protein carbonyl was measured by derivatization of mitochondrial protein with DNPH (0.21% in 2N HCl), followed by removal of underivatized DNPH by extraction with ethanol: ethyl acetate (1:1). The mitochondrial preparation derivatized with DNPH was dissolved in 6N guanidine hydrochloride and read at 370nm (Kamat *et al.*, 2005). Protein thiols were measured by measuring the absorbance of DTNB at 412nm (Moron *et al.*, 1979). Measurement of glutathione peroxidase is based on the degradation of H_2O_2 in the presence of GSH and GSH content was measured as total non-protein sulfhydryl group at 412nm (Moron *et al.*, 1979). Assays of NADPH cytochrome c reductase were carried out by measuring the increase in absorbance of oxidized cytochrome c at 550nm (Kamat *et al.*, 1980). The activity of GSH-S transferase is based upon the rate of increase in the conjugation between GSH and 1-chloro 2,4 dinitrobenzene (CDNB) at 340nm (Mhatre *et al.*, 1983). Experiments were performed to locate SOD activity by activity staining method (Beauhamp and Fridovich, 1971). In all these assays, appropriate sample blanks were taken to ensure that the extract did not interfere with the assay methods.

Radical scavenging assays

The scavenging action of extract for hydroxyl radical was examined by its ability to inhibit ·OH catalyzed deoxyribose oxidation by TBA method, and measuring the resultant malondialdehyde (MDA) spectrophotometrically at 532nm (Joshi *et al.*, 2007). $O_2^{\cdot-}$ is generated in a PMS/NADH system by oxidation of NADH and reduction of nitroblue tetrazolium (NBT). The assay consisted of Tris-HCl buffer 16mM, pH 8.0, 78mM NADH, 50mM NBT, 10mM PMS with Ap extract (10–100µg/ml). Absorbance was measured at 560nm. Ascorbic acid was used as a positive control (Kamat *et al.*, 2005). H_2O_2 measurement is based on the horseradish peroxidase (HRP)-mediated oxidation of phenol red by H_2O_2, resulting in a chromogenic compound having absorbance at 600nm followed by the addition of Ap (10–100µg/ml) (Kamat and Mishra, 2006).

FRAP assay and reducing capacity of Ap

The ferric-tripyridyltriazine (Fe^{2+}-TPTZ) complex was measured at 595nm with and without Ap. L-ascorbic acid was used as standard (Benzie and Strain, 1996). Reducing power of the samples is based on the reduction of Fe (III)® Fe (II). Ap extract (5–100µg/ml) was mixed with 1% potassium ferricyanide at 50°C for 20 min, centrifuged and the clear supernatant was treated with 0.1% $FeCl_3$. Absorbance was measured at 700nm. Increased absorbance reflects increased reducing power (Kamat *et al.*, 2005; Arora *et al.*, 2005b).

Estimation of phytochemicals from Ap

Total phenolic contents are measured based on the reduction of Folin-Ciocalteu reagent by reducing equivalents from phenols. Briefly, Folin-Ciocalteu reagent was mixed equally with sample and mixture was treated with 80% sodium carbonate and the absorbance was measured at 730nm (Kamat and Mishra, 2006). Measurement of total flavonoids content of the extract was carried out by the standard method (Kamat and Mishra, 2006). $NaNO_2$ (5%) was added to the extract mixture, followed by the addition of $AlCl_3.6H_2O$ (10%). Absorbance was taken at 510nm, following the addition of NaOH. Quercetin was used as a positive control.

Evaluation of antioxidant ability of Ap by pulse radiolysis with ABTS radical

Using high-energy electron pulses of 7Mev for 50ns, the pulse radiolysis studies were carried out with ABTS. $ABTS^{\cdot-}$ were generated (by pulse radiolysis) and their decay was measured at 600nm (Scot *et al.*, 1993). The decay of $ABTS^{\cdot-}$ was also measured with known concentration of the extracts. From the measured rate constant value, the ascorbic acid equivalent present in the extract was calculated.

Statistical analysis

The results are expressed as mean ± SEM. Student's t-test was used to make a statistical comparison between the groups. A statistical comparison was done with the radiation alone group vs. Ap and radiation group. The significance levels were set at $p^{\#} < 0.01$, and $p^{*} < 0.001$.

Results

Radioprotective effects of Ap against antioxidant enzymes

The activity of antioxidant enzymes *viz.*, SOD and catalase were studied with various concentrations of Ap (5–100µg/ml) following irradiation of rat liver mitochondria at 450Gy and the results are incorporated in Fig. 16.1. Irradiation of mitochondria caused a significant decrease in the activities of SOD (56%) and catalase (52%), compared to unexposed mitochondria. Presence of the extract during irradiation, however, showed concentration-dependent restoration in the activity of both the enzymes.

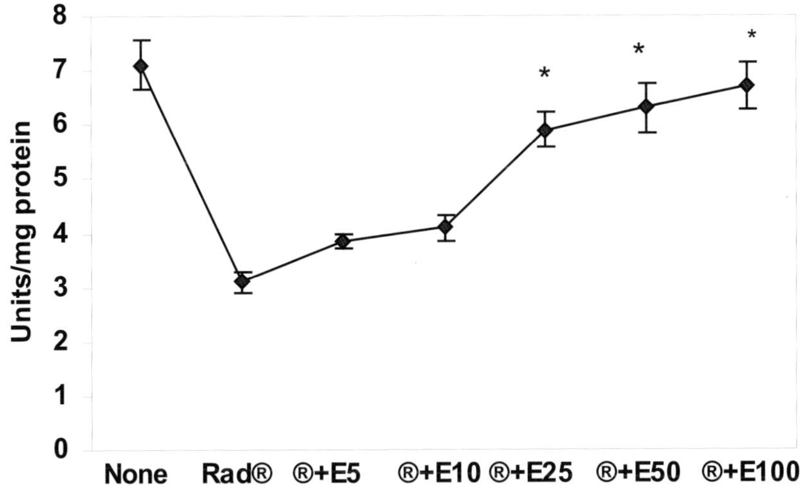

Fig. 16.1A. Radiation-induced depletion of superoxide dismutase (SOD) in rat liver mitochondria and its restoration by aqueous extract of *Andrographis paniculata* (Ap). Mitochondria (2mg protein/ml) were exposed to 450Gy with and without extract (5–100μg / ml). None-unexposed mitochondria, Rad®-mitochondria exposed to 450Gy, ®+E5, ®+E10 ®+E25, ®+E50 and ®+E100 are the respective Ap concentrations added to mitochondria during radiation exposure. The values are mean ± SEM from five experiments. p^{*} <0.001 compared to radiation treatment.

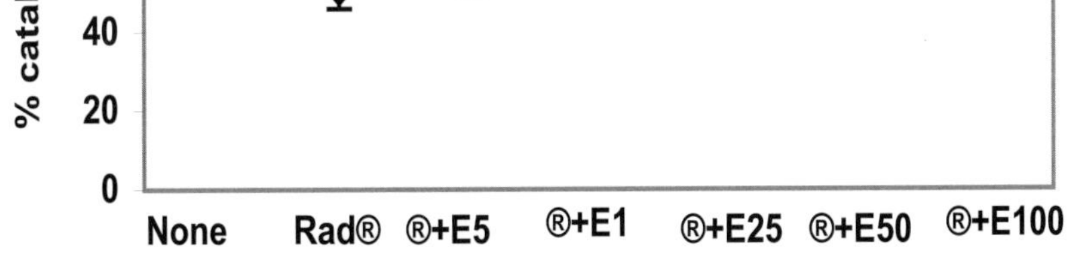

Fig. 16.1B. Radiation-induced depletion of catalase in rat liver mitochondria and its restoration by aqueous extract of *Andrographis paniculata* (Ap). Mitochondria (2mg protein/ml) were exposed to 450Gy, with and without extract (5–100μg /ml). None-unexposed mitochondria, Rad®-mitochondria exposed to 450Gy, ®+E5, ®+E10 ®+E25, ®+E50 and ®+E100 are the respective Ap concentrations added to mitochondria during radiation exposure. The values are mean ± SEM from five experiments. p^{*} < 0.001 compared to radiation treatment.

Restoration in SOD was marginal up to 10μg, but at 25μg/ml it showed significant increase resulting in 83% restoration in the enzyme activity. The protective ability of the extract was found to be maximum at 50μg/ml (Fig. 16.1A).

Hence, this dose was selected to measure radioprotective ability of Ap extract against protein damage. The extract also prevented the decreased activity of catalase. However, the effect was less compared to observed effect with SOD activity. Ap 25μg and 50μg showed only 20% and 30% restoration respectively (Fig. 16.1B).

Radioprotective effects of Ap for protein thiol, GSH and GSH peroxidase

Table 16.1 shows the data on the levels of endogenous thiol scavenging agents, protein thiols, GSH and GSH dependent peroxidase activity, following radiation exposure to mitochondria at 450Gy and prevention by Ap extract. A significant decrease in the levels of these parameters was demonstrated. Protein thiols were decreased by 40.9%, while GSH was decreased by 35.3%. However, *in vitro* addition of Ap (50μg/ml) during radiation showed effective restoration of thiol level and GSH contents. Similarly, the activity of GSH peroxidase was also reduced at 450Gy by 57%, and the extract showed 40% prevention.

Radioprotection by Ap (drug detoxicating antioxidant enzymes and protein oxidation)

Studies were carried out to monitor the modulating effects of Ap against redox sensitive Phase I marker drug metabolism, mediated by NADPH cytochrome c reductase and Phase II marker enzyme, GSH S-transferase at 450Gy. The enzyme activities decreased by (>50%) after irradiation.

Table 16.1. Radiation-induced damage to protein thiols, GSH and Gpx activity in rat liver mitochondria and modulating effects of Ap.

Treatments	Protein thiols (nmoles protein thiols/mg protein)	GSH (μg/mg protein)	Glutathione peroxidase (Units/litre protein)
Unexposed	16.1 ± 1.5	45.6 ± 3.8	32.91 ± 1.3
Radiation 450Gy	9.5 ± 0.75	29.5 ± 2.6	14.08 ± 3.31
Ap50 + 450Gy	15.7 ± 1.1*	43.5 ± 2.9*	27.31 ± 6.1*

Values are mean ± SEM from five experiments $^{*}p < 0.001$ compared with respective radiation treatments.

Different concentrations of Ap were used to find out the effective concentration against these enzymes. We found a concentration-dependent modulating effect of Ap for the activities of both enzymes and maximum restoration was demonstrated at 50μg/ml (data not shown). Therefore, Ap concentration (50μg/ml) was selected to measure the radioprotective effects. Presence of the extract during irradiation could significantly reverse the decrease in

NADPH cytochrome c reductase and GSH transferase levels by 94%. The levels of protein carbonyls, an important index of protein oxidation, showed a nine-fold increase in the levels of protein oxidation following irradiation ($p < 0.001$) and addition of the extract resulted in 68.7% protection (Table 16.2).

Table 16.2. Radiation-induced depletion of NADPH cyt. c reductase, Glutathione-S transferase and enhancement in protein carbonyls and modulating effects of Ap.

Treatments	NADPH cyt. c reductase (n moles cytochrome c reduced/min/mg protein)	Glutathione-S-transferase (nmoles CDNB/min/mg protein)	Carbonyls (nmoles/mg protein)
Unexposed	50 ± 2.3	4.5 ± 0.31	2.52 ± 0.18
Radiation (450Gy)	14.5 ± 1.9	1.9 ± 0.11	18.5 ± 016
450Gy+Ap50	47.5 ± 3.2*	3.9 ± 0.21*	7.5 ± 0.61*

Values are mean ± SEM from five experiments *$p < 0.001$, #$p < 0.01$ compared with respective radiation treatments.

Preventive effects of Ap against SOD enzyme, detected by activity staining method

Detection of SOD enzyme was carried out by activity staining experiments in the mitochondria, exposed to radiation (450Gy) with and without extract (50µg/ml). Irradiation of mitochondria showed a decrease in the intensity of the SOD bands as compared to unexposed control and presence of the extract resulted in effective prevention in the band intensity corresponding to SOD enzyme (Fig. 16.2A).

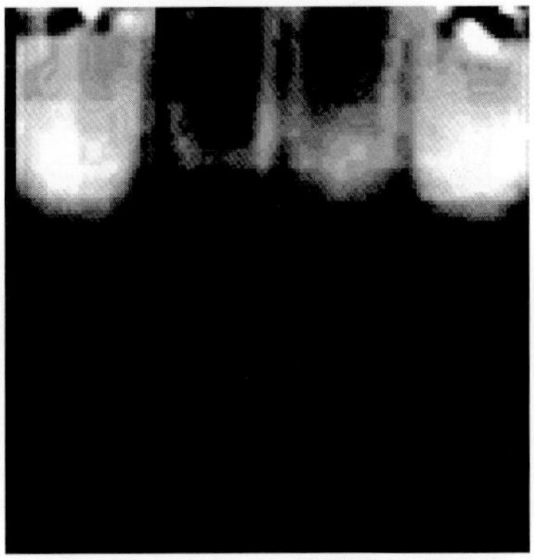

C D GSH Ap

Fig. 16.2A. Restoring effects of *Andrographis paniculata* (Ap) against radiation-induced degradation of SOD enzyme. Mitochondria were exposed to 450Gy and activity staining was performed to check SOD activity. **C**-unexposed mitochondria; **D**-damage (mitochondria exposed to 450Gy); **GSH**-Mitochondria exposed to 450Gy with GSH, 1mM (positive control); **Ap**- Mitochondria exposed to 450Gy with Ap (50µg/ml).

Radical scavenging capacity of Ap extract

The molecular mechanism underlying radioprotective effects of Ap was examined by monitoring scavenging effects of Ap for various ROS. Ap extract (10–200μg/ml) was subjected to radical scavenging assays. A concentration-dependent inhibitory profile was observed for ·OH and superoxide radicals by the extract (Fig. 16.2B). Ap (25 and 50μg/ml) showed 38% and 65% inhibition for ·OH, and 78% and 85% for superoxide radical respectively. Mannitol (182μg/ml for 1mM), a positive control, required 3.64 times higher concentration than Ap (required about 50μg/ml) to show 56% scavenging of ·OH. Mannitol (50μg/ml) showed only 12% scavenging. Ap (10μg/ml) showed 58% scavenging of superoxide radical. Although vit.C showed 60% scavenging effect, it required 17.5 times higher concentration (175μg/ml; 0.1mM) than Ap extract. At 10μg/ml, the scavenging effect of vitamin C was only 8%. Scavenging capacity of the extract for H_2O_2 was found to be concentration dependent and Ap (25μg and 50μg/ml) showed 19% and 31% scavenging respectively (Fig 16.2C). These results demonstrated adequate scavenging ability of Ap for both ·OH and $O_2^{\cdot-}$, and to some extent for H_2O_2.

Measurements of phytoconstituents, reducing power and FRAP of Ap extract

The results on the total phenolic and flavonoids contents and ferric reducing antioxidant power (FRAP) of the extracts are incorporated in Table 16.3. A concentration-dependent increase in the phenolic and flavonoids contents was demonstrated. The phenolic content and flavonoids correlated well. Reducing power as well as FRAP was also enhanced as a function of the extract. All these parameters sufficiently explained antioxidant efficacy of the extract.

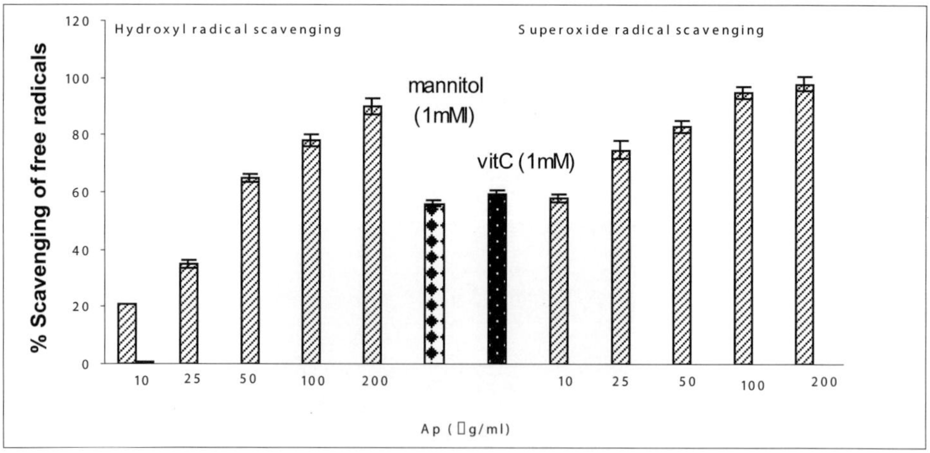

Fig. 16.2B. Restoring effects of *Andrographis paniculata* (Ap) against radiation-induced scavenging ability of Ap for ·OH and $O_2^{\cdot-}$.

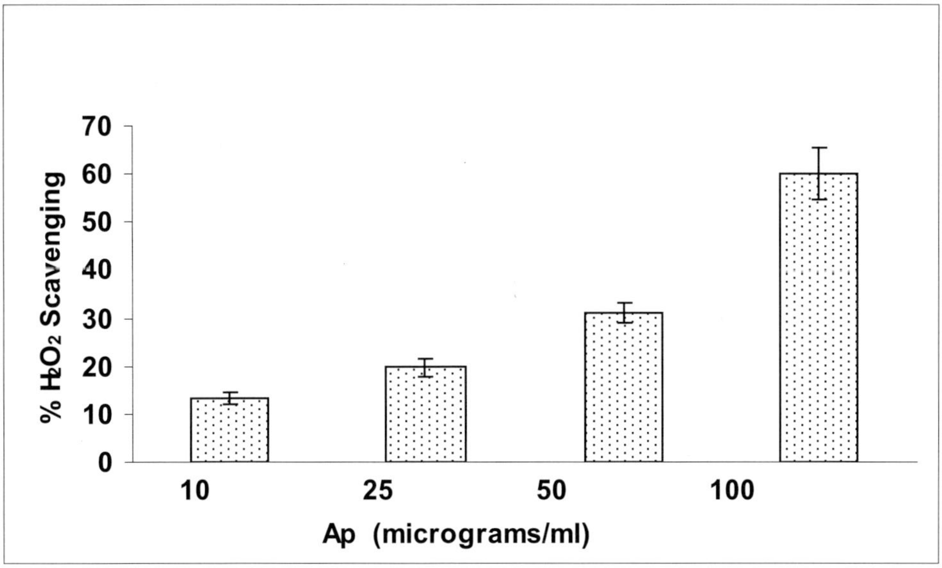

Fig. 16.2C. Restoring effects of *Andrographis paniculata* (Ap) against radiation-induced scavenging ability of Ap for H_2O_2.

Extract Concentration (μg/ml)	Total phenolic contents (equivalent μg gallic acid)	Flavonoid contents (equivalent μg quercetin)	Reducing power at 700nm	Ferric reducing antioxidant power (FRAP) (μg equivalent vitamin C)
Ap10	-	-	0.14	-
Ap25	2.8 ± 0.11	3.14 ± 0.52	0.22	1.4± 0.86
Ap50	5.3 ± 0.15	8.58 ± 1.43	0.28	3.1 ± 0.21
Ap100	10 ± 0.91	14.03± 2.34	0.32	4.68± 0.87
Ap200	25 ± 0.96	29.7 ± 4.95	0.39	not detected

Values are mean ± SEM from four experiments.

Table 16.3. Total phenolic and flavonoid contents, reducing power and ferric reducing antioxidant power of Ap extract.

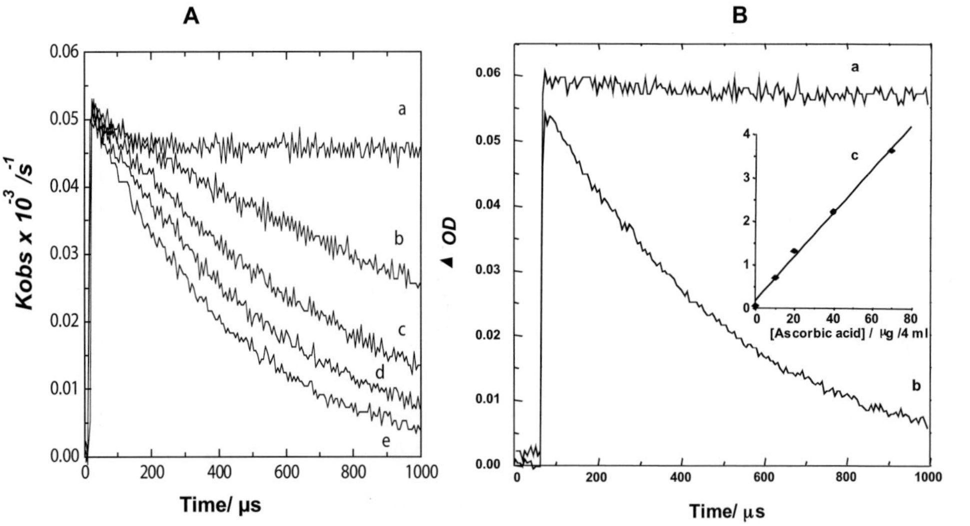

Fig. 16.3. Decay of ABTS radical at 600nm (**A**) in absence (a) and presence of different concentrations of ascorbic acid, b- 10µg/4ml, c- 20µg/4ml, d- 30µg/4ml, e- 40µg/4ml; (**B**) plot of Kobs, over the decay of ABTS radical at 600nm as a function of ascorbic acid concentration.

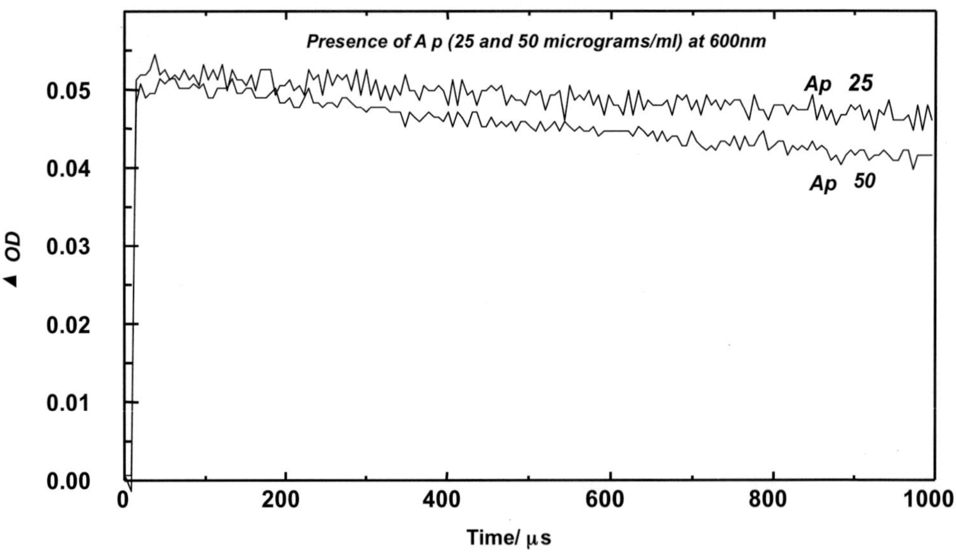

Fig.16.3C. Decay of ABTS radical in the presence of Ap extract.

Pulse radiolysis of Ap Extract

The results on antioxidant capacity of the extract, examined by pulse radiolysis in the presence of ABTS radical, are given in Fig 16.3. The ABTS radicals were generated by pulse radiolysis in the presence of N_2O saturated aqueous solution of N3- and ABTS. ABTS- decays as a function of ascorbic acid concentrations (Fig. 16.3A). The decay constants were measured with known concentration of the extract. The pseudo-first order rate constants (Kobs) increased linearly with ascorbic acid concentrations (Fig. 16.3B). The ascorbic acid equivalent of extract was calculated, and it showed 3.5 acid equivalent of ascorbic for Ap. Decay of ABTS is given in Fig. 16.3C.

Discussion

Radiation exposure to living organisms is known to inflict several damaging effects and these events are mediated by a wide range of ROS, generated by low LET radiations. Adverse impacts of ROS on cellular constituents altering structural and functional relationships of cells leading to increased risk of diseases are well documented (Sonntag, 1987). Among the cell constituents, proteins are the most abundant components interacting readily with free radicals forming long-lived ROS (Alberts, *et al.*, 2002). Evidence has been accumulating to demonstrate that mitochondrial proteins have a key role in oxidative stress and are the initial major targets for the·OH attack (Gebicki and Du, 2004; Gupta *et al.*, 2003). Attack of ·OH is associated with the conversion of several amino acids to corresponding oxyacids, an earliest observable event occurring in oxidative stress (Sonntag, 1987; Davies *et al.*, 1987; Stadman, 2006). As discussed above, radiation exposure brings about several undesirable changes in the cells, including generation of protein radicals, which are known to enhance the activity of DNA binding transcriptional factors and results in genomic instability (Schreck *et al.*, 1991). Due to persistent oxidative stress, the proteins are oxidized, cleaving polypeptide chain/ amino acids, and are further degraded by the proteosome complex or alternatively repaired by antioxidants. Protein oxidation plays a key role also in aging, forming increased oxidized proteins with animal age, ROS generation, decreases in antioxidants and degradation of oxidized proteins. In general, oxidation of proteins brings about severe changes resulting altogether in the loss of cellular integrity and, hence, several physiological disorders (Sies, 1997; Stadman, 2006). Therefore, studying antioxidant effects of natural herbs is of prime importance.

In view of this, the ability of Ap to prevent radiation-induced oxidative damage to proteins was assessed. For this, *in vitro* experiments on rat liver mitochondria at high radiation dose of 450Gy were performed. It is well known that exposure of humans even to a considerably lower dose of γ-radiation causes extensive protein damage. From that perspective, the chosen radiation dose was admittedly too high. However, our *in vitro* experiments with liver mitochondria with lower doses of γ-radiation (75, 150 and 300Gy) did not show significant amount of oxidative damage (Kamat *et al.*, 1999, 2000). Hence, a radiation dose of 450Gy was chosen primarily to demonstrate the capacity of the extract in preventing radiation-induced protein damage. It was observed that presence of the extract during radiation significantly protected against extensive protein damage in liver mitochondria even at high doses of radiation under *in vitro* conditions.

Andrographis paniculata extract (Ap) has been shown to exert multifunctional beneficial effects, including immunomodulating activity (Reyes *et al.*, 2006). Several natural products, including herbs with radioprotective properties, are known to have immunomodulating activity (Motoki *et al.,* 1995; Arora *et al.,* 2005a, 2006a,b, 2008; Amna *et al.,* 2007). It is likely that the immunomodulating activity of Ap may get operative through radioprotective mechanism. The present results demonstrated the radioprotective effect of the extract. The results are highly significant, showing prevention of membrane protein damage, as demonstrated by prevention in protein oxidation, restoration in antioxidant defenses, SOD, catalase, Gpx, protein thiols and drug metabolizing enzymes. Spontaneously generated electrons during wide ranges of physiological reactions or radiation exposure instantly react with oxygen to form $O^{\cdot-}_{2,}$ an important pathological mediator. Though $O^{\cdot-}_2$ itself is not very reactive, it is a precursor of the powerful oxidant, OH (Halliwell and Gutteridge, 1999). In this regard, SOD plays an important role as an early cellular defence by catalyzing the dismutation of $O_2^{\cdot}$. H_2O_2 generated during dismutation of $O^{\cdot-}_2$ is further detoxified by catalase. Therefore, cooperative activity of these two enzymes is a crucial step in the maintenance of redox homeostasis. Our results confirmed the modulating effects of extract against SOD, catalase and GPx indicating possible scavenging of $O^{\cdot-}_2$, H_2O_2 and ·OH. The extract also effectively prevented radiation-induced decline in SOD activity assessed by activity staining experiments

Currently, antioxidant research is mostly focused on phenolics, the major group of antioxidants present in the diet rich in vegetables and fruits. The bioactivity of phenolics may be due to the multifunctional beneficiary roles, including antioxidant behaviour, attributed to their reducing, metal chelating, hydrogen donating capacity, and scavenging of ROS (Rice Evans *et al.,* 1997; Chawla *et al.,* 2005, 2007). Flavonoids with hydroxylyzed B ring and/or unsaturated C ring have been recently suggested as natural proteasome inhibitors and apoptosis inducers, thus providing a molecular basis for the clinically observed cancer-preventive effects of fruits and vegetables (Chen *et al.,* 2006). We have observed increase in phenolics and flavonoids as a function of Ap (25–200μg/ml) concentration (Table 16.3). This is also supported by some studies demonstrating the presence of 12 flavonoids, *viz.,* 5,7,tetramethoxy flavone and 5-hydroxy-7, trimethoxy flavone in the Ap extract (Chen *et al.,* 2006).

The extract has been reported to contain several active components like andrographolide, 14-deoxy-11-oxoandrographolide, 14-deoxyandrographolide and neoandrographolide. Generally, structural perspective chain-breaking antioxidants possess an extensive system of conjugated double bonds that stabilize reaction transients by resonance and are hydrogen donors. The above active components do not have such structural features. Our present study and those by others showed significant scavenging of $O_2^{\cdot-}$ (Kamdem *et al.,* 2002) and ·OH by Ap extract. The extract showed the presence of some xanthones. The presence of hydroxyl group at 2 position of xanthones has been shown as a powerful life-enhancing factor with a wide range of pharmaceutical properties (Dua *et al.,* 2004). Xanthones have also been shown to play a role in prevention of lipid peroxidation reactions (Menkoviz *et al.,* 2002). Our results showed the presence of considerable amount of phenolic/flavonoid contents in Ap, which exhibited an increase in the reduction of Fe (III)®Fe (II) as shown by an increase in absorbance of reducing potential and FRAP, the established indices for antioxidant action. One of the possible reasons for antioxidant effect of Ap may be due to scavenging of ROS. Our results showed concentration-dependent scavenging effects against radiation-induced ROS, $O_2^{\cdot-}$, ·OH, H_2O_2 and the effect was quite potent compared to the standard antioxidants *viz.,* mannitol and vitamin C. ESR signals for DPPH radicals were also scavenged by 50% with Ap at 70μg/ml (data not shown). These observations extensively justify the antioxidative ability of Ap extract.

Pulse radiolysis experiments, carried out with high-energy electron pulses (7Mev, 50ns), demonstrated that pseudo-first order rate constants (Kobs) increase linearly with ascorbic acid concentration. The rate constant was measured and the ascorbic acid equivalent of the extract was calculated. It showed good reactivity for the extract.

Generation of ABTS: **ABTS2- + N3. ® ABTS-. + N3-** ....(I)

Various studies have extensively demonstrated the radioprotective/antioxidant effects of natural products (Uma Devi and Ganasoundari, 1995; Kamat *et al.,* 2000, Kamat and Venkatachalam, 2004; Harikumar and Kuttan, 2004; Baliga *et al.,* 2004; Gupta *et al.,* 2004; Kamat and Mishra, 2006; Harikumar and Kuttan, 2004; Bhattacharya *et al.,* 2005; Arora *et al.,* 2005, 2006a,b, 2008). We attempted to explore the possible radioprotective effect of the extract *Andrographis paniculata* against cellular proteins. We conclude that *Andrographis paniculata* aqueous extract (Ap) is useful in reducing radiation-induced oxidative stress and maintaining redox balance in the normal tissues. Further investigations with the extract appear to be promising for developing a new natural radioprotectant.

Acknowledgements

The authors wish to thank Dr. Hari Mohan of the Radiation Chemistry and Chemical Dynamics Division, Bhabha Atomic Research Center (BARC), Mumbai, India, for his help in pulse radiolysis studies.

References

Aebi, H. (1974) In: *Methods of Enzymatic Analysis*. Berg Meyr, H.V. (ed) Verlag Chemie New York, USA, p. 673.

Alberts, B., Johnson, A., Lewis, J., Raff, M., Roberts, K. and Walter, P. (2002) *Molecular Biology of the Cell*. Garland Science, New York, USA, p. 1463.

Amna, T., Puri, S.C., Gupta, A., Khajuria, A., Arora, R., Spiteller, M. and Qazi, G.N. (2007) Immunomodulatory activity of an extract of the novel endophyte *Entrophosphora infrequens* isolated from *Nothapodytes foetida* (Wight) Sleumer. *Acta Microbiologica et Immunologica Hungarica* 54(3), 237–260.

Arora, R., Chawla, R., Prasad, J., Singh, S., Kumar, R., Sharma, A., Sagar, R.K., Singh, S. and Sharma, R.K. (2005b) Radioprotective activity of *Rhodiola imbricata*– a high altitude medicinal plant. *Molecular and Cellular Biochemistry* 273, 209–223.

Arora, R., Chawla, R., Singh, S., Sagar, R.K., Kumar, R., Sharma, A.K., Singh, S., Prasad, J., Sharma, R.K. and Tripathi, R.P. (2006b) Radioprotection by Himalayan High-altitude Region Plants. In: Sharma, R.K. and Arora, R. (eds) *Herbal Drugs: A Twenty First Century Perspective*, Jaypee Brothers Medical Publishers Private Ltd., Delhi, India, pp. 301–325.

Arora, R., Chawla, R., Singh, S., Sagar, R.K., Kumar, R., Sharma, A., Prasad, J., Singh, S., Gurudatta, G.U. and Sharma, R.K. (2006a) Bioprospection for Radioprotective Molecules from Indigenous Flora. In: Govil, J.N., Singh, V.K. and Bhardwaj, R. (eds) *Recent Progress in Medicinal Plants*, Studium Press, LLC, USA, Vol. 16, pp. 179–219.

Arora, R., Gupta, D., Chawla, R., Sagar, R., Sharma, A., Kumar, R., Prasad, J., Singh, S., Samanta, N. and Sharma, R.K. (2005a) Radioprotection by plant products: present status and future prospects. *Phytotherapy Research* 19, 1–22.

Arora, R., Singh, S., Puri, S.C. and Sharma, R.K. (2008) Himalayan Mayapple (*Podophyllum hexandrum Royle*): Traditional Uses, Clinical Indications and Future Prospects. In: Watson, R.R. and Preedy, V.R. *The Encyclopedia of Herbal Medicine in Clinical Practice*, CABI Publishing, Wallingford, Oxon, United Kingdom, p. 26 (in press).

Baliga, M.S., Jagetia, G. C., Venkatesh, P., Reddy, R. and Ulloor, J.N. (2004) Radioprotective effect of abana, a polyherbal drug following total body irradiation. *British Journal of Radiobiology* 77, 1027–1035.

Beauhamp, C. and Fridovich, I. (1971) Superoxide dismutase: improved assay and an assay applicable to acrylamide gels. *Analytical Biochemistry* 44, 276–287.

Benzie, I. and Strain, J.J. (1996) The ferric reducing ability of plasma (FRAP) as a measure of "antioxidant power" the FRAP assay. *Analytical Biochemistry* 239, 70–76.

Bhattacharya, S., Subramanian, M., Raychowdhury, S., Bauri, A.K., Kamat, J.P., Chattopadhyay, S. and Kumar Bandhopadhyay, S. (2005) Radioprotective property of the ethanolic extract of *Piper betel* leaf. *Journal of Radiation Research* 46,165–171.

Chawla, R., Arora, R., Kumar, R., Sharma, A.K., Prasad, J., Singh, S., Sagar, R.K., Chaudhary, P., Shukla, S., Kaur, G., Sharma, R.K., Puri, S.C., Handa, G., Gupta, V.K. and Qazi, G.N. (2005) Antioxidant activities of fractionated extracts of high altitude *Podophyllum hexandrum* in radiation protection. *Molecular and Cellular Biochemistry* 273, 193–208.

Chawla, R., Arora, R., Singh, S., Sagar, R.K., Sharma, R.K., Kumar, R., Sharma, A., Gupta, M.L., Singh, S., Prasad, J., Khan, H.A., Swaroop, A., Sinha, A.K., Gupta, A.K., Tripathi, R.P. and Ahuja, P.S. (2007) Radioprotective and antioxidant activity of fractionated extracts of berries of *Hippophae rhamnoides*. *Journal of Medicinal Food* 10(1), 101–109.

Chen, L.X., Qu, G.X. and Qiu, F. (2006) Studies on flavonoids of *Andrographis paniculata*. *Zhongguo Zhong Yao Za Zhi* 5, 391–395.

Davies, K.J.A., Lin, S.W. and Pacifici, R. (1987) Protein damage and degradation by oxygen radicals: IV. degradation of denatured protein. *Journal of Biological Chemistry* 262, 9908–9913.

Dua, V.K., Ojha, V.P., Roy, R., Joshi, B.C., Valecha, N., Devi, C.U., Bhatnagar, M.C., Sharma, V.P. and Subbarao, S.K. (2004) Anti–malarial activity of some xanthones isolated from the roots of *Andrographis paniculata*. *Journal of Ethnopharmacology* 95, 247–251.

Gebicki, J.M. and Du, J. (2004) Proteins are major initial cell targets of hydroxyl free radicals. *International Journal of Biochemistry and Cell Biology* 36, 2334–2343.

Gupta, D., Arora, R., Garg, A.P. and Goel, H.C. (2003) Radiation protection of HepG2 cells by *Podophyllum hexandrum* Royale. *Molecular and Cellular Biochemistry* 247, 25–40.

Gupta, D., Arora, R., Garg, A.P., Balas, M. and Goel, H.C. (2004) Modification of radiation damage to mitochondrial system *in vivo* by *Podophyllum hexandrum*: mechanistic aspects. *Molecular and Cellular Biochemistry* 266, 65–77.

Halliwell, B. and Gutteridge, J.M.C. (1997) *Free Radicals in Biology and Medicine*, Oxford University Press, Oxford, UK.

Hari Kumar, K.B. and Kuttan, R. (2004) Protective effect of an extract of *Phyllanthus amarus* against radiation–induced damage in mice. *Journal of Radiation Research* 45, 133–139.

Hidalgo, M.A., Romero, A., Figueroa, J., Cortes, P., Concha, I.I., Hancke, J.L. and Burgos, R.A. (2005) Andrographolide interferes with binding of nuclear factor-kappa β to DNA in HL-60-derived neutrophilic cells. *British Journal of Pharmacology* 144, 680–686.

Joshi, R., Kamat, J.P. and Mukherjee, T. (2007) Free radical scavenging reactions and antioxidant activity of embelin: biochemical and pulse radiolytic studies. *Chemico-Biological Interactions* (in press).

Kamat, J.P. (2006) Peroxynitrite as a potent oxidizing and nitrating agent: A review article. *Indian Journal of Experimental Biology* 44, 436.

Kamat, J.P. and Mishra, K.P. (2006) Herbal Drugs in Radioprotection: Modulation by Phytochemicals. In: Sharma, R.K. and Arora, R. (eds) *Herbal Drugs: A Twenty First Century Perspective*, Jaypee Brothers Medical Publishers Private Ltd, New Delhi, India, pp. 257–267.

Kamat, J.P. and Venkatachalam, S.R. (2004) *Asparagus racemosus* and radioprotection. In: Ramawat, K.G. (ed) *Biotechnology of Medicinal Plants: Vitalizer and Therapeutics*. Scientific Publishing Inc., USA /Oxford/IBH Publishing Company, p.78.

Kamat, J.P., Boloor K.K., Devasagayam, T.P.A. and Kesavan, P.C. (1999) Protection of superoxide dismutase by caffeine in rat liver mitochondria against gamma-irradiation. *Current Science* 77, 286–289.

Kamat, J.P., Boloor, K.K., Devasagayam, T.P.A., Jayashree, B. and Kesavan, P.C. (2000) Differential modification of oxygen-dependent and independent effects of gamma irradiation in rat liver mitochondria by caffeine. *International Journal of Radiation Biology* 76, 1281–1288.

Kamat, J.P., Gopalswamy, U.V., Mohan, H. and Mishra, K.P. (2005) Studies on mechanism of preventive role of diethyldithiocarbamate against oxidative damage. *Indian Journal of Radiation Research* 2, 38–49.

Kamat, J.P., Narurkar, L.M., Mhatre, N.A. and Narurkar, M.V. (1980) Nicotinamide induced microsomal MFO system in rats. *Biochimica et Biophysica Acta* 628, 26.

Kamdem, R.E., Sang, S. and Ho, C.T. (2002) Mechanism of the superoxide scavenging activity of neoandrographolide– a natural product from *Andrographis paniculata* Nees. *Journal of Agricultural and Food Chemistry* 50, 4662–4665.

Menkoviz, N., Foduloovie, K., Bulatovie, V., Alijanziz, I., Juvanis, N., Macura, S., Vajs, V. and Milosavljevis, S. (2002) Xanthones as inhibitors of microsomal lipid peroxidation and TNF alpha induced ICAM–1 expression on human umbilical vein endothelial cells. *Bioorganic Medicinal Chemistry* 10, 3431–3436.

Mhatre, N.A., Kamat, J.P., Narurkar, L.M. and Narurkar, M.V. (1983) Regulation of Hepatic Glutathionine-S-transferase. In: *Proceedings of the Department of Atomic Energy Symposium on Cellular Control Mechanism*, pp. 263–272.

Moron, M.A., Depierre, J.W. and Mannervick, B. (1979) Levels of glutathione, glutathione reductase and glutathione S-transferase activities in rat liver. *Biochimica et Biophysica Acta* 582, 67–68.

Motoki, K., Kobayaski, E., Morita, M., Vchida, T., Akimoto, K., Fushima, H. and Koezuka, Y. (1995) Radioprotective effects of α-galactosylceramides. *Bioorganic Medicinal Chemistry Letters* 5, 2413–2416.

Rades, D., Fehlauer, F., Bajrovic, A., Mahlmann, B., Richter, E. and lberti, W. (2004) Serious adverse effects of amifostine during radiotherapy in head and neck cancer patients. *Radiotherapy Oncology* 70, 261–264.

Rajagopal, S., Kumar, R.A., Deevi, D.S., Satyanarayana, C. and Rajagopalan, R. (2003) Andrographolide, a potential cancer therapeutic agent, isolated from *Andrographis paniculata. Journal of Experimental Therapeutics and Oncology* 3, 147–158.

Rana, A.C. and Avadhoot, Y. (1991) Hepatoprotective effects of *Andrographis paniculata* against carbon tetrachloride–induced liver damage. *Archives of Pharmacal Research* 1, 93–95.

Reyes, B.A., Bautista, N.D., Tanquilut, N.C., Anunciado, R.V., Leung, A.B. and Sanchez, G.C. (2006) Anti-diabetic potentials of *Momordica charantia* and *Andrographis paniculata* and their effects on estrous cyclicity of alloxan-induced diabetic rats. *Journal of Ethnopharmacology* 105,196–200.

Rice Evans, C.A., Miller, N.J. and Paganga, G. (1997) Antioxidant properties of phenolic compounds. *Trends in Plant Sciences* 4, 304–309.

Schreck, R., Rieber, P. and Baeuerle, P.A. (1991) Reactive oxygen intermediates as apparently widely used messengers in the activation of the NF-kβ transcription factor and HIV-1. *The EMBO Journal* 10, 2247–2258.

Scot, S.L., Chen, W.J., Bakac, A. and Espenson, J.H. (1993) Spectroscopic parameters, electrode potentials, acid ionization constants and electron exchange rates of the 2,2′ azinobis (3-ethylbenzothiazoline-6-sulfonate) radicals ions. *Journal of Physical Chemistry* 97, 6710–6714.

Sheeja, K., Shihab, P.K. and Kuttan, G. (2006) Antioxidant and anti-inflammatory activities of the plant *Andrographis paniculata* Nees. *Immunopharmacology and Immunotoxicology* 28, 129–140.

Sies, H. (1997) *Antioxidants in Disease, Mechanisms and Therapy*, Academic Press, New York, USA.

Sonntag, V.C. (1987) *The Chemical Basis of Radiation Biology*, Taylor and Francis, London, UK.

Srivastava, A., Misra, H., Verma, R.K. and Gupta, M.M. (2004) Chemical fingerprinting of *Andrographis paniculata* using HPLC, HPTLC and densitometry. *Phytochemical Analysis* 5, 280–285.

Stadman, E.R. (2006) Protein oxidation and aging. *Free Radical Research* 40, 1250–1258.

Uma Devi, P. and Ganasoundari, A. (1995) Radioprotective effect of leaf extract of Indian medicinal plant *Ocimum santum. Indian Journal of Experimental Biology* 33, 205–209.

Zhou, J., Zhang, S., Choon-Nam, O. and Shen, H.M. (2006) Critical role of pro-apoptotic Bcl-2 family members in andrographolide–induced apoptosis in human cancer cells. *Biochemical Pharmacology* 72, 132–144.

Chapter 17

Mitigation of Deleterious Effects of Ionizing Radiation by Phytoceuticals: Mechanistic Studies With *Centella asiatica*

C.K.K. Nair and Jisha Joy

Introduction

Oxidative free radicals have been implicated in a large number of human ailments such as cardiovascular diseases, neural, liver and degenerative disorders, cancer, diabetes, inflammation and aging, apart from the deleterious consequences and health hazards due to exposures to ionizing radiation. For treating these diseases, preparations from various medicinal plants have been used by the Indian System of Medicine –"*Ayurveda*". The extracts of these medicinal plants contain various bioactive molecules, including vitamins that possess antioxidant and anti-free-radical activity, in addition to other biological activities (Nair *et al.*, 2001, 2003; Weiss and Landauer, 2003). In addition, the extracts of many of these medicinal plants have been found to possess radioprotective properties (Nair *et al.*, 2001; Arora *et al.*, 2005a,b; Arora *et al.*, 2006a,b; Nair, 2006; Maurya *et al.,* 2006).

The present chapter summarizes the radioprotective properties of *Centella asiatica*, an important medicinal plant of the indigenous Ayurvedic system.

Centella asiatica

Centella asiatica (Linn) (Family: Umbelliferae) is an ethnomedicinal plant used in different continents by diverse ancient cultures and tribal groups. In India, it is usually described under the name of 'Mandukaparni' in the Ayurvedic system of medicine (Nadkarni, 1976; Sukh Dev, 2006). Different medical uses are claimed for the plant, the more common being its use as a wound-healing agent (Chopra *et al.,* 1986), which has been attributed to the increased collagen formation and angiogenesis (Shukla *et al.,* 1999). These properties have been ascribed to the active principles, asiatic acid, asiaticoside, madecassic acid and madecassoside. We examined the ability of the hydro-alcoholic extract of this plant with the following aims:

i) to afford protection against gamma radiation-induced damage to plasmid DNA under *in vitro* conditions,
ii) to scavenge free radicals such as DPPH and hydroxyl under *in vitro* conditions and
(iii) to render protection against radiation-induced oxidative stress in animal models under *in vivo* conditions.

Preparation of the *Centella asiatica* extract

Authenticated whole plant parts of *Centella asiatica* L. were obtained from Amala Ayurvedic Hospital and Research Centre, Thrissur, Kerala, India. A known amount of the dry powder was extracted with 70% ethanol. The ethanolic extract was concentrated under vacuum at 45°C and lyophilized to get the extract in powder form with a yield of 17 % under the experimental conditions.

Effect of *Centella asiatica* extract: *In vitro* studies

The ability of *Centella asiatica* extract to protect plasmid (pBR322) DNA against gamma radiation-induced damage was examined, as DNA is the most important target of radiation in living organisms. Since radiation inflicts its deleterious effects through free radicals, the effect of *Centella asiatica* extract on scavenging of free radicals was examined using stable free radical DPPH and Fenton reaction, which generates OH radicals.

Protection of plasmid DNA from gamma radiation-induced damage

Major damage in DNA caused by ionizing radiation is strand breaks. Introduction of strand-breaks results in disappearance of the supercoiled form of plasmid DNA and this can be easily monitored by agarose gel electrophoresis of the DNA. Radiation-induced damage in DNA was determined by irradiating plasmid pBR322 (0.003mg) at 25Gy in the presence and absence of different concentrations of extract up to 10mg/ml. After irradiation, the DNA was electrophoresed in 1% agarose gel with TBE buffer consisting of 0.8mM Tris Borate/2mM EDTA, pH 8.3 (Sambrook *et al.,* 1989; Rajagopalan *et al.,* 2002). The gels were stained with ethidium bromide and DNA bands were photographed and analyzed. Figure 17.1 presents the effect of different concentrations of *Centella asiatica* extract on plasmid pBR322 DNA against gamma radiation-induced strand breaks at 25Gy.

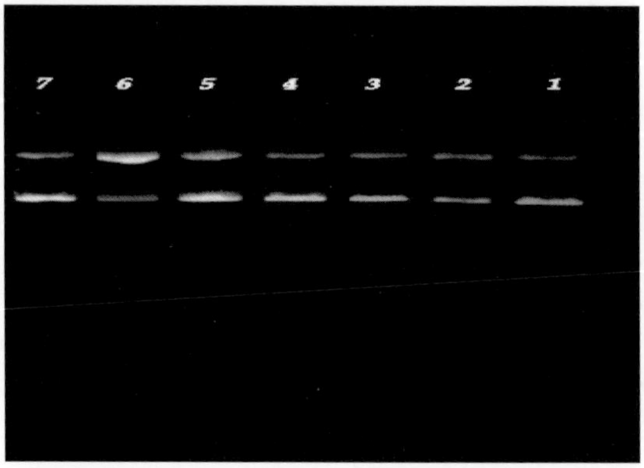

Fig. 17.1. Effect of different concentrations of *Centella asiatica* extract on plasmid pBR322 DNA against gamma radiation-induced strand breaks at 25Gy. Lane**1**- 0Gy, 10mg/ml *C. asiatica* extract; Lane **2**- 25Gy, 2mg/ml *C. asiatica* extract; Lane **3**- 25Gy, 4mg/ml *C. asiatica* extract; Lane **4**- 25Gy, 6mg/ml *C. asiatica* extract; Lane **5**- 25Gy, 10mg/ml *C. asiatica* extract; Lane **6**- 25Gy control; Lane **7**- 0Gy control.

Exposure to gamma radiation leads to DNA strand breaks resulting in relaxation of plasmid DNA from super coiled (ccc) form to open circle (oc) form. It can be seen that the *Centella asiatica* extract reduced ionizing radiation-induced disappearance of ccc form of the plasmid DNA significantly. It was seen that a large portion of the ccc form of plasmid DNA disappeared or got converted into open circular form at a dose of 25Gy (Figure 17.1; lane 6). From the data presented in the figure it can be realized that the extract protects plasmid DNA in a concentration-dependent manner from gamma radiation-induced strand breaks.

Scavenging of DPPH free radicals by *Centella asiatica* extract

DPPH (2,2-diphenyl-1-picryl-hydrazyl) is a stable free radical and has been used as a model free radical compound to evaluate the effectiveness of antioxidants. The free radical scavenging activity of the extract was determined by the method of Gadow and co-workers with some modifications (Gadow *et al.*, 1997). Methanolic solution of DPPH (63.4μM) was incubated at ambient temperature with various concentrations of the extract and absorbance was measured using a spectrophotometer at 515nm (Gandhi and Nair, 2004). The percent inhibition of DPPH reduction (decolourization) was calculated according to the formula:

$$\% \text{ of inhibition} = \frac{A_o - A20}{A_o} \times 100$$

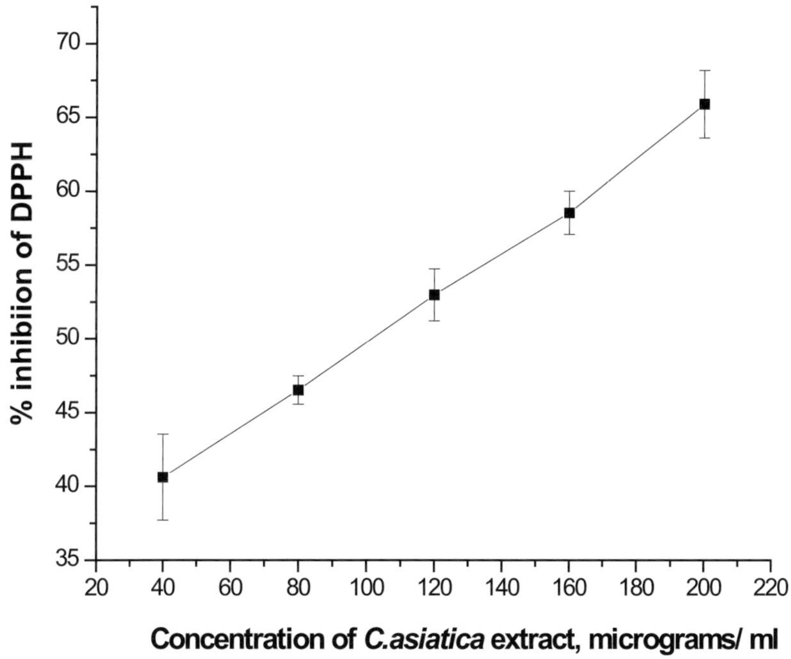

Fig. 17.2. Scavenging of DPPH free radicals by *Centella asiatica* extract.

The data on inhibition of DPPH free radical by various concentrations of *Centella asiatica* extract is presented in Fig. 17.2. It is discernible from the figure that a concentration of 110 micrograms per ml of the extract would bring about 50% inhibition of stable free radical DPPH under *in vitro* conditions.

Scavenging of hydroxyl radicals by *Centella asiatica* extract

Hydroxyl radicals induce oxidative damage to deoxyribose to form malondialdehyde (MDA)-like substance that reacted with thiobarbituric acid (TBA) to give a chromogen with absorbance maxima 532nm. Scavengers of OH radicals protect deoxyribose from radical-mediated oxidative damage. Hydroxyl radicals can be generated by Fenton reaction with Fe^{3+}/ascorbate and EDTA/H_2O_2 systems. The reaction mixture (1.0ml) contained deoxyribose (3mM), FeC13 (0.1mM), EDTA (0.1mM), H_2O_2 (1mM), ascorbic acid (0.1mM), phosphate buffer (20mM; pH 7.4) and various concentrations of the extract. The reaction mixture was incubated at 37°C for 1 hour, followed by addition of 1ml TBA reagent (0.375% thiobarbituric acid, 0.025N HCl, 15% trichloroacetic acid and 6mM EDTA). It was then heated at 90°C for 30 minutes and cooled to room temperature and A_{532} was determined (Elizabeth and Rao, 1990).

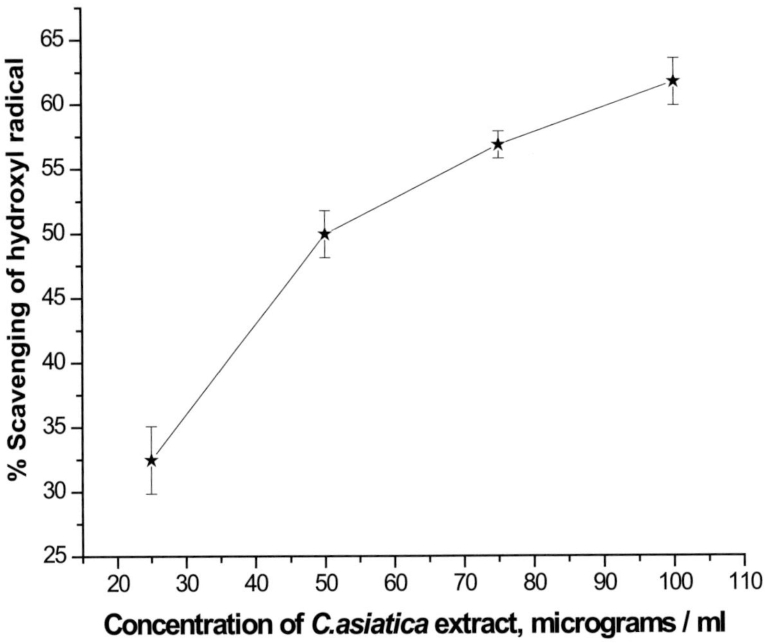

Fig.17.3. Protection of deoxyribose from hydroxyl radicals (Fenton reaction)_due to scavenging of the radicals by *Centella asiatica* extract.

The results presented in Figure 17.3 revealed that OH radicals generated by Fenton reaction caused oxidative damage to deoxyribose and this damage was prevented by *Centella asiatica* extract. Protection to deoxyribose by the *Centella asiatica* extract resulted from scavenging of the OH radicals by the extract in a concentration-dependent manner. Nearly fifty per cent of the OH radicals formed under the reaction conditions were scavenged by the extract at a concentration of 50 micrograms per ml.

In vivo studies in mice to evaluate the effect of *Centella asiatica* extract on radiation-induced oxidative stress

In vivo experiments were carried out in mice. Four to six weeks old male Swiss albino mice (20–25g) were purchased from the Small Animal Breeding Station of Kerala Agricultural University, Mannuthy, Thrissur, Kerala, India, and kept under standard conditions of temperature and humidity in the centre's Animal House Facility. The animals were provided with standard mouse chow (Sai Durga Feeds and Foods, Bangalore, India) and water *ad libitum*. All animal experiments were conducted strictly adhering to the guidelines of Institutional Animal Ethics Committee (IAEC). The toxicity of the extract in mice was first determined by oral feeding. Maximum tolerance dose (MTD) was investigated on the basis of acute toxicity and even a dose of 500mg/kg body weight did not show any toxic symptoms (Goel *et al.*, 2001; Shobi and Goel, 2001). We administered the extract orally at a dose of 100mg/kg body weight for the experiments. The effect of the extract on radiation-induced oxidative stress was investigated by examining:

(i) peroxidative damage in various tissues,
(ii) biochemical parameters of antioxidant status such as levels of reduced glutathione, glutathione peroxidase, superoxide dismutase in animal tissues and
(iii) deleterious effects on haemopoietic system such as hemoglobin levels and blood cell counts.

Peroxidative damage in various tissues

Measurement of lipid peroxidation

Damage to membranes, in cells and tissues, resulting from exposure to gamma radiation can be assessed in terms of peroxidation of membrane lipids (Buege and Aust, 1978). Following exposure of Swiss Albino mice to a sub-lethal dose of whole-body gamma radiation (4Gy) and administration of extract of *Centella asiatica*, the animals were sacrificed by cervical dislocation and various tissues were excised and homogenates were prepared and reacted with thiobarbituric acid and TCA at 90°C for 30 minutes. The reaction mixture was cooled and centrifuged at 10,000 × g for 10 minutes. The amount of thiobarbituric acid reacting substances (TBARS) in the supernatant was estimated by measuring the absorption at 532nm. The lipid peroxidation values are expressed as nano moles of malonyldialdehyde (MDA) per mg protein. Protein in the tissue was calculated by the method of Lowry *et al.* (1951).

Table 17.1 presents the data on lipid peroxidation in different tissues following whole-body gamma irradiation and administration of *Centella asiatica* extract. Whole-body exposure to gamma radiation increased lipid peroxidation in various tissues. Administration of the extract prevented this radiation-induced lipid peroxidation in these tissues.

Table 17.1. Effect of *Centella asiatica* extract on lipid peroxidation of various tissues exposed to whole-body gamma radiation (4Gy). The values are expressed as mean ± S.D.

Treatment	nMoles of MDA/mg protein *(Liver)*	nMoles of MDA/mg protein *(Brain)*	nMoles of MDA/mg protein *(Kidney)*	nMoles of MDA/mg Protein *(Spleen)*	nMoles of MDA/mg protein *(G.I. mucosa)*
Unirradiated control (0Gy)	1.62 ± 0.14	2.64 ± 0.19	0.99 ± 0.13	1.19 ± 0.28	1.38 ± 0.18
Irradiated control (4Gy)	3.12 ± 0.28	4.97 ± 0.33	1.91 ± 0.11	1.85 ± 0.17	3.07 ± 0.26
Centella asiatica (4Gy)	2.11 ± 0.31	1.65 ± 0.17	1.24 ± 0.17	1.33 ±0.162	1.53 ± 0.146

Biochemical parameters of antioxidant status

Determination of tissue reduced glutathione (GSH)

GSH is the major non-protein sulfhydryl reducing agent, which plays an important role in detoxification reactions. It also participates in the reaction that destroys H_2O_2, organic peroxides and free radicals. GSH reacts with DTNB and gets reduced to a yellow-coloured complex which has an absorption maximum at 412nm (Moron *et al.,* 1979). To quantitate GSH, proteins were precipitated from aliquots of tissue-extracts with cold TCA and the protein free supernatant was reacted with DTNB at pH 8.0 in phosphate buffer. After the reaction, the yellow chromophore formed was measured at 412nm. From the standard graph, the quantity of GSH in tissue extracts was calculated and expressed in terms of nanomoles per milligram protein. The values of GSH in different tissues, following whole-body radiation and administration of *Centella asiatica* extract, are presented in Table 17.2. It can be seen from the data that radiation decreased the GSH content in tissues of the liver and brain and administration of *Centella* extract prevented the radiation-induced depletion of GSH in these tissues.

Table 17.2. Effect of *Centella asiatica* extract on levels of GSH in liver and brain tissues of whole-body irradiated (4Gy) mice. The values are expressed as mean ± S.D.

Treatment	GSH (n Moles/ mg protein)	
	Liver	Brain
Unirradiated Control (0Gy)	26.82 ± 2.073	18.01 ± 1.64
Irradiated Control (4Gy)	21.02 ± 1.77	15.027 ± 1.73
Centella asiatica (4Gy)	27 ± 1.77	19.49 ± 1.35

Determination of tissue glutathione peroxidase (Gpx) activity

The main role of GPx is in defence against oxidative damage. The biochemical function of GPx is to reduce lipid hydroperoxides to their corresponding alcohols and to reduce free H_2O_2 to H_2O in presence of GSH. During the reaction, reduced glutathione gets converted to the oxidized form. Levels of GPx in tissue were measured by reacting excess GSH and H_2O_2 in presence of GPx and quantitating the unreacted GSH using DTNB as described above (Hafeman *et al.*, 1974).

$$2GSH + H_2O \longrightarrow GSSG + 2H_2O$$

The GPx activity was calculated and expressed in terms of units per milligram protein.

Table 17.3. Effect of *Centella asiatica* extract on levels of GPx in liver and brain tissues of whole-body irradiated (4Gy) mice. The values are expressed as mean ± S.D.

Treatment	GPx (U /mg protein)	
	Liver	Brain
Unirradiated Control (0Gy)	57.7 3 ± 2.57	36.19 ±2.27
Irradiated Control (4Gy)	28.66 ± 2.84	27.75 ± 2.33
Centella asiatica (4Gy)	42.712 ± 2.59	38.51 ± 1.70

Table 17.3 presents the data on GPx levels in liver and brain tissues of whole-body irradiated (4Gy) mice treated with *Centella asiatica* extract. The extract showed significant protective effect on GPx levels in brain tissues. *Centella asiatica* extract also protects against radiation-induced decrease of GPx in liver and brain of whole-body irradiated mice.

Estimation of tissue superoxide dismutase (SOD) activity

SOD is an intracellular enzyme that catalyses the conversion of superoxide anion to molecular oxygen and hydrogen peroxide. The ability of the enzyme to inhibit the reduction of nitroblue tetrazolium (NBT) by superoxide (O2-), which is generated by the reaction of photo-reduced riboflavin with oxygen, was estimated by the method of McCord and Fridovich (1969).

Enzyme activities were calculated from the inhibition of reduction using a standard curve constructed by using varying amounts of homogenate. One unit of enzyme activity is defined as the amount of enzyme required for giving 50% inhibition of the reduction of NBT and expressed as units/mg protein. Table 17.4 presents the levels of SOD in tissues of liver and brain of whole-body irradiated mice administered *Centella asiatica* extract. SOD levels in liver and brain decreased following exposure to whole-body gamma radiation and administration of the extract protected the tissues from this radiation-induced decrease.

Table 17.4. Effect of *Centella asiatica* extract on levels of SOD in liver and brain tissues of whole-body irradiated (4Gy) mice. The values are expressed as mean ± S.D.

Treatment	SOD(U/mg protein)	
	Liver	Brain
Unirradiated Control (0Gy)	6.90 ± 0.38	4.095 ± 0.64
Irradiated Control (4Gy)	4.26 ± 0.17	2.64 ± 0.30
Centella asiatica (4Gy)	6.56± 0.39	3.64 ± 0.44

Effect of *Centella* extract on radiation-induced alterations in hematopoetic system

Exposure of mammalian organisms to whole-body ionizing radiation produces a characteristic sequence of acute symptoms depending upon the radiation dose (Bond *et al.*, 1965). Haemopoietic syndrome occurs at sub-lethal doses [below LD $_{50/(30)}$], characterized by depression in bone marrow cells, severe anaemic conditions, platelet depletion etc. In the present study we have estimated the haemoglobin levels and monitored the blood cell count in animals following administration of the extract and whole-body radiation exposure. It can be seen in Table 17.5 that in whole-body exposed animals, the Hb levels decreased by about 40% and administration of *Centella asiatica* extract prior to radiation exposure prevented this radiation-induced decrease of haemoglobin.

It is also discernible from Table 17.5 that radiation significantly lowered the total leukocyte counts in irradiated mice. Administration of hydroalcoholic extract of *Centella asiatica* extract was found to prevent this lowering of cell counts.

Table 17.5. Effect of *Centella asiatica* extract on haemoglobin (Hb) levels and blood cell counts in whole-body gamma irradiated mice. The values are expressed as mean ± S.D.

Treatment	Hb g/ dL	WBC/ mm^3
Irradiated control (4Gy)	8.53 ± 1.102	4766.675 ± 292.97
Centella asiatica (4Gy)	11.84 ± 0.776	6816.667 ± 596.51
Unirradiated control (0Gy)	13.46 ± 0.844	8866.667 ± 160.73

Discussion and Conclusion

It has been reported that *Centella asiatica* contains many antioxidant molecules like carotenoids, ascorbic acid and terpenoids (Padma *et al.,* 1998). These antioxidants can scavenge the free radicals produced during irradiation, thus protecting the organism against free radical-induced cytotoxic and genotoxic effects, especially in the radiosensitive gastro intestinal system. GSH is one of the most abundant tripeptide non-enzymatic biological antioxidant which protects cells against reactive oxygen species (ROS). Its functions are concerned with the removal of free radical species such as hydroxyl, superoxide radicals, alkoxy radicals and maintenance of membrane protein thiols. GSH is also a very important constituent of detoxification pathways.

In recent years there has been an increased awareness and interest in the area of radiation and human health. The general public has considered radiation as an enigma and the use of radiation for therapeutic and other uses has always been associated with some scepticism. Currently ionizing radiation is being used in a large number of therapeutic, industrial and other applications, apart from generation of nuclear power and developing new varieties of high-yielding crops and enhancing the storage-period of food materials. Ionizing radiation causes cellular lethality through induction of damage to vital targets such as DNA and membranes. An ionizing event can result in a variety of DNA damage scenarios leading to mutagenesis, cell death etc. The degree of cell damage induced by radiation depends on numerous factors, including the radiation dose, its scheduled administration, the stage of the cell within the cell cycle, the levels of cellular antioxidant defence system and the availability of oxygen in the tissues. The rapidly dividing tissues such as cells of the haematopoetic system are more prone to radiation-induced damage.

Radiotherapy is one of the dominant and effective modes of treatment of cancer, which is in fact one of the leading causes of morbidity and mortality in several populations of the world, and there is a need to protect normal cells from deleterious effects of radiation during radiotherapy. Recent reports indicate the possibility of cancer induction due to exposure of humans to radiation during the medical diagnostic X-rays and CT-scans. Space travel is another area where there is a risk of exposure to ionizing radiation (Wilson, 2000).

Centella asiatica extract protected various tissues of mice from damage resulting from exposure to whole-body gamma radiation and offered excellent protection to haemopoietic system in experimental animals exposed to 4Gy gamma radiation. Since *Centella* extract is non-toxic up to doses of 500mg/kg body weight in animal studies it could be a safe radioprotector for human applications.

References

Arora, R., Chawla, R., Prasad, J., Singh, S., Kumar, R., Sharma, A., Sagar, R., Singh, S. and Sharma, R.K. (2005b) Radioprotective activity of *Rhodiola imbricata*– a high altitude medicinal plant. *Molecular and Cellular Biochemistry* 273, 209–223.

Arora, R., Chawla, R., Singh, S., Sagar, R.K., Kumar, R., Sharma, A.K., Singh, S., Prasad, J., Sharma, R.K. and Tripathi, R.P. (2006b) Radioprotection by Himalayan High-altitude Region Plants. In: Sharma, R.K. and Arora, R. (eds) *Herbal Drugs: A Twenty First Century Perspective*, Jaypee Brothers Medical Publishers Pvt. Ltd., Delhi, India, pp. 301–325.

Arora, R., Chawla, R., Singh, S., Sagar, R.K., Kumar, R., Sharma, A., Prasad, J., Singh, S., Gurudutta, G.U. and Sharma, R.K. (2006a) Bioprospection for Radioprotective Molecules from Indigenous Flora. In: Govil, J.N., Singh, V.K. and Bhardwaj, R. *Recent Progress in Medicinal Plants*, Studium Press, LLC, USA, Vol. 16, pp. 179–219.

Arora, R., Gupta, D., Chawla, R., Sagar, R., Sharma, A., Kumar, R., Prasad, J., Singh, S., Samanta, N. and Sharma, R.K. (2005a) Radioprotection by plant products: Present status and future prospects. *Phytotherapy Research* 19, 1–22.

Bond, V., Fliedner, T.M. and Archambeau, J.O. (1965) *Mammalian Radiation Lethality: A Disturbance in Cellular Kinetics*, Academic Press, Inc., New York, USA.

Buege, A.J. and Aust, S.D. (1978) Microsomal lipid peroxidation. *Methods in Enzymology* 52, 302–310.

Chopra, R.N., Chopra, I.C., Handa, K.L. and Kapur, L.D. (1986) *Indigenous Drugs of India,* UN Dhar and Sons Private Limited, Calcutta, India, p.145.

Elizabeth, S. and Rao, M.N.A. (1990) Oxygen radical scavenging activity of curcumin. *International Pharmacopoeia* 58, 237–240.

Gadow, A., Joubert, E. and Hansmann, C.F. (1997) Comparison of the antioxidant activity of aspalathin with that of other plant phenols of roobios tea (*Aspalathus linearis*) alpha tocopherol, BHT and BHA. *Journal of Agricultural and Food Chemistry* 45, 632–638.

Gandhi, N.M. and Nair, C.K.K. (2004) Radiation protection by diethyldithiocarbonate. Protection of membrane and DNA *in vitro* and *in vivo* against γ-radiation. *Journal of Radiation Research* 45,175–180.

Goel, H.C., Arora, R., Shobi, V. and Mathew, T.L. (2001) A process for preparation of a behavioural radioprotective herbal extract. Indian Patent Office, New Delhi, India (Patent no. 194325).

Hafeman, D.G., Sundae, R.A. and Houster, W.G. (1974) Effect of dietary selenium on erythrocyte and liver glutathione peroxidase in rat. *Journal of Nutrition* 104, 580–588.

Lowry, O.H., Rosebrough, N.J., Lewis, F.A. and Randall, R.J. (1951) Protein measurement with the Folin phenol reagent. *Journal of Biological Chemistry* 193, 262–275.

Maurya, D.K., Devasagayam, T.P.A. and Nair, C.K.K. (2006) Some novel approaches for radioprotection and the beneficial effect of natural products. *Indian Journal of Experimental Biology* 44, 93–114.

McCord, J.M. and Fridovich, I. (1969) Superoxide dismutase, an enzymatic function for erythrocuperin. *Journal of Biological Chemistry* 244, 6049–6055

Moron, M.A., De Pierre, J.W. and Mannerwick, B. (1979) Levels of glutathione, glutathione reductase and glutathione-s-transferase activities in rat liver. *Biochimica et Biophysica Acta* 582, 67–68.

Nadkarni, A.K. (1976) *Indian Materia Medica.* 3rd revised and enlarged edition. Vol. 1. Popular Prakashan Private Limited, Bombay, India, p. 1319.

Nair, C.K.K. (2006) Screening Methodologies for Effective Radioprotectors. In: Sharma, R.K. and Arora, R. (eds) *Herbal Drugs: A Twenty First Century Perspective,* Jaypee Brothers Medical Publishers Pvt. Ltd., New Delhi, India, pp. 237–245.

Nair, C.K.K., Devi, P.U., Shimanskaya, R., Kunugita, N., Murase, H., Gu, Y.H. and Kagiya, T.V. (2003) Water soluble vitamin E (TMG) as a radiopotector. *Indian Journal of Experimental Biology* 41, 1365–1371.

Nair, C.K.K., Parida, D.K. and Nomura, T. (2001) Radioprotectors in radiotherapy. *Journal of Radiation Research* 42, 21–37.

Padma, P.R., Bhuvaneswari, V., Silambuchelvi, K. (1998) The activities of enzymatic antioxidants in selected green leaves. *Indian Journal of Nutrition and Diet* 35 (1), 1–3.

Rajagopalan, R., Khalida, W., Huilgol, N.G., Kagiya, V.T. and Nair, C.K.K. (2002) Inhibition of gamma radiation induced DNA damage in plasmid pBR322 by TMG, a water soluble derivative of vitamin E. *Journal of Radiation Research* 43, 153–159.

Sambrook, K.J., Fritsch, J. and Maniatis, T. (1989) *Molecular Cloning: A Laboratory Manual*, Vol I, 2nd ed. Cold Spring Harbor Laboratory Press, New York, USA.

Shobi, V. and Goel, H.C. (2001) Protection against radiation-induced conditioned taste aversion by *Centella asiatica. Physiology and Behaviour* 73, 19–23.

Shukla, A., Rasik, A.M. and Dhawan, B.N. (1999) Asiaticoside-induced elevation of antioxidant levels in healing wounds. *Phytotherapy Research* 13(1), 50–54.

Sukh Dev (2006) *A Selection of Prime Ayurvedic Plant Drugs. Ancient-Modern Concordance.* Anamaya Publishers, New Delhi, India, p. 501.

Weiss, J.F. and Landauer, M. (2003) Protection against ionizing radiation by antioxidants and phytochemicals. *Toxicology* 189, 1–20.

Wilson, J.W. (2000) Overview of radiation environments and human exposures. *Health Physics* 79, 470–494.

SECTION V

CANCER THERAPY: THE ROLE OF RADIOSENSITIZERS OF NATURAL ORIGIN

Chapter 18

The Radiosensitizing Effects of L-Canavanine

David R. Worthen and Peter A. Crooks

Introduction

L-Canavanine (L-2-amino-[4-guanidinooxy] butanoic acid, L-CAV) is a naturally occurring L-arginine (L-ARG) analogue produced by a variety of leguminous plants (Rosenthal, 1977; Fig. 18.1). L-CAV, the δ-oxa analogue of L-ARG, differs fundamentally from L-ARG in terms of its functional group chemistry. L-ARG has a strongly basic terminal guanidino group with a pKa of approximately 12.5. Thus, in aqueous biological systems, L-ARG residues are almost exclusively protonated and positively charged, thereby serving as important sites for ion-ion and ion-dipole interactions. The exo-form of the L-ARG guanidino tautomer is generally favoured (Soman *et al.,* 1985). In contrast, the guanidinooxy moiety of L-CAV is much less basic. With a pKa near 7.05, L-CAV residues are much less likely to be protonated, and the more rigid guanidinooxy endo-tautomer predominates. These functional group differences have a profound effect on the relative biological effects of L-ARG and L-CAV in many living systems. A substrate for arginyl-tRNA synthetase, L-CAV may be rapidly incorporated in place of L-ARG into newly synthesized proteins in a wide variety of organisms. In addition to adversely affecting the synthesis, stability and cellular transport of proteins in general, L-CAV incorporation may also lead to the formation of non-functional proteins in particular. These non-functional proteins may occur as structural and functional defects, including morphological and developmental aberrations, altered protein conformation and structure, and impaired enzymatic activity, as well as decreased cellular tolerance to heat, radiation and other stressors (Rosenthal, 1977; Laszlo and Li, 1993; Mattei *et al.,* 1992). L-CAV may also exert deleterious effects on DNA replication, RNA synthesis and on RNA translation into protein (Rosenthal, 1977). L-CAV may seriously compromise the stability and function of histones and heat shock proteins, critical, L-ARG-rich cationic chaperones for proteins and other important cellular macromolecules (Chiu and Oleinick, 1997).

L-ARG undergoes arginase [E.C.3.5.3.1]-catalyzed hydrolytic deguanidination to produce L-ornithine (L-ORN), an important precursor to several polyamines, and urea (Rosenthal, 1977; Fig. 18.2). L-CAV is also an arginase substrate, and is analogously cleaved to give urea and L-canaline (L-CAN), a structural analogue of L-ORN with important biological activities of its own. L-CAN may competitively inhibit the cellular transport of L-ORN and other compounds, and acts as a strong nucleophile for many aldehydic and ketone-containing biomolecules. L-CAN is also a potent inhibitor of pyridoxal phosphate (PLP)-dependent enzymes, including the rate-limiting enzyme in polyamine biosynthesis, ornithine decarboxylase [ODC] (Worthen *et al.,* 1996a; Fig. 18.3).

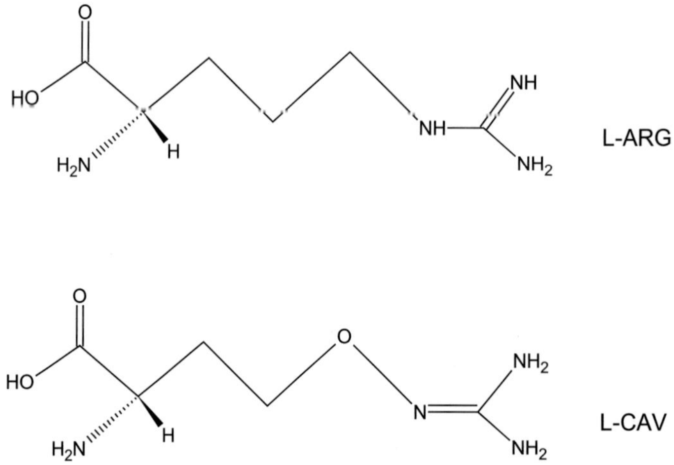

Fig. 18.1. Structures of L-arginine (L- ARG) and L-canavanine (L-CAV).

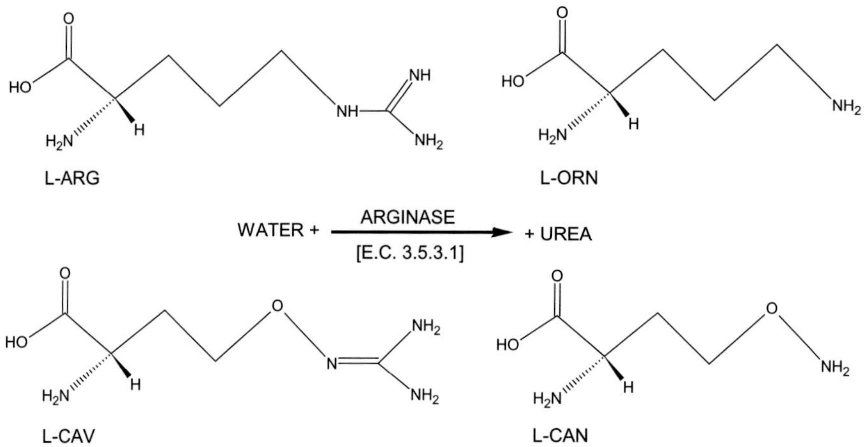

Fig. 18.2. Arginase [E.C. 3.5.3.1]-catalyzed hydrolytic conversion of L-arginine (L-ARG) to L-ornithine (L-ORN), and L-canavanine (L-CAV) to L-canaline (L-CAN).

The diverse biological activities of L-CAV and L-CAN have led to a number of investigations evaluating their potential as anticancer agents. The antineoplastic activity of L-CAV has been examined in a number of studies, either as a single agent or in combination with standard chemotherapeutic agents (Worthen *et al.,* 1998; Bence *et al.,* 2003).

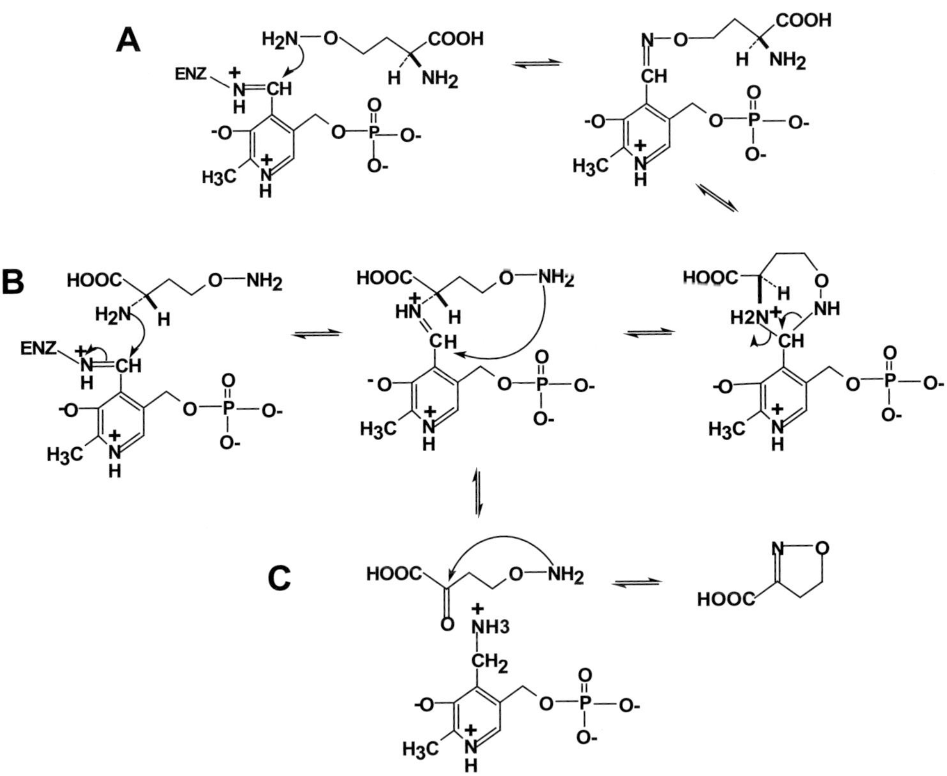

Fig. 18.3. Proposed mechanisms for L-CAN interactions with pyridoxal phosphate (PP)-dependent enzymes, including direct nucleophilic attack on the enzyme-bound co-factor by the L-CAN aminooxy moiety, forming a stable oxime (A); an initial alpha-amino-PP imine formation, followed by an intramolecular aminooxy reaction to produce a stable oxime (B); final inactivation of deaminated, α-keto-L-CAN via an intramolecular condensation of L-CAN to a non-nucleophilic, cyclic product (C). (Reprinted from Worthen *et al.* (1996a) with permission from the American Chemical Society).

In addition, the capacity of L-CAV to sensitize both parental and multi-drug resistant human tumour cells to a variety of standard antineoplastic drugs has been determined (Worthen *et al.*, 1998). L-CAN also shows promise as a novel anticancer agent. The antineoplastic activity of L-CAN and certain derivatives has been evaluated in several studies. It has been suggested that the mechanisms underlying the observed cytotoxic activity of L-CAN may differ from those of the parent compound, L-CAV, giving rise to the possibility of a single parent drug (L-CAV) ultimately exerting multiple, independent antineoplastic effects (NaPhuket *et al.*, 1999). This extensive body of work clearly demonstrates the cytotoxicity and potential utility of L-CAV and L-CAN as antineoplastic and chemosensitizing agents.

Along with chemotherapy, biologics and surgery, ionizing radiation is an important therapeutic modality for the management of many human cancers (Sekhar *et al.*, 2007). In light of the diverse biological activities of L-CAV, many of which may be implicated in the mode of action and efficacy of radiation therapy, L-CAV is being considered as a radiosensitizing agent for the treatment of cancer. A modest number of studies have been conducted in this area in order to assess the potential utility and biological mechanisms of L-CAV as a radiosensitizer. The rational, directed development of L-CAV as a radiosensitizing agent requires a thorough examination of the proposed mechanisms underlying its biological effects in order to appreciate the demonstrated and potential radiosensitizing effects of the molecule.

A review of the radiosensitizing activity of L-CAV might be approached in a number of ways. Clearly, L-CAV manifests profound biological activity in many systems with important implications on the mechanism of action and anticancer effects of ionizing radiation. An analysis of these biological activities and their relationship to radiosensitization might facilitate more focused research efforts based upon individual biochemical and pharmacological pathways. Accordingly, this chapter provides a review of the effects of L-CAV on a number of systems incumbent on the biological effects of ionizing radiation, as well as a summary of published research describing the radiosensitizing effects of L-CAV in living tissues.

L-ARG antagonism and non-functional proteins

As a potent L-ARG antagonist, L-CAV competes with L-ARG for cellular uptake, transport and incorporation into proteins in many organisms. L-CAV inhibits L-ARG uptake in diverse living systems, thereby depriving tissues of this essential amino acid. L-CAV may also significantly decrease the bioavailability of other basic amino acids, including L-lysine (L-LYS) and L-histidine (Michelangeli and Vargas, 1994). L-CAV is extensively incorporated into newly synthesized proteins in place of L-ARG, resulting in a variety of structural and functional defects (Rosenthal, 1977). L-CAV incorporation may disrupt the tertiary and quaternary structure of critical cellular proteins, resulting in the formation of so-called non-functional proteins (Rosenthal *et al.,* 1989). These non-functional proteins may include cellular transporters, enzymes, metabolic proteins and a host of other macromolecules critical for normal cell function (Rosenthal, 1977; Rosenthal, 1991). In addition to causing the formation of non-functional proteins, canavanyl proteins incorporating L-CAV may also be degraded faster than native proteins, thereby further compounding the increased metabolic stress and cellular energy demands induced by L-CAV (Ballard and Knowles, 1977).

Since L-CAV may be incorporated into proteins throughout an organism, including those proteins necessary for radiation resistance and repair, its potential as a radiosensitizing agent is of particular interest. It has long been recognized that basic amino acids and basic proteins may serve as endogenous radioprotective molecules (Efimov *et al.,* 1969). Thus, the replacement of L-ARG by L-CAV may be particularly deleterious to irradiated tissues. L-CAV induced disruption of certain tissue-specific proteins, such as caveolin-1, an integral membrane protein believed to mediate radiation resistance in pancreatic adenocarcinoma, may one day be developed into an important therapeutic modality (Cordes *et al.,* 2007).

Other promising targets for L-CAV-mediated radiosensitization include histones and heat shock proteins.

Histones

Histones, the primary protein components of chromatin, comprise a group of nuclear proteins that are critical to DNA assembly, compaction and stabilization (Villar-Garea and Imhof, 2006). Histone structure is evolutionarily conserved across a vast number of organisms, a characteristic that underscores their enormous biological importance, while suggesting that deleterious mutations in their structure or function would seriously impair critical cellular processes. These basic proteins interact with DNA through dispersion forces, helix-dipoles, hydrogen bonds and salt bridges between DNA phosphate oxygen atoms and the basic, cationic functional groups of L-LYS and L-ARG. These basic functionalities are also required for histone water solubility, membrane association and intracellular transport. The complexation of histones with mammalian DNA protects DNA from radiation-induced strand breakage (Chiu and Oleinick, 1997). Certain histone subtypes bind to segments of DNA that contain double strand breaks, thereby marking damaged cellular DNA for repair (Cerna *et al.*, 2006). Structurally modified histones that have undergone post-translational modification or acylation may regulate cell cycle distribution and cytostasis in specific ways (Karagiannis *et al.*, 2004).

Since DNA is a primary target for the cytotoxic effects of radiation, which include double strand DNA breaks, compounds that induce deleterious changes in histone structure, function and biochemical modification have been investigated as radiosensitizing agents. Numerous compounds that alter histone structure and subsequent chromatin formation sensitize tissues to radiation damage while concurrently disrupting or inhibiting normal cellular DNA repair processes, thereby amplifying radiation-induced tumour cell death (Flatmark *et al.*, 2006). The inclusion of L-CAV in yeast media completely blocked histone incorporation into yeast chromatin (Childs *et al.*, 1971). L-CAV is also a powerful antagonist of L-ARG incorporation into mammalian histone proteins, blocking L-ARG incorporation into histones in a dose-dependent manner (Ackermann *et al.*, 1965). Since the cationic L-ARG guanidino moiety is critical to the salt bridging, protein solubilization and electrostatic interactions required for chromatin structure, it is plausible that L-CAV incorporation may seriously compromise DNA-histone interactions. Notably, L-CAV pre-treatment significantly curtails functional histone synthesis and increases cellular vulnerability to treatment by DNA lesion-inducing chemicals in some living organisms (Subramani *et al.*, 1983). More extensive analysis of the rate, extent and schedule-dependent effects of L-CAV exposure on histone synthesis, structure and function may lead to the development of effective, histone-targeted L-CAV radiosensitizing therapies.

Heat shock proteins (HSP)

Heat shock proteins (HSP) serve as molecular chaperones for other cellular proteins (Csermely and Yahara, 2003). HSP play a prominent role in facilitating precise protein folding, and they regulate protein transport across membranes in living organisms. HSP expression is increased when cells are exposed to heat, chemicals, radiation and other stressors (Kumar *et al.*, 2005a,b; Kumar *et al.*, 2006). In order to overcome the protective role of HSP during this so-called stress response, thereby sensitizing cells to radiation and other insults, various methods have been explored to inhibit HSP synthesis and function. Cell radiosensitization has been accomplished by treating tumour cells with nanomolar concentrations of 17-AAG, an inhibitor of HSP 90, followed by X-ray irradiation (Russell *et al.*, 2003). Exposure to 17-AAG reduced the expression of three cellular proteins associated with radiosensitivity, and cellular radiosensitivity increased between 1.3- and 1.7-fold. Notably, this HSP inhibitor had no effect on normal human fibroblasts, and appeared to act exclusively on transformed tissues.

Agents that inhibit HSP synthesis and function may serve as effective radiosensitizing agents. L-CAV has a profound effect on HSP synthesis and function, and may serve as an effective radiosensitizing compound through HSP-related mechanisms. Like a number of amino acid analogues, L-CAV induces the protective heat shock response in a variety of organisms (Olsen *et al.*, 1983). However, unlike other stressors, L-CAV fails to induce thermotolerance, despite this increase in HSP expression. Human melanoma cells treated with L-CAV became significantly more sensitive to the cytotoxic effects of heat than untreated controls (Mattei *et al.*, 1992). The induction of the synthesis of apparently non-functional HSP was dose- and time-dependent. L-CAV treatment rendered melanoma cells more sensitive to heat, and disrupted cellular control and feedback mechanisms responsible for the regulation of HSP synthesis.

The phenomenon of L-CAV-induced HSP synthesis with a concurrent loss of thermotolerance has been observed in a number of other tissue types. Chinese hamster fibroblasts treated with L-CAV were unable to mount a protective thermotolerant response, following heat or sodium arsenite treatment (Laszlo *et al.*, 1993). Here, HSP synthesis was again upregulated, but their protective function upon cellular exposure to heat or radiation was absent. L-CAV exposure increased thermosensitivity in parental Chinese hamster fibroblasts, but not in heat-resistant variants (Li *et al.*, 1985). The authors suggest that L-CAV induces the formation of non-functional HSP, and conclude that heat-resistant cells already contain a high level of constitutive, functional HSP, thereby mitigating the deleterious effects of L-CAV on thermotolerance. Thus, L-CAV pre-treatment may be the most effective means of inducing thermal- and radiosensitization in target tissues.

Nucleic acid synthesis

The nuclear material of tumour cells is a primary target for ionizing radiation and many alkylating, radical generating and intercalating chemotherapeutic drugs. Cellular DNA and RNA are particularly sensitive to these therapies, and compounds that alter nucleic acid structure, synthesis and repair are important candidates for development as radiosensitizing agents. L-CAV inhibits DNA and RNA synthesis in a variety of organisms, and its incorporation into nuclear proteins may result in extended inhibition of DNA synthesis long after L-CAV exposure is withdrawn (Naha, 1981). L-CAV significantly inhibits DNA and RNA synthesis in mammalian, microbial, insect and plant tissues (Miller *et al.*, 1974; Bell, 1974; Weaks *et al.*, 1973). L-CAV rapidly blocked DNA synthesis, as measured by tritiated thymidine incorporation, in Novikoff hepatoma cells (Doyle *et al.*, 1983). L-CAV treatment also inhibited RNA and DNA synthesis in mouse and hamster embryonic cells, an effect that was reversible when additional L-ARG was added to the culture medium (Hare, 1969). Rather than a non-specific inhibitory mechanism, the author suggested that DNA synthesis was inhibited by the formation of an abnormal canavanyl protein that is required for DNA synthesis.

The same observation was made in *Escherichia coli*, where reduced DNA synthesis was attributed to the formation of non-functional canavanyl proteins that are otherwise vital to the normal replication of the bacterial genome. These effects on nucleic acid synthesis underscore the potential of L-CAV as an anticancer and radiosensitizing agent, and they may also have applications as inhibitors of cellular repair and resistance mechanisms (Naha, 1981).

Polyamines

The polyamines comprise a group of nitrogen-rich biomolecules that play a critical role in molecular condensation and transport, as well as in nucleic acid and DNA complexation, stabilization and in other biochemical pathways (Warters *et al.*, 1999). Important polyamines include putrescine, produced by the ODC-catalyzed decarboxylation of L-ORN, cadaverine, the decarboxylated backbone of L-lysine, agmatine, the decarboxylated metabolite of L-ARG, and spermidine and spermine, produced by the successive substitution of putrescine (Fig. 18.4). At physiologic pH, these basic, protonated polyamines are cationic, and thereby form strong, polyvalent electrostatic interactions with anionic biomolecules, including the polyphosphate backbone of DNA. The interaction of polyamines with nucleic acids is critical to cell growth and differentiation in many organisms, and may strongly influence the effects of ionizing radiation in living tissues (Wallace *et al.*, 1986; Warters *et al.*, 1999).

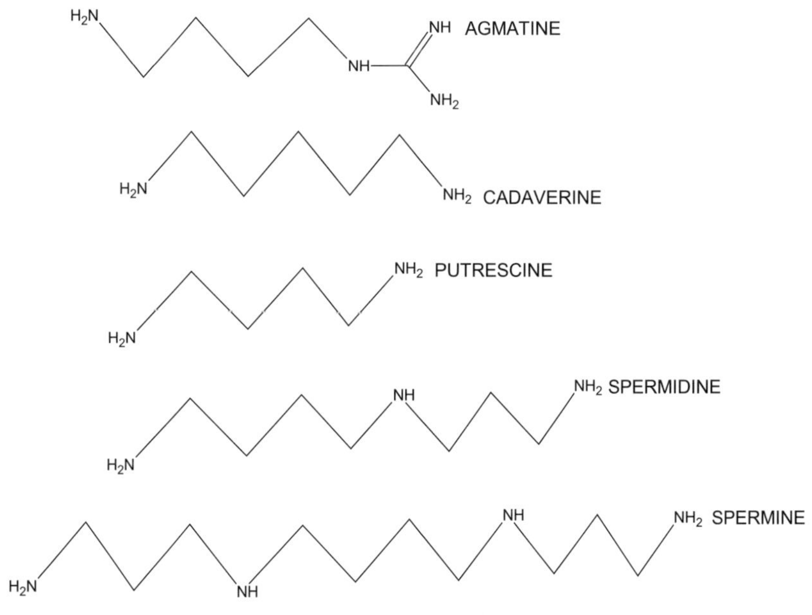

Fig. 18.4. Representative polyamines include putrescine, spermidine, spermine, cadaverine and agmatine.

The polyamines putrescine and spermine protect human cell DNA against double-stranded DNA breaks induced by ionizing radiation (Warters *et al.*, 1999). Tightly bound in the nucleus, spermine associates avidly with nuclear chromatin at physiologically relevant concentrations, and may be critical to promoting the resistance of cellular DNA to radiation-induced double strand breaks. The polyamines may also serve as important scavengers of radiation-induced radicals, highly reactive compounds with unpaired electrons which might otherwise damage DNA, membrane lipids and other critical biomolecules. In light of these observations, researchers have evaluated compounds that interfere with the production, transport and function of polyamines as potential radiosensitizers.

Cellular polyamine depletion, either by reduced precursor availability, impaired transport, blockade of polyamine biosynthesis, or reaction with sequestering molecules, may induce radiosensitization in both normal and neoplastic tissues. In cultured HeLa cells, depletion of cellular polyamines by treatment with polyamine synthesis inhibitors significantly increases cellular sensitivity to X-ray induced damage and disrupts the normal cell cycle (Snyder *et al.,* 1994). Defects in repair mechanisms and lengthened phase-delay in the relatively radiosensitive G2 cell phases may underlie this radiosensitizing effect. CAL 18 breast carcinoma cells treated with difluoromethylornithine (DFMO), like L-CAN an ODC inhibitor, were significantly more sensitive to ionizing radiation than were untreated controls (Courdi *et al.,* 1986). DFMO-induced polyamine depletion enhanced the radiation response and profoundly inhibited radiation-induced potentially lethal damage (PLD) repair after only a 1 hour exposure to the compound.

L-CAV and its arginase metabolite L-CAN inhibit polyamine synthesis, transport and function in a variety of tissues. As such, L-CAV may exert polyamine-related radiosensitizing effects in tumours through these polyamine-related mechanisms. L-CAV inhibits both ornithine- and agmatine-related polyamine pathways (Schwartz *et al.,* 1997). L-CAN is a potent inhibitor of pyridoxal phosphate (PLP)-dependent enzymes, including the rate-limiting enzyme in polyamine biosynthesis, ODC (Worthen *et al.,* 1996a). Notably, L-CAV inhibits polyamine synthesis during the critical, radiosensitive M phase in some organisms, which may further enhance radiation-induced damage (Bagni *et al.,* 1981). In addition to inhibiting polyamine biosynthesis, L-CAV interferes with polyamine transport and efflux in mammalian cells, and adversely affects polyamine acetylation and metabolism (Wallace *et al.,* 1986). Polyamine supplementation in polyamine-depleted tissues may reverse radiosensitization, and many tumour types exhibit increased polyamine influx activity (Snyder *et al.,* 1994). Thus, polyamine transport inhibition by L-CAV and L-CAN may further augment radiation therapy by interfering with this potentially protective adaptation. Since L-CAV and L-CAN appear to operate by independent, yet complementary, mechanisms, they may prove to be versatile radiosensitizers via their diverse effects on polyamine biochemistry (Bence *et al.,* 2002; NaPhuket *et al.,* 1999).

Other biological effects

In addition to its effects on macromolecular synthesis, stability and function, L-CAV has additional characteristics that could impact radiosensitivity and cytotoxicity in tumour cells. An inhibitor of constitutive and inducible nitric oxide synthase, L-CAV might alter tumour blood flow and perfusion, perhaps restricting tumour blood supply and inducing a potentially toxic environment containing elevated levels of highly reactive superoxide (Pass *et al.,* 1996). In addition to L-CAN, L-CAV may be converted to other potentially toxic metabolites *in vivo* by the action of amino acid oxidases. L-CAV undergoes oxidation to form hydroxyguanidine and vinylglyoxylate, a highly electrophilic substance that may react with cellular macromolecules, further disrupting normal function (Hollander *et al.,* 1989). These secondary cytotoxic effects may improve response to radiation therapy in a number of ways. Further, the toxic effects of L-CAV appear to be selective for rapidly proliferating cells (Naha *et al.,* 1980). Indeed, L-CAV appears to induce apoptotic cell death in human

Jurkat T leukemia cells through a highly specific, cytochrome c-independent caspase-3 activation pathway (Jang *et al.,* 2002).

L-Canavanine as a Radiosensitizing Agent

Practical examination of the profound and diverse biological activities exerted in transformed tissues by L-CAV has led to important studies on its effects as an adjunct to standard cancer therapies. L-CAV is equally cytotoxic to both parental and multi-drug resistant (MDR) cancer cells, which are often refractory to chemotherapy and resistant to radiation. L-CAV significantly increases cellular sensitivity to standard chemotherapeutic drugs, in particular, to those that alter DNA structure and function, as well as to those that may act as radical generators (Worthen *et al.,* 1998). Since ionizing radiation induces DNA damage and generates reactive radicals *in vivo*, the radiosensitizing effects of L-CAV in living tissues have been similarly explored.

Compounds such as L-CAV, which inhibit DNA synthesis and repair and also interfere with normal cellular metabolism and structure, are known to induce radiosensitization. At minimally cytotoxic levels, L-CAV was nevertheless an efficient radiosensitizer in the human colonic tumour cell line, HT-29 (Green and Ward, 1983). These researchers assessed L-CAV-induced radiosensitization in cells cultured in media containing serial ratios of L-CAV: L-ARG (range 2.5–20:1). Radiosensitization was evaluated in cancer cells exposed to L-CAV for various periods of time (12 to 48 hours), both before and after irradiation at 150 rads/min from a ^{137}Cs generator (total dose 150 to 900 rads). Notably, as compared to untreated controls, more than 90% of DNA synthesis was inhibited in cells exposed to L-CAV for a mere 6 hours.

Maximum enhancement of killing by irradiation was noted in HT-29 tumour cells grown in 100mM/ml L-CAV for 48 hours. Increasing L-CAV exposure time did not alter this radiosensitizing effect, while incubation with L-CAV for shorter time periods reduced L-CAV radiosensitization. L-CAV radiosensitization was also dose-dependent. Treatment of HT-29 tumour cells with lower concentrations of L-CAV resulted in correspondingly lower enhancements of cell death. Irradiation was significantly more lethal to cells treated with L-CAV both before and after irradiation. This is an important finding, as it suggests that L-CAV may sensitize existing and newly synthesized cellular components and processes to the effects of irradiation through a number of mechanisms. In addition, L-CAV may impair the rate, accuracy and efficiency of post-irradiation repair mechanisms, thereby reducing tumour resistance to radiation.

Concurrent cell proliferation experiments in both HT-29 colon tumour cells and in mitogenically transformed human lymphocytes revealed that L-CAV preferentially exerts its lethal effects in rapidly proliferating cells, while a summary of previous literature indicates that L-CAV is preferentially cytotoxic to tumour cells *in vivo*. These observations suggest the possibility of selective toxicity for neoplastic tissues, as well as those healthy tissues already known to be affected by standard cancer therapies. Accordingly, L-CAV therapy might be appropriately monitored in the clinic by including standard side effect profiles and clinical endpoints in therapeutic protocols. The authors note that since non-proliferating cells still synthesize and presumably incorporate L-CAV into nascent proteins, L-CAV radiosensitization may be most relevant to those processes involved with cellular proliferation.

These encouraging findings demonstrated the clear dose-, time- and schedule-dependent radiosensitizing properties of L-CAV in human tumour cells. Possible mechanisms underlying L-CAV radiosensitization and synergy were further explored in two human pancreatic tumour cell lines: PANC-1 and MIA PaCa-2 (Bence *et al.*, 2003). In this comprehensive study, the *in vitro* effects of L-CAV on cell cycle distribution were evaluated across serial L-CAV concentrations and exposure times. The synergistic effects of L-CAV in combination with ionizing radiation on killing tumour cells were examined using median-effect and combination index analysis.

Many effective radiosensitizing agents cause cells to accumulate in the G2/M phase of the cell cycle, where DNA and associated structures may be particularly vulnerable to radiation-induced strand breaks and radical generation. It was hypothesized that an appreciation of the effect of L-CAV on cell cycle distribution might thereby offer mechanistic insight into its radiosensitizing activity. In this study, human pancreatic cancer cells were grown in media containing L-CAV (0.5 to 3mM) for 24, 48 or 72 hours. Propidium iodide staining with flow cytometry analysis permitted quantification of the percentage of viable cells that were in the G1, S or G2/M phases of the cell cycle.

Exposure to the lower L-CAV concentrations for 24 or 48 hours did not significantly alter cell cycle distribution. However, MIA PaCa-2 cells exhibited enhanced accumulation in the G2/M phase after exposure to 3mM L-CAV for 24 and 48 hours (25.5 and 35.8 percent, respectively). Similarly, PANC-1 cells accumulated in the G2/M phase after incubation with 3mM L-CAV for 24 and 48 hours (37.6 and 38.1 percent, respectively). Longer L-CAV exposure produced more dramatic results. The percentage of MIA PaCa-2 cells in the G2/M phase increased 2-fold after a 72 hour exposure to 2mM L-CAV; a 4-fold increase was observed after analogous incubation in 3mM L-CAV. PANC-1 cells manifested a 6-fold increase in the percentage of cells in the G2/M phase after a 72-hour exposure to 3mM L-CAV. A concomitant decrease in the percentage of pancreatic cells in the less radiosensitive G0/G1 phase was noted with these changes.

Important control experiments were conducted in order to determine whether the G2/M accumulation was due to L-CAV itself, as opposed to reduced L-ARG uptake into the cells due to transport competition by L-CAV. Notably, pancreatic cancer cells cultivated for as long as 72 hours in an L-ARG-free environment did not display an increase in G2/M phase accumulation. Cell cycle accumulation was not affected by exposure to other L-ARG analogues, including D-ARG, L-homoarginine, which, like L-CAV, can also be incorporated into proteins in place of L-ARG, L-indospicine or the L-CAV derivative, L-CAN. The authors concluded that the G2/M accumulation was an effect attributable to L-CAV itself, rather than to non-specific L-ARG antagonism, amino acid transport inhibition or non-specific replacement of L-ARG incorporation into proteins. L-CAV appears to be a specific, time- and concentration-dependent inducer of cell cycle accumulation in the vulnerable G2/M phase.

A concurrent study on the radiosensitizing activity of L-CAV in these human pancreatic cancer cell lines was conducted using a standardized synergistic approach based upon the concentration of L-CAV in the incubation medium as a function of radiation dose. Here, the cells were incubated for 72 hours with serial doses of L-CAV, such that the final L-CAV (mM): radiation (Gy) ratio ranged from 1:1 to 1:40. After a 72-hour incubation with the appropriate concentration of L-CAV, the cells were irradiated, grown another 72 hours in L-CAV-free media, and then assayed for viability. In order to more precisely assess the interaction between L-CAV and radiation treatments, as well as determine synergism versus antagonism, both median effect and combination index analyses were employed to evaluate the results of the combined L-CAV radiation viability experiments.

As compared to controls, both PANC-1 and MIA PaCa-2 cells were significantly more sensitive to the cytotoxic effects of radiation after exposure to L-CAV for 72 hours. L-CAV was found to be synergistic with radiation at all of the L-CAV: radiation dose ratios tested. For a 72h exposure, the ratio of L-CAV (mM): radiation (Gy) exposure that was most effective at killing tumour cells was 1:10. Even at a corresponding dose ratio of 1:40, cytotoxic synergy between L-CAV and radiation treatments was observed. In pancreatic cancer cells, the synergistic effects of combined L-CAV and radiation exposure appeared to be schedule-dependent. Cells exposed to L-CAV 72 hours before irradiation were significantly more sensitive to radiation. In contrast, those cells treated with L-CAV 72 hours after irradiation, and those incubated with L-CAV both 72 hours before and after irradiation, were, at best, only modestly sensitized to the effects of ionizing radiation. This study suggests that the radiosensitizing effects of L-CAV may be more closely associated with cell cycle alterations, and consequent accumulation in the more radiosensitive G2/M cell phase, than with deleterious modulation of post-irradiation DNA and cellular repair mechanisms.

These schedule-dependent radiosensitization effects are consistent with those previously observed in other cancerous tissues, including CFPAC-1, another human pancreatic adenocarcinoma cell line, as well as in the human uterine sarcoma cell line MESSA and its multi-drug resistant (MDR) variant, Dx-5 (Worthen *et al.,* 1996b). In this study, cancer cells were incubated in media containing 250μM L-CAV for 72 hours, then returned to L-CAV-free media for 24 hours before irradiation. Cell viability was then determined 72 hours post-irradiation (total dose 125 to 2000 rads from a ^{137}Cs source). As determined by trypan dye exclusion, tritiated thymidine uptake and total protein analyses, L-CAV treatment did not by itself significantly affect cell viability. However, as compared to irradiated controls not previously exposed to L-CAV, all three lines of L-CAV-treated cancer cells were some 20 to 53 percent more sensitive to the lethal effects of radiation. Here, the radiosensitivity of cancer cells increased steadily and significantly with an increasing radiation dose. The magnitude of this increase in radiosensitivity was consistent in both parental and MDR cancer cells. This effect is clinically relevant, as MDR tumours resistant to chemotherapy are often resistant to radiation treatment as well. As was observed in the study reported by Bence *et al.* (2003), L-CAV radiosensitization was noted only when cancer cells were cultured with L-CAV before irradiation. While both 48- and 72-hour L-CAV treatment with either 100 or 250μM L-CAV before irradiation increased radiosensitivity, analogous post-irradiation treatment did not significantly affect cell viability or radiation lethality.

Conclusions

L-CAV is a promising radiosensitizing agent for the treatment of cancer. This L-ARG analogue exerts potent biological effects in diverse living systems, and may increase tissue sensitivity to ionizing radiation in a number of ways. L-CAV preferentially sensitizes cancer tissues to ionizing radiation at non-cytotoxic doses, and appears to act primarily by altering pre-irradiation cell cycling. These radiosensitizing effects are exerted in numerous tissue types, including parental and multi-drug resistant cancer cells. In addition, unlike standard radiation and chemotherapeutic treatments, L-CAV does not appear to induce multi-drug resistance in tumour cells, a characteristic that may allow continuous or multiple L-CAV treatments while maintaining therapeutic effectiveness (Worthen *et al.*, 1998).

Although the influence of L-CAV on tumour cell sensitivity to ionizing radiation, chemotherapeutic drugs and therapeutic resistance is clear, this important natural product still affords many opportunities for fundamental mechanistic and therapeutic studies. L-CAV may induce cellular accumulation in the G2/M phase, but the pathways underlying this effect are unknown. L-CAV sensitizes tumour cells to intercalating and radical-generating chemotherapeutics, as well as to ionizing radiation, therapeutic modalities primarily associated with nucleic acid damage, immobilization and scission.

However, the complicated and simultaneous influence of L-CAV exposure on DNA and RNA synthesis, polyamine biochemistry and histone and heat shock protein function render the elucidation of this nucleic acid sensitization mechanism profoundly complex. Nor is L-CAV without potential toxicity. In addition to the general antimetabolic effects of L-CAV and its arginase metabolite, L-CAN, L-CAV treatment may be immunotoxic to certain patients, a particularly untoward effect in persons already beset with neoplastic disease (Bence *et al.*, 2002). The development and optimization of effective L-CAV dosing regimens is an unexplored area, as the pharmacokinetic, pharmacodynamic and metabolic characterization of this compound in humans has not been extensively described. The design and evaluation of less toxic, more effective L-CAV analogues may be warranted; likewise, the design and evaluation of L-CAV pro-drugs and pharmaceutical formulations for optimizing L-CAV bioavailability and tumour delivery. This promising radiosensitizing agent offers many opportunities for basic and applied research, and the elucidation of its mechanisms of action will provide significant insight into the directed development of adjuncts for the effective radiation therapy of cancer.

References

Ackermann, W.W., Cox, D.C. and Dinka, S. (1965) Control of histone and DNA synthesis with canavanine, puromycin and polio virus. *Biochemical and Biophysical Research Communications* 19, 745.

Bagni, N., Torrigiani, P. and Barbieri, P. (1981) Effect of various inhibitors of polyamine synthesis on the growth of *Helianthus tuberosus*. *Medical Biology* 59, 403–409.

Ballard, F.J. and Knowles, S.E. (1977) Increased degradation rates of canavanine-containing proteins in hepatoma cells. *Intracellular Protein Catabolism* 43–48.

Bell, D. (1974) Effect of canavanine on herpes simplex virus replication. *Journal of General Virology* 22, 319–330.

Bence, A.K., Adams, V.R. and Crooks, P.A. (2003) L-canavanine as a radiosensitization agent for human pancreatic cancer cells. *Molecular and Cellular Biochemistry* 244, 37–43.

Bence, A.K., Worthen, D.R., Adams, V.R. and Crooks, P.A. (2002) The antiproliferative and immunotoxic effects of L-canavanine and L-canaline. *Anticancer Drugs* 13, 313–320.

Cerna, D., Camphausen, K. and Tofilon, P.J. (2006) Histone deacetylation as a target for radiosensitization. *Current Topics in Developmental Biology* 73, 173–204.

Childs, E.A., Ayres, J.C. and Koehler, P.E. (1971) Differentiation in *Aspergillus flavus* as influenced by L-canavanine. *Mycologia* 63, 181–184.

Chiu, S.M. and Oleinick, N.L. (1997) Radioprotection against the formation of DNA double-strand breaks in cellular DNA but not native cellular chromatin by the polyamine spermine. *Radiation Research* 148, 188–192.

Cordes, N., Frick, S., Brunner, T.B., Pilarsky, C., Grutzmann, R., Sipos, B., Kloppel, G., McKenna, W.G. and Bernhard, E.J. (2007) Human pancreatic tumour cells are sensitized to ionizing radiation by knockdown of caveolin-1. *Oncogene* 1–12.

Courdi, A., Milano, G., Bouclier, M. and Lalanne, C.M. (1986) Radiosensitization of human tumour cells by alpha-difluoromethylornithine. *International Journal of Cancer* 38, 103–107.

Csermely, P. and Yahara, I. (2003) Heat Shock Proteins. In: Keri, G. and Toth, I. (eds) *Molecular Pathomechanisms and New Trends in Drug Research*, Taylor and Francis, London, pp. 67–75.

Doyle, D.G., Bratton, G.R. and Natelson, S. (1983) Effect of canavanine and 2,3-dimercapto-1-propanol (British antilewisite) on the proliferation of Novikoff hepatoma cells in the presence of concanavalin A. *Clinical Physiology and Biochemistry* 1, 305–317.

Efimov, M.L., Yakovleva, S.S. and Ismailov, B.I. (1969) Histochemical characteristics of radioresistant and radiosensitive tumours. *Voprosy Onkologii* 15, 94–96.

Flatmark, K., Nome, R.V., Folkvord, S., Bratland, A., Rasmussen, H., Ellefsen, M.S., Fodstad, O. and Ree, A.H. (2006) Radiosensitization of colorectal carcinoma cell line by histone deacetylase inhibition. *Radiation Oncology* 1, 25.

Green, M.H. and Ward, J.F. (1983) Enhancement of human tumor cell killing by L-canavanine in combination with gamma-radiation. *Cancer Research* 43(9), 4180-4182.

Hare, J.D. (1969) Reversible inhibition of DNA synthesis by the arginine analog canavanine in hamster and mouse cells *in vitro*. *Experimental Cell Research* 58, 174.

Hare, J.D. (1970) Nuclear alterations in mammalian cells induced by L-canavanine. *Journal of Cell Physiology* 75, 129.

Hollander, M.M., Reiter, A.J., Horner, W.H. and Cooper, A.J.L. (1989) Conversion of canavanine to α-keto-γ-guanidinooxybutyrate and to vinylglyoxylate and 2–hyrdoxyguanidine. *Archives of Biochemistry and Biophysics* 270, 698–713.

Jang, M.H., Jun, D.Y., Rue, S.W., Han, K.H., Park, W. and Kim, Y.H. (2002) Arginine antimetabolite L-canavanine induces apoptotic cell death in human Jurkat T cells via caspase–3 activation regulated by Bcl-2 or Bcl-xL. *Biochemical and Biophysical Research Communications* 259, 283–288.

Karagiannis, T.C., Smith, A.J. and El Osta, A. (2004) Radio- and chemo-sensitization of human erythroleukemic K562 cells by the histone deacetylase inhibitor Trichostatin A. *Helvetica*

Journal of Nuclear Medicine 7, 184–191.
Kumar, R., Singh, P.K., Arora, R., Sharma, A., Prasad, J., Sagar, R.K., Singh, S. and Sharma, R.K. (2005a) Radioprotection by *Podophyllum hexandrum* in the liver of mice: A mechanistic approach. *Environmental Toxicology and Pharmacology* 20, 326–334.
Kumar, R., Singh, P.K., Arora, R., Sharma, A.K. and Sharma, R.K. (2006) Modulation of Apoptosis-inducing Protein in Mouse Kidney by *Podophyllum hexandrum*. In: Govil, J.N., Singh, V.K. and Bhardwaj, R. *Recent Progress in Medicinal Plants*, Studium Press, LLC, USA, Vol. 16, 221–236.
Kumar, R., Singh, P.K., Sharma, A., Prasad, J., Sagar, R., Singh, S., Arora, R. and Sharma, R.K. (2005b) *Podophyllum hexandrum* extract provides radioprotection by modulating the expression of proteins associated with apoptosis role in radioprotection. *Biotechnology and Applied Biochemistry* 42(1), 81–92.
Laszlo, A. and Li, G.C. (1993) Effect of amino acid analogs on the development of thermotolerance and on thermotolerant cells. *Journal of Cell Physiology* 154, 419.
Li, G.C. and Laszlo, A. (1985) Amino acid analogs while inducing heat shock proteins sensitize CHO cells to thermal damage. *Journal of Cell Physiology* 122, 91–97.
Mattei, E., Damasi, D., Mileo, A.M., Delpino, A. and Ferrini, U. (1992) Stress response, survival, and enhancement of heat sensitivity in a human melanoma cell line treated with L-canavanine. *Anticancer Research* 12, 757.
Michelangeli, C. and Vargas, R.E. (1994) L-canavanine influences feed intake, plasma basic amino acid concentrations and kidney arginase activity in chicks. *Journal of Nutrition* 124, 1081–1087.
Miller, G.C. and Consigli, R.A. (1974) Effect of canavanine on cultured mouse cells. *Proceedings of the Society of Experimental Biology and Medicine* 146, 549–553.
Naha, P.M. (1981) Cytotoxicity of L-canavanine *in vitro*. *Anticancer Research* 1, 275–278.
Naha, P.M., Silcock, J.M. and Fellows, L. (1980) Reappraisal of L-canavanine as an antitumour agent. *Cell Biology International Reports* 4, 155–166.
NaPhuket, S.R., Trifonov, L.S., Yu, C., Worthen, D.R., Crooks, P.A., Rosenthal, G.A. and Freeman, J.W. (1999) *Drug Development Research* 47, 170–177.
Olsen, A.S., Triemer, D.F. and Sanders, M.M. (1983) Dephosphorylation of S6 and expression of the heat shock response in *Drosophila melanogaster*. *Molecular and Cellular Biology* 3, 2017–2027.
Pass, M.A., Arab, H., Pollitt, S. and Hegarty, M.P. (1996) Effects of the naturally occurring arginine analogs indospicine and canavanine on nitric oxide mediated functions in aortic endothelium and peritoneal macrophages. *Natural Toxins* 4, 135–140.
Rosenthal, G.A. (1977) The biological effects and mode of action of L-canavanine, a structural analog of L-arginine. *Quarterly Review of Biology* 52, 155–178.
Rosenthal, G.A. (1991) The biochemical basis for the deleterious effects of L-canavanine. *Phytochemistry* 30, 1055–1108.
Rosenthal, G.A., Reichhart, J.M. and Hoffman, J.A. (1989) L-Canavanine incorporation into vitellogenin and macromolecular conformation. *Journal of Biological Chemistry* 264, 13693.
Russell, J.S., Burgan, W., Oswald, K.A., Camphausen, K. and Tofilon, P.J. (2003) Enhanced cell killing induced by the combination of radiation and the heat shock protein 90 inhibitor 17-allylamino-17-demethoxygeldanamycin: a multitarget approach to radiosensitization. *Clinical Cancer Research* 9, 3749–3755.
Schwartz, M., Altman, A., Cohen, Y. and Arzee, T. (1997) Inhibition of polyamine biosynthesis by L-canavanine and its effect on meristematic activity, growth and development of *Zea mays* roots. *Israel Journal of Plant Sciences* 45, 23–30.
Sekhar, K.R., Sonar, V.N., Muthusamy, V., Sasi, S., Laszlo, A., Sawani, J., Horikoshi, N., Higashikubo, R., Bristow, R.G., Borrelli, M.J., Crooks, P.A., Lepock, J.R., Roti Roti, J.L. and Freeman, M.L. (2007) Novel chemical enhancers of heat shock increase thermal radiosensitization through a mitotic catastrophe pathway. *Cancer Research* 67, 1–7.
Snyder, R.D. and Schroeder, K.K. (1994) Radiosensitivity of polyamine-depleted HeLa cells and

modulation by the aminothiol WR-1065. *Radiation Research* 137, 67–75.
Soman, G., Hurst, M.O. and Graves, D.J. (1985) Chemical modification of canavanine with p-nitrophenylglyoxal. Factors influencing the chemistry and reactivity of alpha-dicarbonyl-guanidino reactions. *International Journal of Peptide and Protein Research* 25, 517–525.
Subramani, V., Siddiq, E.A. and Palanichamy, K. (1983) Effective use of base-specific chemical mutagens in rice. *Indian Journal of Genetics and Plant Breeding* 43, 44–53.
Villar-Garea, A. and Imhof, A. (2006) The analysis of histone modifications. *Biochimica et Biophysica Acta* 1746, 1932–1939.
Wallace, H.M. and Keir, H.M. (1986) Factors affecting polyamine excretion from mammalian cells in culture. Inhibitors of polyamine biosynthesis. *FEBS Letters* 194, 60–63.
Warters, R.L., Newton, G.L., Olive, P.L. and Fahey, R.C. (1999) Radioprotection of human cell nuclear DNA by polyamines: radiosensitivity of chromatin is influenced by tightly bound spermine. *Radiation Research* 151, 354–362.
Weaks, T.E. and Hunt, G.E. (1973) Effects of canavanine on protein and nucleic acid synthesis in canavanine resistant and sensitive species of higher plants. *Physiologia Plantarum* 29, 421–424.
Worthen, D.R., Chien, L., Tsuboi, C., Gosland, M.P. and Crooks, P.A. (1996b) Modulation of drug sensitivity and the PGP-mediated multi-drug resistance by the naturally-occurring L-arginine analog L-canavanine. *Proceedings of the 212th American Chemical Society National Meeting*, Orlando, Florida, USA, August 25–29, 1996.
Worthen, D.R., Chien, L., Tsuboi, C.P., Mu, X.Y., Bartik, M.M. and Crooks, P.A. (1998) L-canavanine modulates cellular growth, chemosensitivity and P-glycoprotein substrate accumulation in cultured human tumour cell lines. *Cancer Letters* 132, 229–239.
Worthen, D.R., Ratliff, D.K., Rosenthal, G.A., Trifonov, L. and Crooks, P.A. (1996a) Structure-activity studies of L-canaline-mediated inhibition of porcine alanine aminotransferase. *Chemical Research in Toxicology* 9, 1293–1297.

Chapter 19

Withaferin A – A Phytosteroid of Promise for Tumour Sensitization in Cancer Therapy

P. Uma Devi

Introduction

Radiotherapy is the most common non-surgical method of solid tumour therapy. Even though several new therapeutic modalities like hormone therapy, immune therapy and gene therapy have been introduced, radiotherapy still continues to be the major treatment option for localized as well as many metastatic tumours. However, the therapeutic outcome is often less than expected because of the normal tissue toxicity, which limits the safely applicable radiation dose. A major reason for the failure of radiotherapy is considered to be the presence of hypoxic areas, which make the tumour radioresistant. Hypoxic cells are about three times more resistant to radiation damage than normally oxygenated cells (Hall, 1994). The therapeutic benefit can be increased if the radiosensitivity of the tumour cells is increased. Hyperbaric oxygen (HBO) breathing was tried to increase the oxygenation of hypoxic tumours, thus increasing their sensitivity to radiation. In spite of encouraging clinical results, HBO treatment did not get wide acceptance because of technical problems and discomfort to the patients.

Some chemicals have the property of enhancing the radiation response of tumour cells; they are generally called chemical radiosensitizers. Adams in the 1970s demonstrated that metronidazole, a 5-nitroimidazole compound used in the treatment of intestinal infections, increased the radiosensitivity of hypoxic cells, without affecting the response of normal oxygenated cells. When injected before radiation metronidazole produced a higher tumour regression, giving a sensitizer enhancement ratio (ER, the ratio of the radiation doses to produce the same amount of cell killing in the absence and in the presence of the drug) of 1.4 (Adams, 1977). Even though this drug showed good radiosensitization of hypoxic tumours in the clinical trials, daily administration with fractionated radiotherapy produced unacceptable neurotoxicity. Misonidazole, the 2-nitroimodazole derivative of metronidazole, was a more effective hypoxic cell sensitizer (ER = 1.6) and less toxic than its parent compound (Adams *et al.*, 1979).

Clinical trials with misonidazole gave encouraging results in some types of head and neck cancers (Overgaard *et al.*, 1989). However, severe peripheral neuropathy on daily administration was contraindicative (Coleman *et al.*, 1994). Second and third generation drugs were synthesized and tested. Some of them (e.g. etanidazole, sanazole) showed good hypoxic cell sensitization in the laboratory (Hall, 1994; Uma Devi *et al.*, 2000), but none of them has been accepted for routine clinical use. Several bioreductive compounds, which get metabolized to form active cytotoxic products inside the cells, were found to be preferentially toxic to hypoxic cells, while presence of oxygen in the cells reduced their activity. The most promising in this category, tirapazamine (SR-4233), produced significant sensitization of hypoxic cells *in vitro* and *in vivo*, with low drug toxicity (Brown, 1993). But clinical trials showed severe muscular cramps in the patients (Hercsher *et al.,* 1994).

A tumour sensitizer, to be of value in clinical radiotherapy, ideally should have the following properties:

1. Act preferentially on the tumours, without sensitizing the normal cells.
2. Able to diffuse through relatively non-vascular tissue to reach the hypoxic cells.
3. Metabolized slowly and to products that are non-toxic to normal tissues.
4. Easy to administer, preferably by oral route.
5. Affordable by all patients– low cost.

More than three decades of research has not produced a single radiosensitizer that has the ideal properties or is safe to use with fractionated radiotherapy. The earlier researchers had selected mostly synthetic compounds for the studies, based on their favourable structural properties, but the same structures were also responsible for their prohibitive toxicity. Recent years have witnessed a shift in interest to the natural products and phytochemicals as possible sources of non-toxic tumour sensitizers and research on these products, especially medicinal plants, has been on a steep rise during the last decade.

The medicinal plant-derived drugs have the following advantages:

a. These plants have been in therapeutic use for many centuries and, therefore, they are more readily accepted.
b. Most of them are orally effective and hence easy to administer.
c. Since they are made from indigenous sources, the drugs can be manufactured at low cost, and made available at affordable price.
d. Since they are part of nature and useful for the plant life, they are expected to show little or no systemic toxicity.

Investigations on the plant withanolide, Withaferin A, have revealed that this compound has promising tumour radiosensitizing properties and low normal tissue toxicity.

Withaferin A

Withaferin A is a steroidal lactone (Fig. 19.1) found in the leaves and roots of *Withania somnifera* (Sanskrit: Ashwagandha; Family: Solanaceae). *Withania somnifera* is an evergreen shrub with diverse medicinal properties, mentioned in the ancient medical treatise of India, Charaka Samhita. Because of the wide range of its medicinal uses, it is also known as the "Indian ginseng". In the traditional Indian systems of medicine, Ayurveda and Siddha, Ashwagandha root preparations are used as nerve tonic, aphrodisiac, immune stimulator and anti-stress agent, and in the treatment of consumption, senile debility, rheumatism, emaciation in children and as diuretic and narcotic (Kirtikar and Basu, 1991). The pharmacological and therapeutic properties of the plant have been attributed to the presence of withanolides, the most important of which is Withaferin A (Uma Devi, 1996). Withaferin A, from the leaves of the Israeli variety of Withania plant, was demonstrated to have growth inhibitory effects on cancer cells *in vitro* (Shohat *et al.*, 1970; Fuska *et al.*, 1984) and experimental tumours *in vivo* (Shohat *et al.*, 1967).

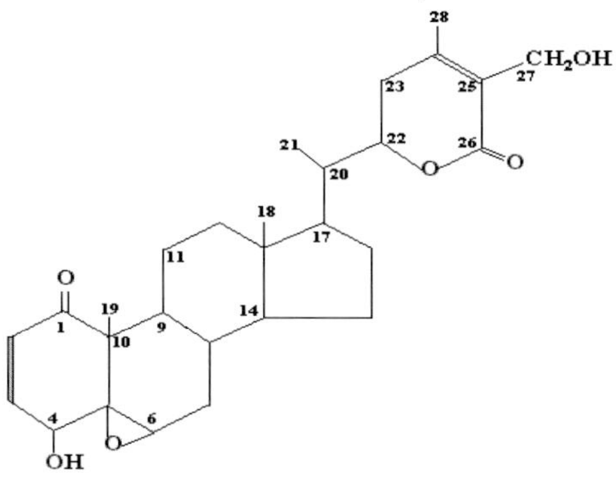

Fig. 19.1. Withaferin A.

Radiosensitizing effect of Withaferin A

Uma Devi and co-workers were the first to report the radiosensitizing effect of Withaferin A. They showed that Withaferin A (WA) has a weak tumour inhibitory effect, but treatment before irradiation produced significant radiosensitization of Ehrlich ascites carcinoma *in vivo* (Uma Devi *et al.*, 1995; Sharada *et al.*, 1996) and V79 Chinese hamster cells *in vitro* (Uma Devi *et al.*, 1996). The *in vivo* tumour growth inhibitory effect increased with increase in dose of Withaferin A from 10mg/kg to 60mg/kg body weight, but the optimum single dose which could be used with radiation, without noticeable toxicity, was 40mg/kg body weight. This compound gave an ED_{50} of 33mg/kg body weight for Ehrlich ascites carcinoma *in vivo* (Uma Devi *et al.*, 1995), while its LD_{50} for mouse is ~80mg/kg body weight (Sharada *et al.*, 1996). In mice bearing Ehrlich ascites carcinoma, Sharada *et al.* (1996) demonstrated that i.p. injection of Withaferin A before local irradiation of the tumour significantly increased the mean survival time of the mice. Multiple doses of Withaferin A produced better radiosensitization than single dose and the effect increased with increase in drug per dose. Maximum tumour sensitization was obtained when 30mg/kg Withaferin A was administered for 3 consecutive days, followed by radiotherapy after the third dose (Sharada *et al.*, 1996).

In order to examine its radiosensitizing effect on solid tumours, further investigations were carried out using the transplanted mouse tumours, fibrosarcoma and B16F1 melanoma.

B16F1 melanoma and fibrosarcoma solid tumours were developed in adult C57BL mice and Swiss albino mice, respectively, by injecting 5×10^5 on the dorsal skin and tumours of $100 \pm 10 mm^3$ were used for treatment. Withaferin A was injected intraperitoneally (i.p.) in acute or fractionated daily doses before local gamma irradiation of the tumour.

Hyperthermia was included to study the effect of trimodality combination of Withaferin A, radiation and hyperthermia (immediately after irradiation). Treatment response was studied by measuring tumour regression and growth delay and animal survival. Experiments using acute dose treatments showed that i.p. injection of 30 or 40mg/kg Withaferin A 1h before local irradiation of the tumour produced significant increase in the radiation response of both fibrosarcoma and B16F1 melanoma (Kamath *et al.,* 1999; Uma Devi *et al.,* 2000). Oral administration also produced significant tumour sensitization, even though a higher dose (60mg/kg) was needed to bring about the same level of sensitization as that obtained by i.p. injection (Kamath *et al.,* 1999). Mice also tolerated higher doses by oral route (LD_{50} = 130mg/kg vs. 80mg/kg i.p.). Melanoma showed a lower response than fibrosarcoma. This is expected, as melanomas are demonstrated to contain high hypoxic fractions (Rockwell and Moulder, 1990). In the above studies on mouse tumours, 30–50Gy radiation was given as a single dose, and the best effect in terms of tumour growth delay was obtained for both tumours when 40mg/kg WA was combined with 30Gy radiation. But such high doses are not feasible in human patients.

Fractionation of radiation dose to 10Gy daily for 5 days produced significantly higher growth delay and complete regression than single dose treatment in melanoma, while the difference between acute and fractionated irradiation was not pronounced in fibrosarcoma (Uma Devi and Kamath, 2003). The higher response of melanoma to fractionated radiotherapy can be explained on the basis of the high hypoxic fraction in this tumour; reoxygenation between fractions will lead to increased cell killing with each fraction. Reoxygenation of tumour cells and repair of normal cells during the interval between fractions are the two main components of the 4 R's contributing to the better tumour control and normal tissue sparing effect of conventional radiation dose fractionation (Withers, 1975). Injection of 30mg/kg WA before 10Gy for 5 days resulted in a synergistic increase in complete regression and growth delay of both fibrosarcoma and melanoma. This effect was further enhanced when one dose of hyperthermia (43°C for 30 min) was given locally to the tumour immediately following the first radiation fraction in a trimodality treatment (Uma Devi and Kamath, 2003). However, even 10Gy per fraction may be too high for daily treatment in humans.

Conventional dose fractionation in radiotherapy of cancer patients uses 2Gy daily, locally to the tumour, 5 days a week for 4–6 weeks. This schedule was followed in a combined modality regimen to see if Withaferin A can enhance the response of the radioresistant tumour melanoma to conventional radiation dose fractionation. Radiation, as a single modality, was given for 4 weeks (total dose = 40Gy), while, in bimodality combination, 15mg/kg WA was injected 1h before each fraction of radiation for 3 weeks (total dose = 30Gy) Hyperthermia (43ºC; 30min) was given locally to the tumour once a week after the first radiation dose for three weeks in trimodality combination. Tumour growth delay increased from 32 days with radiation alone for 4 weeks to 49 days with Withaferin A + radiation for 3 weeks; mean survival time of mice increased from 53 days with RT alone to 77 days with WA + RT. Addition of hyperthermia once a week to the 3 weeks WA + RT schedule further enhanced the tumour growth delay (69 days) and mean survival time (105 days; Table 19.1). None of the animals showed any drug-related toxicity when WA was combined with the other modalities (Uma Devi and Guruprasad, 2007). Kamath (2000) has shown that 16mg/kg body weight of Withaferin A, given daily for 30 days to mice did not result in any significant systemic toxicity. These results indicate that by combining a non-toxic dose of WA with conventional fractionated radiotherapy, it is possible to reduce the therapeutic dose of radiation, consequently reducing the normal tissue reactions.

Table 19.1. Response of B16F1 melanoma to fractionated Withaferin A (WA) + radiation (RT) fractionated dose (5 days a week) treatment with or without hyperthermia (HT, 43°C for 30 min once a week).

Treatment	Treatment duration (weeks)	Tumour growth delay (days)	Mean survival time (days)
Control	-	-	40
WA (15 mg/kg) × 5	3	11	58
HT (43°C; 30') × 1	3	13	51
RT (2Gy) × 5	4	32	53
WA + RT	3	49	77
WA + RT + HT	3	69	105

Mechanism of radiosensitization by Withaferin A

Sensitization can be brought about by different mechanisms–enhancing the oxidative activity, depleting the cellular antioxidants, inhibiting DNA repair, forming DNA adducts and inhibiting DNA synthesis, etc. Recent studies using Ku70 and rad54 deficient chicken B-lymphocytes have indicated that WA inhibits DNA repair, leading to higher cell kill (Uma Devi *et al.*, 2007). How it is brought about is not clear, but cell cycle arrest in the synthetic phase by WA has been reported (Uma Devi *et al.,* 1996). Withaferin A was found to reduce the glutathione (GSH) level and cellular antioxidant enzyme activity in transplanted mouse tumours (Kamath, 2000), which may have some role in its radiosensitizing effect. Glutathione, in addition to being a strong antioxidant, also promotes biochemical repair by acting as a co-enzyme (Revesz, 1985). Depletion of GSH in tumour cells by Withaferin A will not only enhance oxidative damage, but also inhibit DNA repair, leading to increased cell death. Withaferin A has also been demonstrated to enhance radiation induced cell kill by apoptosis in the B16F1 mouse melanoma (Uma Devi and Guruprasad, 2007). Pharmacokinetic studies in mice have shown that Withaferin A gets accumulated in the tumour within 1h after i.p. administration, while uptake in the normal tissues is less efficient (unpublished observation). This is relevant to cancer therapy, as it can lead to preferential sensitization of tumours.

Conclusions

Pre-clinical data indicate that the plant withanolide, Withaferin A, is an effective tumour radiosensitizer, and is safe to use with daily doses of fractionated radiotherapy. Clinical studies are needed to establish its utility in cancer patients. If proved clinically, it can lead to the indigenous development of a non-toxic and low cost radiosensitizer for human application.

References

Adams, G.E. (1977) Hypoxic Cell Sensitizers for Radiotherapy. In: Becker, F.F. (ed), *Cancer,* Vol. 6, Plenum Press, New York, pp.181–223.

Adams, G.E., Clark, E.D. and Flockhart, I.R. (1979) Structure–activity relationship in the development of hypoxic cell sensitizers. II. Cytotoxicity and therapeutic ratio. *International Journal of Radiation Biology* 35, 151–160.

Brown, J.M. (1993) SR-4233 (tirapazamine): a new anticancer drug exploiting hypoxia in solid tumours. *British Journal of Cancer* 67, 1163–1170.

Coleman, C.N., Beard, C.L. and Hlatky, L. (1994) Biochemical modifiers. In: Mauch, P.M. and Loeffler, J.S. (eds) *Radiation Oncology: Technology and Biology*, W. B. Saunders, Philadelphia, pp. 56–89.

Fuska, J., Fuskova, A., Rosazza, J.P. and Nicholas, A.W. (1984) Novel cytotoxic and antitumour agents. IV. Withaferin A: relation of its structure to the *in vivo* cytotoxic effects on P-388 cells. *Neoplasma* 31, 31–36.

Hall, E.J. (1994) *Radiobiology for the Radiologist.* 4th ed., J.B. Lippincott, Philadelphia, USA.

Hercsher, L.L., Krishna, M.C. and Cook, J.A. (1994) Protection against SR-4233 (tirapazamine) aerobic cytotoxicity by the metal chelators desferrioxamine and tiron. *International Journal of Radiation Oncology, Biology and Physics* 30, 879–885.

Kamath, R. (2000) Studies on the radiosensitizing effect of Withaferin A and hyperthermia on two mouse tumours. Ph.D. Thesis, Manipal Academy of Higher Education, Manipal, India.

Kamath, R., Rao, B.S.S. and Uma Devi, P. (1999) Response of a mouse fibrosarcoma to Withaferin A and radiation. *Pharmaceutical and Pharmacological Communication* 5, 287–291.

Kirtikar, K.R. and Basu, B.D. (1991) In: Blatter, E., Caind, J.F. and Bhaskar, K.S. (eds) *Indian Medicinal Plants*, Periodical Experts Book Agency, Delhi, p. 1774.

Overgaard, J., Sand-Hansen, H. and Anderson, A.P. (1989) Misonidazole combined with split-course radiotherapy in the treatment of invasive carcinoma of larynx and pharynx. Final report from the DAHANCA 2 study. *International Journal of Radiation Oncology Biology and Physics* 16, 1065–1068.

Revesz, L. (1985) The role of endogenous thiols in intrinsic radioprotection. *International Journal of Radiation Biology* 47, 361–368.

Rockwell, S. and Moulder, J.E. (1990) Hypoxic fraction in human tumours xenografted into mice: a review. *International Journal of Radiation Oncology, Biology and Physics* 19, 197.

Sharada, A.C., Solomon, F.E. and Uma Devi, P. (1996) Antitumour and radiosensitizing effects of Withaferin A on mouse Ehrlich ascites carcinoma *in vivo. Acta Oncologica* 35, 95–100.

Shohat, B., Gitter, S., Abraham, A. and Lavie, D. (1967) Antitumour activity of Withaferin A. *Cancer Chemotherapy Reports* 51, 271–276.

Shohat, B., Gitter, S. and Lavie, D. (1970) Effect of Withaferin A on Ehrlich ascites tumour cells–cytological observations. *International Journal of Cancer* 5, 244–252.

Uma Devi, P. (1996) *Withania somnifera* Dunal (Ashwagandha): potential plant source of a promising drug for cancer chemotherapy and radiosensitization. *Indian Journal of Experimental Biology* 34, 927–932.

Uma Devi, P., Akagi, K., Ostapenko, V., Tanaka, Y. and Sugahara, T. (1996) Withaferin A: a new radiosensitizer from the Indian medicinal plant *Withania somnifera. International Journal of Radiation Biology* 69, 193–197.

Uma Devi, P. and Guruprasad, K. (2007) Enhancement of response of mouse B16F1 melanoma to fractionated radiotherapy by Withaferin A and hyperthermia. *Acta Oncologica* (communicated).

Uma Devi, P. and Kamath, R. (2003) Radiosensitizing effect of Withaferin A combined with hyperthermia on mouse fibrosarcoma and melanoma. *Journal of Radiation Research* 44, 1–6.

Uma Devi, P., Kamath, R. and Rao, B.S.S. (2000) Radiosensitization of a mouse melanoma by Withaferin A: *in vivo* studies. *Indian Journal of Experimental Biology* 38, 432–435.

Uma Devi, P., Nagaratnam, A. and Rao, B.S.S. (2000) *Introduction to Radiation Biology.* Chapter 10. B. I. Churchill Livingstone, New Delhi, pp. 131–150.

Uma Devi, P., Sharada, A.C. and Solomon, F.E. (1995) *In vivo* growth inhibitory and radiosensitizing effects of Withaferin A on mouse Ehrlich ascites carcinoma. *Cancer Letters* 95, 189–193.

Uma Devi, P., Utsumi, H., Takata, M. and Takeda, S. (2007) Inhibition of repair of X–ray induced DNA damage in DT40 cells by Withaferin A. *Indian Journal of Experimental Biology* (communicated).

Withers, H.R. (1975) The Four R's of Radiotherapy. In: Lett, J.T. and Alder, H. (eds) *Advances in Radiation Biology.* Vol. 5. John Wiley and Sons, New York, USA, pp. 241–272.

Chapter 20

The Radiosensitizer Hypericin as Adjuvant Therapy in the Treatment of Central Nervous System Tumours

Toba Niazi and William T. Couldwell

Introduction

Hypericin, a polycyclic aromatic compound (Fig. 20.1), can be found in certain insects, fungi, protozoa and more than 350 species of the plant genus Hypericum, family Hypericaceae (Delaey *et al.,* 2003). This compound is particularly concentrated in the species *Hypericum perforatum*, also known as St. John's Wort. Hypericin effectively crosses the blood-brain barrier, owing to its lipophilic nature, and has been administered as an antidepressant for more than five decades with salutary effects (Couldwell *et al.,* 1994).

Ingestion of hypericin-containing plants by grazing animals was noted to cause a unique phenomenon called hypericism, a condition of severe sensitivity to light (Miccoli *et al.,* 1998). As a result, over the last decade there has been renewed interest in hypericin as a photosensitizing and radiosensitizing agent and in its applications to different disease states, particularly viral and neoplastic diseases. The ubiquitous availability of the St. John's Wort plant worldwide and the relatively convenient and inexpensive extraction of this compound make it particularly attractive for use (Couldwell *et al.,* 1994).

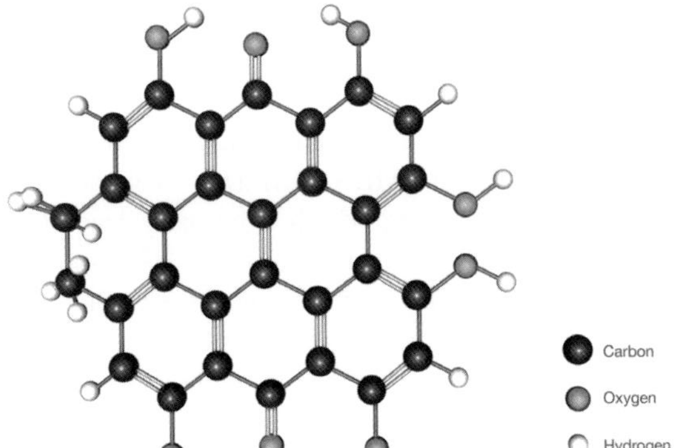

Fig. 20.1. The hypericin molecule.

Hypericin and photodynamic therapy

The photochemical properties of hypericin are best exploited in the frame of photodynamic therapy/treatment (PDT) (Berlanda *et al.,* 2006). PDT is based on the administration of a photosensitizing agent, in this case hypericin, and subsequent excitation by irradiation with light in the visible range matching the absorption spectrum of the compound (the maximum absorption spectrum of hypericin is approximately 595nm) (Delaey *et al.,* 2003). Photosensitizers acquire their antitumour activity when excited by light and are relatively inactive in the absence of light. In the presence of oxygen, an absolute requirement for PDT, the hypericin generates superoxide radicals, which, in turn, produce reactive oxygen species (ROS) such as peroxide and hydroxyl radicals in a Type I photochemical reaction or singlet oxygen molecules in a Type II photochemical reaction (Berlanda *et al.,* 2006). In concert with the production of the ROS and singlet oxygen molecules, there is an ability to produce a photogenerated pH drop through an intramolecular proton transfer in the excited state, which is likely to result in acidification of cells in solvent and those *in vivo* when exposed to hypericin (Miccoli *et al.,* 1998).

These ROS are also involved in other signal pathways within the cell, including membrane lipid peroxidation and inhibition of protein kinase C (PKC) and other growth factor-stimulated protein kinases. The subcellular localization of hypericin is mainly limited to the lipid membranes; however, hypericin has also been reported to accumulate extensively within the endoplasmic reticulum, Golgi apparatus, lysosomes and mitochondria (Theodossiou *et al.,* 2006).

Hypericin has not been shown to have any genotoxicity or toxicity in animals or humans, with the exception of hypericism, and is considered a relatively safe clinical photodynamic antiviral and antitumour agent (Miccoli *et al.,* 1998). Hypericin has a distribution half-life of 2 hours and an elimination half-life of 38.5 hours in humans with a method for rapid and specific measurement of the drug by high-performance liquid chromatography, lending it to be a compound that can be safely and easily administered and measured (Couldwell *et al.,* 1994). Upon photoactivation, hypericin also exhibits a bright red fluorescence that is helpful in clinical treatment paradigms.

Antiviral applications of hypericin

AZT (3'-azido2',3'dideoxythymidine), a nucleoside analogue that inhibits reverse transcription of HIV, was one of the first antiretroviral agents shown to prolong the survival and improve the quality of life of those afflicted with human immunodeficiency virus (HIV) and acquired immunodeficiency syndrome (AIDS). Efficacy of most antiretroviral drugs depends on frequent administration, however, which eventually leads to severe toxicity to the patient. Severe toxicity and rapidly resistant strains of HIV limited AZT's clinical usefulness and led to further research into a milder, non-toxic alternative. In the late 1980s, hypericin gained attention for its antiretroviral properties in two murine retroviral systems, Friend Virus (FV) and LP-BM5 murine immunodeficiency virus. Administration of hypericin to murine models infected with FV precluded the onset of acute FV-associated erythroleukemia and subsequent development of splenomegaly (Lavie *et al.,* 1989). In those animals afflicted with LP-BM5, administration of a single 10–50μg dose of hypericin per mouse prevented the development of significant viremia and minimized disease (Meruelo *et al.,* 1988; Lavie *et al.,* 1989). This success suggested that this herbal substance might be used in the fight against HIV without the toxic effects seen in other treatment paradigms.

In pioneering studies in which they treated HIV-infected cell lines with hypericin, Lavie *et al.* (1989) and Degar *et al.* (1992) found that hypericin effectively inhibits release of reverse transcriptase (RT) but does not affect total mRNA levels or expression of the antigens within the cells. Hypericin did not inhibit reactions catalyzed by commercially purified RT from murine viruses, indicating that suppression of the RT activity does not occur via direct action on the enzyme itself. Instead, these results suggest that hypericin directly inactivates the virions or interferes with the assembly or shedding of the virus particles. More recently, Darbinian-Sarkissian *et al.* (2006) identified a novel protein, p27 (SJ), present in St. John's Wort that suppresses transcription of the HIV genome in primary culture of microglia and astrocytes. This has the potential to control HIV transcription and replication in cells associated with HIV infection within the central nervous system.

The antiviral activity of hypericin also appears to involve interaction of hypericin with cell membranes, since its known red fluorescence enabled localization to surface membranes (Meruelo *et al.,* 1988). This is further suggested because normal laboratory lighting conditions were required for dramatic inhibition of HIV in the cell lines studied. In those HIV-infected cell lines treated with hypericin in the dark, the antiretroviral effect was substantially diminished (Degar *et al.,* 1992). In fact, hypericin's demonstrated potent antiretroviral activity is enhanced more than 100-fold in the presence of light (Meruelo *et al.,* 1988).

Unfortunately, Phase I clinical trails were not successful in patients who were HIV positive because of severe phototoxicity (Gulick *et al.,* 1999). In addition, most patients afflicted with HIV are on a medical cocktail of antiretroviral therapy and protease inhibitors. Unfortunately, hypericin was shown to increase the metabolism of many of the protease inhibitors and non-nucleoside RTs because of the activation of cytochrome P450 within the liver. Thus, the most beneficial application of hypericin in the treatment of HIV has not yet been defined. Promising applications must be further tailored to today's HIV-positive patient on cocktail therapy.

Antitumoural applications of hypericin

Because of many challenges associated with successful treatment of malignant gliomas (World Health Organization grades 3 and 4) (Zhang *et al.,* 1996), radiation therapy after surgical resection remains the most common treatment modality to fight the disease process. Unfortunately, the tumour cells render themselves resistant to the radiation therapy because of either an acquired defect or one that is intrinsically present, and as such are a major cause of tumour recurrence. Therefore, much effort has been focused on increasing cellular sensitivity to radiation therapy. These tumours have very high protein kinase C (PKC) activity when compared with non-transformed glia, and this high level of activity is strongly correlated with the growth rates of these tumour cells *in vitro* (Couldwell *et al.,* 1991; Couldwell *et al.,* 1992).

PKC represents a family of closely related phospholipid-activated protein kinases that modify the function of other proteins by phosphorylation at their specific serine/threonine sites. They are known to play a role in a wide variety of cellular processes, including proliferation, differentiation, and programmed cell death (i.e., apoptosis) (Zhang *et al.,* 1995). At least 12 distinct isoforms of PKC have been elucidated to date, and several have been found to be expressed in malignant glioma cell lines (Zhang *et al.,* 1997). Surgically resected frozen human glioma specimens also display elevated PKC activity within the same range that is seen in the glioma cell lines (Couldwell *et al.*, 1994). Pituitary adenoma cell lines have also been shown to have very high levels of PKC activity, with levels even more heightened in invasive adenoma cell lines. This may suggest that PKC activity may also have a functional role in the rate of growth and invasive properties of this cell line as well (Hamilton *et al.*, 1996). Because hypericin and pseudohypericin have been identified as relatively specific inhibitors of PKC (Takahashi *et al.*, 1989), hypericin, which has greater antiretroviral activity than pseudohypericin, has been proposed for use in treatment of malignant gliomas.

It has been shown that inhibition of PKC activity reduces DNA synthesis and cell growth in malignant glioma cell lines. Fuks *et al.* (1994) demonstrated that PKC-mediated basic fibroblast growth factor protection of endothelial cells against radiation induces apoptosis, suggesting that PKC activation may constitute a generic mechanism of radiation resistance in affected cell lines. As a negative regulator of apoptosis, high levels of PKC in malignant glioma cell lines would work in favour of metastases and invasion (Couldwell *et al.*, 1994).

The molecular structure of hypericin is similar to that of calphostin C, a relatively specific inhibitor of PKC (Zhang *et al.*, 1995). In fact, Takahashi *et al.* (1989) have shown that hypericin is a potent PKC inhibitor by interacting with the regulatory domain of the PKC enzyme. Hypericin differs from the classic PKC inhibitors such as staurosporine and tamoxifen, which competitively block the ATP catalytic site. Instead, hypericin inhibits PKC by interacting directly with the regulatory domain of the enzyme and displays more selectivity in its inhibitory actions than classic PKC inhibitors, resulting in fewer global side effects (Hamilton *et al.*, 1996).

Given the high activity of PKC in glioma cell lines, antitumour agents such as hypericin targeted against PKC hold promise for antitumoural therapy. The correlation of the PKC system with cell growth, invasion and apoptosis suggests that the PKC system is involved with the control of these processes and that the inhibition of PKC can provide a potential avenue for clinical tumour growth inhibition in tumour cell lines. Research involving PDT with hypericin indicates that there is an inhibitory effect on glioma growth that is increased by 13% by visible light, indicating that light may facilitate the antitumour effect *in vitro* (Couldwell *et al.*, 1994). Hypericin was also found to inhibit the growth of glioma cell lines and pituitary adenoma cell lines in a dose-related manner, with longer periods of incubation with hypericin yielding more inhibition of cell growth and promotion of apoptosis (Couldwell *et al.*, 1994; Hamilton *et al.*, 1996). Low dosages of hypericin caused little or no growth inhibition in both established and low-passage malignant glioma cell lines, but nonetheless it did enhance the radiation-induced cell killing in these cells, suggesting a potent role for hypericin as a radiosensitizer for malignant gliomas (Zhang *et al.*, 1996).

Researchers have demonstrated immediate evidence of skin shrinkage around the tumour after PDT with hypericin in a P388 mouse lymphoma tumour model with subsequent microvascular occlusion of the tumour and skin surrounding the area of tumour with eschar formation within four days after treatment (Chen *et al.*, 2001). Further microscopic analysis of the tumours confirmed necrosis at the area of eschar formation, but tumour regrowth in the periphery of the irradiated area was also apparent. The treatment involved 1mg/kg, 5mg/kg or 20mg/kg hypericin administration with subsequent irradiation of the tumour at 1 or 24 hours after dosage. The results from this study showed that mice treated with PDT 1 hour after a 1mg/kg dosage had the lowest tumour concentration of hypericin but the highest plasma drug level. This was shown to be the most effective treatment protocol in these mice with the most evidence of microvascular damage, and it increased survival time of these mice considerably in comparison with the other groups tested (Chen *et al.*, 2001). This study confirmed that the efficacy of hypericin-mediated PDT is highly dependent on the circulating drug level rather than the tumour drug concentration.

Similar results were seen with hypericin-induced PDT in the neuroblastoma cell line SK-N-SH. Neuroblastoma is the most common extracranial solid tumour arising from the nervous system in children, and overall survival rates for patients diagnosed with advanced disease are grim (Zhang *et al.*, 1995). Human neuroblastoma tissues also possess very high levels of PKC activity. Treatment of the neuroblastoma cell line with hypericin-induced PDT resulted in a dose-dependent inhibition of cell growth with associated apoptosis as already described with the other cell lines (Zhang *et al.*, 1995).

The cellular events that increase the radiosensitivity and contribute to the phototoxicity of hypericin-treated cell lines are multifactorial in nature. As alluded to earlier, multiple actions of the drug are due to the initial effects of the production of superoxide radicals and singlet oxygen species that damage the cell membrane and organelles with resultant acidification of the cells and direct cytotoxicity on tumour cells. The overall intracellular effect is one of depleting intracellular adenosine triphosphate, acidifying the cell milieu, and promoting synthesis of thromboxanes, prostaglandins and leukotrienes (Du *et al.*, 2003). These highly reactive products may induce vascular damage via direct effects on the endothelial cells, platelets and red blood cells, with subsequent hypoxia and nutritional deprivation, thereby initiating tumour necrosis via tissue ischemia (Chen *et al.*, 2001; Du *et al.*, 2003). Both the direct tumour cell killing and the secondary vascular effect have been shown to contribute to the antitumoural activity of hypericin (Chen *et al.*, 2002).

As previously mentioned, the inhibitory effect of hypericin on the PKC system renders the tumour cells more radiosensitive by depletion of PKC. Depletion of PKC also renders the cell more susceptible to apoptosis, as is evidenced by extensive degradation of genomic DNA into discrete oligonucleosomal fragments in those cell lines treated by hypericin-induced PDT (Zhang *et al.*, 1995). There also seems to be an association between hypericin-induced PDT and levels of zinc within the cell. There is an association between increased apoptosis and a decrease in intratumoural zinc levels. It appears that a decrease in zinc levels triggers apoptosis mainly through activation of a caspase-3 activation leading to apoptosis (Thong *et al.*, 2006). Activation of PKC has also been shown to be involved in cell motility and therefore associated with invasion and metastases. Hypericin has been shown to significantly inhibit human malignant glioma cell invasion *in vitro* in concentrations previously shown to inhibit PKC activity in cultured glioma cells (Zhang *et al.*, 1997). By inhibiting migration of cell lines after exposure to hypericin, this lends further credence to the use of hypericin to avoid the crucial processes of metastases and invasion that are inherent in malignant gliomas (Zhang *et al.*, 1997).

The metabolism of tumour cells is directly affected by hypericin-induced PDT as well, given that these cells have some of the highest rated aerobic metabolism recorded. By reducing the ability for the cell to perform aerobic metabolism and therefore causing energy deprivation, growth inhibition is obtained (Miccoli *et al.*, 1998). In human glioma cells treated with hypericin-induced PDT, there was a notable inhibition in the hexokinase bound to mitochondria with associated drop in internal pH, in a light- and dose-dependent manner (Miccoli *et al.*, 1998). Glucose-dependent energy required for glioma metabolism is regulated by hexokinase, which catalyzes the phosphorylation of glucose for its entry into glycolysis.

Malonaldehyde (MDA), a product derived from oxidative degradation of unsaturated phospholipids and cholesterol, is an established marker of lipid peroxidation, and is increased in cell lines exposed to hypericin-induced PDT. The peroxidation of lipids is known to disrupt membrane structure and induce loss of function leading to cell death, adding another means by which hypericin may have its effect on cell death (Du *et al.*, 2003).

There also seems to be a synergistic interaction between hypericin-mediated PDT and hyperthermia, with hyperthermia further contributing to the cell membrane damage (Chen *et al.*, 2002). In the RIF-1 murine tumour model, it appears that hypericin-induced PDT induced sublethal damage to some tumour cells. These cells may be able to survive in the absence of subsequent hyperthermia by reversing sublethal cellular damage. However, when hypericin-induced PDT is followed by hyperthermia, the combined cellular damage becomes irreversible, triggering a synergistic effect on tumour cell demise (Chen *et al.*, 2002). This synergistic effect is only seen if hyperthermia is applied directly after PDT. Apoptotic cell death was increased at a 15-hour interval with the two modalities, whereas necrosis was seen as an immediate effect. Why a different mode of cell death is seen at two different time frames has yet to be elucidated (Chen *et al.*, 2002).

Angiogenesis of tumour vasculature is a major means by which tumour recurrence occurs, and vascular endothelial growth factor (VEGF) is the means by which much of the angiogenic process occurs. Hypericin-induced PDT offers another means of action of ROS directed against the vasculature leading to tissue ischemia and subsequent necrosis. In a study using a mouse fibrosarcoma model, it was shown that hypericin-induced PDT resulted in lowered serum VEGF levels, and that this correlated with a prolonged survival rate compared with controls (Thong *et al.*, 2006). Therefore, hypericin also works in this other modality where lowered serum VEGF levels prevent the formation of new tumour vasculature and tumour regrowth.

Conclusions

Despite the advances that have been made in the realm of conventional surgical, radiotherapeutic and chemotherapeutic modalities, the prognosis of those afflicted with malignant gliomas remains poor, with the median survival of patients harbouring glioblastoma multiforme of less than 12 months (Zhang *et al.*, 1996). Radiotherapy continues to be the mainstay of treatment after surgical resection or biopsy; however, the resistance of glioma cells against radiation represents a formidable treatment challenge. An important goal in malignant glioma management and treatment is to elucidate novel treatment modalities that would help to increase the cellular sensitivity to radiotherapy and, therefore, prolong patient survival. Hypericin has been found to enhance the radiosensitivity of malignant glioma cells, through its intrinsic properties as well as its enhanced antitumoural activity when used in a frame of PDT. Hypericin-induced PDT not only increases sensitization to radiation but also has many other effects on photoactivated tumour cells, thereby potentiating its efficacy against tumour cells. Its mechanism of action is definitely multifactorial and many avenues have yet to be elucidated. One such pathway may be in the area of hypericin-induced PDT in conjunction with hyperthermia against malignant glioma cell line. The association of VEGF levels and hypericin treatment also is an exciting realm in terms of targeting tumour angiogenesis. The ability to synthesize hypericin safely and cheaply also renders it as a good pharmacological compound for widespread use. Thus, we should foster and encourage the current pathways and continue to conduct clinical trials and further our knowledge of the nature of action of this compound in hopes of alleviating and palliating neoplastic diseases.

References

Berlanda, J., Kiesslich, T., Oberdanner, C.B., Obermair, F.J., Krammer, B. and Plaetzer, K. (2006). Characterization of apoptosis induced by photodynamic treatment with hypericin in A431 human epidermoid carcinoma cells. *Journal of Environmental Pathology Toxicology and Oncology* 25(1–2), 173–188.

Chen, B., Roskams, T. and de Witte, P.A. (2002). Enhancing the antitumoural effect of hypericin-mediated photodynamic therapy by hyperthermia. *Lasers in Surgery and Medicine* 31(3), 158–163.

Chen, B., Zupko, I. and de Witte, P.A. (2001). Photodynamic therapy with hypericin in a mouse P388 tumour model: vascular effects determine the efficacy. *International Journal of Oncology* 18(4), 737–742.

Couldwell, W.T., Antel, J.P. and Yong, V.W. (1992). Protein kinase C activity correlates with the growth rate of malignant gliomas: Part II. Effects of glioma mitogens and modulators of protein kinase C. *Neurosurgery* 31(4), 717–724.

Couldwell, W.T., Gopalakrishna, R., Hinton, D.R., He, S., Weiss, M.H., Law, R.E. and Apuzzo, M.L. (1994). Hypericin: a potential antiglioma therapy. *Neurosurgery* 35(4), 705–709.

Couldwell, W.T., Uhm, J.H., Antel, J.P. and Yong, V.W. (1991). Enhanced protein kinase C activity correlates with the growth rate of malignant gliomas *in vitro*. *Neurosurgery* 29(6), 880–886; discussion 886–887.

Darbinian-Sarkissian, N., Darbinyan, A., Otte, J., Radhakrishnan, S., Sawaya, B.E., Arzumanyan, A., Chipitsyna, G., Popov, Y., Rappaport, J., Amini, S. and Khalili, K. (2006). p27(SJ), a novel protein in St John's Wort, that suppresses expression of HIV-1 genome. *Gene Therapy* 13(4), 288–295.

Degar, S., Prince, A.M., Pascual, D., Lavie, G., Levin, B., Mazur, Y., Lavie, D., Ehrlich, L.S., Carter, C. and Meruelo, D. (1992). Inactivation of the human immunodeficiency virus by hypericin: evidence for photochemical alterations of p24 and a block in uncoating. *AIDS Research and Human Retroviruses* 8(11), 1929–1936.

Delaey, E., Zupko, I., Chen, B., Derycke, A., van Laar, F., De Vos, D. and De Witte, P. (2003). Comparison of hexamethylhypericin and tetrabromohypericin to hypericin for their *in vivo* efficacy as PDT tools. *International Journal of Oncology* 23(2), 519–524.

Du, H.Y., Olivo, M., Tan, B.K. and Bay, B.H. (2003). Hypericin-mediated photodynamic therapy induces lipid peroxidation and necrosis in nasopharyngeal cancer. *International Journal of Oncology* 23(5), 1401–1405.

Fuks, Z., Persaud, R.S., Alfieri, A., McLoughlin, M., Ehleiter, D., Schwartz, J.L., Seddon, A.P., Cordon–Cardo, C. and Haimovitz-Friedman, A. (1994). Basic fibroblast growth factor protects endothelial cells against radiation-induced programmed cell death *in vitro* and *in vivo*. *Cancer Research* 54(10), 2582–2590.

Gulick, R.M., McAuliffe, V., Holden-Wiltse, J., Crumpacker, C., Liebes, L., Stein, D.S., Meehan, P., Hussey, S., Forcht, J. and Valentine, F.T. (1999). Phase I studies of hypericin, the active compound in St. John's Wort, as an antiretroviral agent in HIV-infected adults. AIDS Clinical Trials Group Protocols 150 and 258. *Annals of Internal Medicine* 130(6), 510–514.

Hamilton, H.B., Hinton, D.R., Law, R.E., Gopalakrishna, R., Su, Y.Z., Chen, Z.H., Weiss, M.H. and Couldwell, W.T. (1996). Inhibition of cellular growth and induction of apoptosis in pituitary adenoma cell lines by the protein kinase C inhibitor hypericin: potential therapeutic application. *Journal of Neurosurgery* 85(2), 329–334.

Lavie, G., Valentine, F., Levin, B., Mazur, Y., Gallo, G., Lavie, D., Weiner, D. and Meruelo, D. (1989). Studies of the mechanisms of action of the antiretroviral agents hypericin and pseudohypericin. *Proceedings of the National Academy of Sciences of the United States of America* 86(15), 5963–5967.

Meruelo, D., Lavie, G. and Lavie, D. (1988). Therapeutic agents with dramatic antiretroviral activity and little toxicity at effective doses: aromatic polycyclic diones hypericin and pseudohypericin. *Proceedings of the National Academy of Sciences of the United States of America* 85(14), 5230–5234.

Miccoli, L., Beurdeley–Thomas, A., De Pinieux, G., Sureau, F., Oudard, S., Dutrillaux, B. and Poupon, M.F. (1998). Light-induced photoactivation of hypericin affects the energy metabolism of human glioma cells by inhibiting hexokinase bound to mitochondria. *Cancer Research* 58(24), 5777–5786.

Takahashi, I., Nakanishi, S., Kobayashi, E., Nakano, H., Suzuki, K. and Tamaoki, T. (1989). Hypericin and pseudohypericin specifically inhibit protein kinase C: possible relation to their antiretroviral activity. *Biochemical and Biophysical Research Communications* 165(3), 1207–1212.

Theodossiou, T.A., Noronha–Dutra, A. and Hothersall, J.S. (2006). Mitochondria are a primary target of hypericin phototoxicity: synergy of intracellular calcium mobilisation in cell killing. *International Journal of Biochemistry & Cell Biology* 38(11), 1946–1956.

Thong, P.S., Watt, F., Ren, M.Q., Tan, P.H., Soo, K.C. and Olivo, M. (2006). Hypericin-photodynamic therapy (PDT) using an alternative treatment regime suitable for multi-fraction PDT. *Journal of Photochemistry and Photobiology B Biology* 82(1), 1–8.

Zhang, W., Anker, L., Law, R.E., Hinton, D.R., Gopalakrishna, R., Pu, Q., Gundimeda, U., Weiss, M.H. and Couldwell, W.T. (1996). Enhancement of radiosensitivity in human malignant glioma cells by hypericin *in vitro*. *Clinical Cancer Research* 2(5), 843–846.

Zhang, W., Law, R.E., Hintona, D.R. and Couldwell, W.T. (1997). Inhibition of human malignant glioma cell motility and invasion *in vitro* by hypericin, a potent protein kinase C inhibitor. *Cancer Letters* 120(1), 31–38.

Zhang, W., Law, R.E., Hintona, D.R., Su, Y. and Couldwell, W.T. (1995). Growth inhibition and apoptosis in human neuroblastoma SK–N–SH cells induced by hypericin, a potent inhibitor of protein kinase C. *Cancer Letters* 96(1), 31–35.

Chapter 21

Radiosensitizing Activity of the Indian Medicinal Plant *Tinospora cordifolia* Miers Ex h\Hook f & Thoms in Tumour-Bearing Mice

Ganesh Chandra Jagetia

Introduction

The term "Cancer" is derived from the Latin word "Cancer" and the Greek word, "Karkinos", both meaning crab. It signifies the ominous nature of the disease, as the present treatment regimens still remain unsatisfactory for many types of cancer though a significant progress has been made in the last 2–3 decades. The overall incidence of cancer is globally rising, so as to say the "burden" of cancer is increasing. However, "burden" is a concept without formal definition; although it reflects the quantum of a particular disease in a community, it can be more precisely expressed in terms of prevalence or incidence. Prevalence is defined as the proportion of a population alive at a given time that is suffering from cancer. Incidence refers to the new cases of cancer that occur in a defined population and the number of such new cases detected in a specific period of time.

Ancient medical wisdom of India known as "Ayurveda" is not confined to medicine only, it deals with the whole subject of life and its various ramifications. The unique and most important quality of Ayurveda is its health-oriented approach and is not just disease-oriented. Cancer, as we know it today, is not described in classic Ayurvedic texts. But, we come across descriptions of certain diseases that have remarkable similarities with modern interpretations of cancer. Arbuda is a parallel term to cancer found in Ayurvedic treatises and is derived from the root verb, "Arb", that means to kill or hurt. Arbuda denotes any kind of fleshy protuberance in a broad sense. The study of the profound description of modern oncology together with the analysis of the information in terms of Ayurvedic principles and practices would certainly provide us with new vistas for effective interactions and, thereby, improve our clinical wisdom. It needs a special mention here that the basic approach of Ayurveda is integrative and not analytical and hence the modern study parameters, which mostly adopt analytical and reductionist methodology, may not be found sufficiently effective in bringing out the desired results. The intuitive logic and functional observations will have to play a major role in interpreting empirical results. Cancer, a comparatively rare disease in the past, has now become a fairly common ailment. The etiology of cancer still remains largely unknown and so is the cure.

Radiation therapy is an important modality of cancer treatment and approximately

60% of all cancer patients are treated using this mode of treatment. Radiotherapy causes secondary cancers after the remission of primary cancer, which typically occurs several years later. It is also accompanied by other secondary diseases such as pneumonitis and radiation fibrosis. Radiation therapy is associated with both acute and delayed disturbances in nutritional status. However, radiotherapy does not distinguish between neoplastic and normal healthy cells surrounding tumour tissue; as a result, the latter also receive significant radiation dose and get damaged during cancer cure.

A long-standing paradigm in radiation biology has been that many effects induced by ionizing radiation (IR), including its carcinogenic effects and ability to kill cancer cells, are the result of DNA damage arising from the actions of IR in cell nuclei, especially interactions of IR and its products with nuclear DNA (Illiakis, 1991; Goodhead, 1994). Consistent with this view, IR undoubtedly can damage DNA by directly ionizing DNA itself and by indirect processes in which DNA reacts with numerous radiolytic reactive products, e.g., $^{\bullet}OH$, $H^{\bullet}$, $O_2^{\bullet-}$ and H_2O_2, that are generated in aqueous fluid surrounding DNA. Several attempts have been made to increase the effect of ionizing radiation by combining hypoxic cell sensitizers with radiation. Although successful in an experimental setup, clinical success was never achieved. Over recent years, many attempts have been made to combine chemotherapy with radiation for improved results of malignant tumours. This combination proved beneficial in solid neoplastic disorders in randomized clinical trials. The most frequently used combination of chemotherapeutic agents includes cis-dichlorodiammine-platinum (II), 5-fluorouracil, mitomycin C, paclitaxel, docetaxel, gemcitabine, topotecan, irinotecan, crytophycins, camptothecin and combretastatin A-4 (Carde and Laval, 1995; Creane *et al.,* 1999; Dunton *et al.,* 2002; Giocanti *et al.,* 1993; Masunaga *et al.,* 2001; Matsuura *et al.,* 2000; McGinn *et al.,* 2002; Milas *et al.,* 1999; Murata *et al.,* 2001; Owen *et al.,* 2002; Rauth *et al.,* 1983; Rockwell and Kennedy, 1979; Teicher *et al.,* 2000; Tishler *et al.,* 1992; Turner and Pearcey, 1996; Van Belle *et al.,* 1994; Vickers, 2002; Vokes *et al.,* 1992). Combination of chemotherapy and radiotherapy has been successful in many cases, however, the toxic side effects of combination therapy are severe. Therefore, newer approches are required to reduce the toxic effects of combination regimens, which shall reduce the side effects and confer optimum therapeutic gains with good quality of life.

Tinospora cordifolia Miers, belonging to family Menispermaceae, is commonly known as guduchi or giloe. It is widely used in the Ayurvedic system of medicine for its general tonic, anti-inflammatory, anti-arthritic, anti-allergic, anti-malarial, antidiabetic and aphrodisiac properties (Sukh Dev, 2006). The radiomodulatory properties of *Tinospora cordifolia* have been reported by Goel *et al.,* (2001). The alkaloids, most of them belonging to the isoquinoline group, appear to be the active constituents of the plant. So far, berberine, palmatine, tembertarine, magniflorine, choline and tinosporin are reported from its stem (Sarma and Khosa, 1993; Sukh Dev, 2006). The preliminary studies on the stem extracts of *Tinospora cordifolia* have shown promising response in cultured HeLa cells, where various extracts of guduchi were found to reduce the cell survival in a dose-dependent manner. However, dichloromethane (methylene chloride) extract produced the maximum cytotoxic effect *in vitro* and was found to be non-toxic (Jagetia *et al.,* 1998, 2002; Jagetia and Rao, 2006a). Our earlier study on the anticancer activity of dichloromethane extract of guduchi had conclusively shown that 50mg/kg b. wt. dichloromethane extract arrested tumour growth in mice transplanted with Ehrlich's ascites carcinoma (Jagetia and Rao, 2006b). The present work describes the radiomodulatory activity of dichloromethane extract of *Tinospora cordifolia* in mice bearing Ehrlich ascites carcinoma.

Materials and Methods

Collection and extraction of plant material

The stems of *Tinospora cordifolia* (Willd.) Miers ex Hook. F. & Thoms (Family, Menispermaceae) were collected, cleaned, shade dried and coarsely powdered with the help of a ball mill. The plant material was exhaustively extracted sequentially with petroleum ether (60-80°C). Chloroform and dichloromethane, using a Soxhlet continuous extraction apparatus. The final dichloromethane extracts (henceforth TCE) were concentrated *in vacuuo* and dried under reduced pressure. An approximate yield of 1.2% w/w was obtained.

Animal care and handling

The animal care and handling were done according to the guidelines set by the WHO (World Health Organization, Geneva, Switzerland), INSA (Indian National Science Academy, New Delhi, India) and the "Guide for the Care and Use of Laboratory Animals" (NIH publication #86-23, revised in 1985). Ten- to twelve-week-old female Swiss albino mice weighing 30 to 36g were selected from an inbred colony, maintained under controlled conditions of temperature (23±2°C), humidity (50±5%) and light (12h of light and dark, respectively). The animals had free access to the sterile food and water. Four animals were housed in a polypropylene cage containing sterile paddy husk (procured locally) as bedding throughout the experiment.

Tumour model

Ehrlich ascites carcinoma (EAC) procured from the Advanced Center for Treatment and Research in Cancer (ACTREC), Mumbai, India, was maintained and propagated by serial transplantation intraperitoneally in an aseptic environment. 10^6 viable EAC cells were injected intraperitoneally into each animal under aseptic conditions and the day of tumour inoculation was considered as day 0.

Preparation of drug and mode of administration

TCE was dissolved in 100μl of ethanol and diluted with normal sterile saline (SPS). containing 1% carboxymethyl cellulose (CMC). Both SPS and TCE were administered intraperitoneally.

Selection of optimum TCE dose

The dose of TCE was selected following the standard protocol recommended by the Drug Evaluation Branch, Drug Research and Development, National Institutes of Health (NIH), USA (Geran *et al.*, 1972). Twenty-four hours after tumour inoculation, the animals were divided into the following two groups: (i) *SPS + irradiation*: The animals of this group received 0.3 to 0.36ml/kg b. wt. SPS one hour before exposure to 6Gy hemi-body γ-radiation, followed by single injection (once daily) for another eight consecutive days after irradiation. (ii) *TCE + irradiation*: The animals of this group were injected with 0, 25, 30, 40, 50 and 100mg/kg b. wt. TCE one hour before exposure to 6Gy hemi-body

γ-radiation, followed by single injection of each dose of TCE (once daily) for another eight consecutive days after irradiation.

Evaluation of the radiosensitizing effect of TCE

A separate experiment was conducted to ascertain the optimum dose of radiation to obtain highest tumour killing effect in conjunction with 30mg/kg TCE. The animals were divided into the following two groups: (i) *SPS + irradiation*: The animals of this group received 0.3 to 0.36ml/ kg b. wt. of SPS one hour before exposure to 0, 1, 2, 4, 6 or 8Gy hemi-body γ-radiation, followed by single injection (once daily) for another eight consecutive days after irradiation. (ii) *TCE + irradiation*: The animals of this group were injected with 30mg/kg b. wt. TCE one hour before exposure to 0, 1, 2, 4, 6 or 8Gy hemi-body γ-radiation followed by single injection of TCE once daily, for another eight consecutive days after irradiation.

Effect of altered administration regimen

Pre-irradiation administration

A separate experiment was carried out to evaluate the radiosensitizing activity of altered schedule of administration of 30mg/kg b. wt. TCE, where the animals were divided into the following groups:

SPS + irradiation

Twenty-four hours after EAC inoculation, the animals received 0.3 to 0.36ml/kg b. wt. of SPS, one hour before exposure to 6Gy of hemi-body gamma irradiation.

TCE + irradiation (A)

This group of animals was injected with 30mg/kg b. wt. of TCE, one hour before exposure to 6Gy of hemi-body gamma irradiation, followed by single injection (once daily) for another three consecutive days after irradiation.

TCE + irradiation (B)

This group of animals was injected with 30mg/kg b. wt. of TCE, one hour before exposure to 6Gy of hemi-body gamma irradiation, followed by single injection of TCE (once daily) for another six consecutive days after irradiation.

Post-irradiation administration

A separate experiment was carried out to evaluate the radiosensitizing activity of 30mg/kg b. wt. TCE after irradiation, by dividing the animals into the following groups:

Irradiation + SPS

This group of animals received 0.3 to 0.36ml/kg b. wt. of SPS immediately after exposure to 6Gy of hemi-body gamma irradiation.

Irradiation + TCE (C)

This group of animals received 30mg/kg b. wt. of TCE, within five minutes of exposure to 6Gy of hemi-body gamma irradiation, followed by a single administration of TCE (once daily) for another three consecutive days after irradiation.

Irradiation + TCE (D)

This group of animals received 30mg/kg b. wt. of TCE, within five minutes of exposure to 6Gy of hemi-body gamma irradiation, followed by a single administration of TCE once daily for another six consecutive days after irradiation.

Irradiation

Prostrate, immobilized (achieved by inserting cotton plugs in the restrainer) and unanaesthetized tumour-bearing animals of all experiments were restrained in a specially designed well-ventilated perspex box and their lower half of the body, below rib cage (hemi-body) was then exposed to 0 (sham-irradiation), 2, 4, 6 or 8Gy gamma radiation as the case may be. A batch of ten animals was irradiated at a dose rate of 2Gy/min at source to animal distance (midpoint) of 78.9cm, using a Telecobalt therapy source (Theratron, Atomic Energy Agency, Ontario, Canada).

After the last administration of TCE, the animals were monitored regularly for body weight changes, signs of toxicity and mortality. The weights of animals were recorded every third day up to 30 days after tumour inoculation in all the groups. Thirty three percent of drug related deaths or a weight loss of 5 g per mouse was considered as an index of toxicity. The animal survival was monitored daily up to 120 days, since the survival of animals up to 120 days is roughly equivalent to 5 years survival in man (Nias, 1990). Animals, which survived after 60 days were considered to be long time survivors (LTS). The tumour response was assessed on the basis of median survival time (MST), average survival time (AST) and tumour free survival. MST and AST were calculated from the animals dying within 120 days and those surviving 120 days were excluded from these calculations. Statistical analysis of the data was carried out by Student's t-test at 95% confidence limit. The calculations for MST, AST, IALS and IMLS were determined as per National Cancer Institute (NCI) guidelines (Geran *et al.*, 1972).

Biochemical analyses

A separate experiment was performed to estimate glutathione, glutathione-S-transferase and lipid peroxidation in the tumour cells exposed to 6Gy gamma radiation. The animals were inoculated with tumour cells as described above and the tumour was allowed to grow for six days so as to get a reasonable volume for aspiration of cells. On the seventh day, the tumour-bearing animals were divided into the following groups:

SPS + irradiation

The animals of this group received 0.3 to 0.36ml/kg b. wt. of SPS one hour before 6Gy irradiation and subsequently once daily for another six days.

TCE + irradiation

The animals of this group were administered with a single injection of 30mg/kg b. wt. of TCE one hour before 6Gy irradiation and subsequently once daily for another six days.

Four animals from each group were sacrificed at 1.5, 3, 6, 9 12, 15, 18 or 24h after irradiation. The tumour cells were aspirated in an aseptic condition, washed with SPS and counted under an inverted microscope (Ernst Leitz, GmbH, Wetzlar, Germany). One million cells were sonicated (Virsonic, Virtis, NY, USA) and processed for the estimation of glutathione, glutathione-S-transferase and lipid peroxidation as described below.

Glutathione (GSH)

Glutathione concentration was measured by the method of Moron *et al*., (1979). Briefly, proteins were precipitated by 25% TCA, centrifuged and the supernatant was collected. The supernatant was mixed with 0.2 M sodium phosphate buffer pH 8.0 and 0.06mM DTNB and incubated for 10 minutes at room temperature. The absorbance of the sample/s was read against the blank at 412nm in a UV-Visible double beam spectrophotometer (Shimadzu UV-260, Shimadzu Corp, Tokyo, Japan) and the GSH concentration was calculated from the standard curve.

Glutathione-S-transferase (GST)

The cytosolic glutathione-S-transferase (GST) activity was determined spectrophotometrically at 37°C according to the procedure of Habig *et al.* (1974). The reaction mixture contained 2.7ml of 100mM phosphate buffer (pH 6.5), and 0.1ml of 30mM CDNB. After preincubating the reaction mixture at 37°C for 5 min, the reaction was initiated by the addition of 0.1ml of supernatant (S) and the absorbance was followed for 5 min at 340 nm in a UV-Visible double beam spectrophotometer. Reaction mixture without the enzyme was used as blank. The GST activity has been expressed as mM of GSH-CDNB conjugate formed/min/10^6 cells. The activity measured was multiplied by 10 to get total activity per million cells.

Lipid peroxidation (LPx)

Lipid peroxidation was estimated by the method of Beuege and Aust (1972). Briefly, the homogenate was mixed with TCA-TBA-HCl and heated for 15 min in a boiling water bath. After centrifugation, the absorbance was recorded at 535 nm using a UV-Visible double beam spectrophotometer. The lipid peroxidation has been expressed as TBARS in nmol per 10^6 cells. Lipid peroxidation in the samples was determined against the standard curve of MDA (malonaldehyde).

Statistical analyses

The statistical analyses were performed using GraphPad Prism 2.01 statistical software (GraphPad Software, San Diego, CA, USA). The statistical significance between the treatments was determined using "Z" test for survival studies (Abramowitz and Stegun, 1972), whereas Student's 't' test was used for biochemical analyses. A p value of < 0.05 was considered statistically significant. All the data are expressed as mean±SEM (standard error of the mean).

Results

Selection of optimum TCE dose

Mice transplanted with Ehrlich ascites carcinoma (EAC) did not show spontaneous tumour regression. The tumour-bearing mice showed a constant weight gain and increase in the tumour volume due to tumour cell multiplication and growth of EAC cells (Fig. 21.1) where MST and AST were 17 days, respectively (Table 21.1).

Table 21.1. Effect of various doses of TCE in combination with 6Gy of gamma irradiation on the survival of tumour-bearing mice.

TCE dose (mg/kg)	Sham-irradiation*							
	Survival (days)				*Percent survivors (days)*			
	MST	IMLS	AST	IALS	30	60	90	120
SPS	17.5	-	16.6	-	0	0	0	0
TCE 25	21.5	13.1	20.5	14	0	0	0	0
TCE 30	26	36.8	25.7	43.3	0	0	0	0
TCE 40	35.5	86.8	34.9	94.3	91.7	66.7	0	0
TCE 50	55.5	192.1	56.4	213.3	100 [a]	100 [a]	100 [a]	33.3 [b]
TCE 100	26	36.8	25.7	43.3	0	0	0	0
	Irradiation							
SPS	25.2	-	24.7	-	0	0	0	0
TCE 25	36.2	46.6	34.5	39.7	0	0	0	0
TCE 30	45.6	160.6	43.7	163.2	100 [a]	91.7 [a]	50 [b]	16.7
TCE 40	42.1	78.9	41.3	76.1	75 [b]	41.7	16.7	0
TCE 50	36.4	44.4	34.3	38.9	33.3	16.7	0	0
TCE 100	29.7	17.8	27.8	12.5	0	0	0	0

*a = $p < 0.0001$; b = $p < 0.05$ (when compared with SPS); * sham-irradiation = 0Gy*

Exposure of tumour-bearing mice to a single dose of 6Gy hemi-body gamma irradiation resulted in an increase in the MST and AST approximately by 8 days and the animals did not survive beyond 28 days post-irradiation (Table 21.1). Treatment of EAC-bearing mice with TCE, followed by an exposure to a single dose of 6Gy gamma irradiation resulted in a TCE dose-dependent rise in the MST and AST. The greatest effect was observed for the animals receiving 30 or 40mg/kg TCE and 6Gy irradiation, where the MST increased by approximately 21 and 17 days, respectively, over the SPS + irradiation group. Similarly, the AST was also elevated by 19 and 16 days for 30 and 40mg/kg TCE + irradiation, respectively (Table 21.1). The least effect was observed for 100mg/kg b. wt. TCE + irradiation, where MST and AST were found to be approximately 30 and 28 days, respectively, which was marginally higher than the SPS + irradiation group. The highest radiomodifying effect was observed for 30mg/kg b. wt. TCE, which resulted in approximately 92% long-term survivors up to 60 days ($p < 0.0001$), followed by a 50% survival until 90 days ($p < 0.05$). This dose also showed approximately 17% survivors beyond 120 days (Table 21.1). Moreover, 30mg/kg TCE also did not induce any toxic effects in the form of debility and loss of body weight. Therefore, 30mg/kg was considered as an optimum dose and further studies were carried out using this dose.

Treatment of EAC-bearing mice with 25, 30, 40, 50 or 100mg/kg b. wt. TCE, before exposure to 6Gy hemi-body radiation, arrested the tumour weight gain, indicating inhibition of tumour cell proliferation and growth (Fig. 21.1). Administration of 100mg/kg TCE before irradiation was accompanied by toxic side effects like ruffling of hair, sluggishness, weight loss and lacrimation, followed by death. None of the animals survived beyond day 9 post-tumour inoculation (Fig. 21.1).

Radiosensitizing effect

Irradiation of EAC-bearing mice caused a radiation dose-dependent reduction in the tumour growth as evidenced by a dose-dependent increase in MST, AST, IMLS and IALS. However, irradiation failed to prolong the life span of EAC-bearing mice beyond 30 days (Table 21.2). Administration of 30mg/kg TCE before irradiation caused a radiation dose-dependent remission of tumour up to 6Gy, as evidenced by an increase in AST and MST by approximately 20 days. This was reflected in an increased IMLS and IALS, respectively (Table 21.2). The combination of 30mg/kg TCE with 6Gy resulted in LTS as well as survivors up to 120 days, while this effect was not discernible for the combination treatment with other irradiation doses. An increase in irradiation dose up to 8Gy actually resulted in a decline in MST and AST, indicating that this combination was not as effective in reducing the tumour growth as TCE + 6Gy irradiation (Table 21.2). Therefore, a combination of 30mg/kg TCE with 6Gy irradiation was considered the best and remaining studies were performed using this regimen.

Effect of altered administration regimen

Pre-irradiation administration

Exposure of animals to 6Gy hemi-body gamma radiation increased MST and AST up to 25 days, respectively, when compared with SPS + sham-irradiation (Table 21.1). An increase in MST and AST by seven days was observed when TCE was given one hour

before 6Gy irradiation, followed by another three-day treatment. Similarly, when the TCE treatment time was extended up to six days post-irradiation, the MST and AST increased by approximately 7 days, when compared to TCE + irradiation (A) and showed a drastic increase in MST and AST by 15 and 14 days, respectively, when compared to the SPS + irradiation group (Table 21.3).
Treatment of animals with 30mg/kg b. wt. of TCE one hour before exposure to 6Gy of hemi-body gamma irradiation and further treatment of TCE once daily for another six consecutive days post-irradiation also showed an increase in the life span of EAC-bearing mice as more LTS were observed. Almost 17% of EAC-bearing mice showed tumour free survival up to 120 days, when compared with other groups (Table 21.3).

Post-irradiation administration

Hemi-body irradiation of animals (6Gy) increased MST and AST up to 25 days, respectively, when compared with SPS + sham-irradiation (Table 21.1). An increase in MST and AST by approximately nine days was observed when TCE was given immediately after 6Gy irradiation, followed by another three-day treatment. Similarly, extension of TCE treatment to another six days caused an increase in MST and AST approximately by six days when compared to irradiation + TCE (C) and showed a drastic increase in MST and AST approximately by 16 and 14 days respectively in animals that received a total of seven doses of TCE after irradiation, when compared to the irradiation + SPS group (Table 21.3).

Treatment of animals with 30mg/kg b. wt. of TCE within five minutes of exposure to 6Gy of hemi-body gamma irradiation and further treatment with TCE once daily for another six consecutive days also showed a greater increase in the life span of EAC-bearing mice as revealed a greater number of LTS as well as survivors up to 90 days (approximately 17%) when compared with other groups (Table 21.3).

Biochemical analyses

Glutathione (GSH)

The GSH concentration showed a slight fluctuation with time in EAC-bearing mice treated with SPS + irradiation. Treatment of 30mg/kg. b. wt. TCE showed a marginal decline in GSH concentration in EAC cells and the pattern was similar to that of SPS + irradiation group. Administration of TCE before 6Gy irradiation showed a steady decline in GSH concentration that reached a nadir at 15h post-irradiation and elevated thereafter. However, GSH concentration was significantly lower even up to 24h post-irradiation, when compared to TCE alone and SPS + irradiation groups (Fig. 21.2a).

Glutathione-S-transferase (GST)

Exposure of EAC animals to 6Gy gamma radiation showed a slight decline in the GST activity up to 24 h post-irradiation. A similar effect was observed after TCE treatment alone, which was statistically non-significant in comparison to SPS + irradiation group. Treatment of EAC mouse with TCE before irradiation resulted in a steady decline in the GST activity and a maximum reduction was observed at 12h post-irradiation.

Table 21.2. Effect of 30mg/kg of TCE in combination with different doses of □-irradiation on the survival of tumour-bearing mice.

Dose (Gy)	Treatment															
	SPS + Irradiation								TCE + Irradiation							
	Survival (days)				Percent survivors (days)				Survival (days)				Percent survivors (days)			
	MST	IMLS	AST	IALS	30	60	90	120	MST	IMLS	AST	IALS	30	60	90	120
0	17.5	-	16.6	-	0	0	0	0	25.2	-	24.7	-	0	0	0	0
2	22.3	27.4	21.1	27.1	0	0	0	0	24.8	41.7	23.3	40.4	0	0	0	0
4	23.7	34.4	21.5	29.5	0	0	0	0	26.5	51.4	27.7	66.9	16.7	0	0	0
6	25.2	44	24.7	48.8	0	0	0	0	45.6	160.6	43.7	163.2	100[a]	91.7[a]	50[b]	16.7
8	28.4	62.3	30.3	82.5	0	0	0	0	19.5	11.4	15.3	-	33.3	0	0	0

[a] = $p < 0.0001$; [b] = $p < 0.05$ (when compared with SPS)

Table 21.3. Effect of altered treatment regimen on the radiation-induced regression of tumour in mice treated with TCE.

Treatment		Survival (days)				Percent survivors (days)			
		MST	IMLS	AST	IALS	30	60	90	120
SPS + Irradiation		25.2	-	24.7	-	0	0	0	0
TCE + Irradiation	A	33	57.1	32.1	68.9	66.7	33.3	16.7	0
	B	40.5	92.8	39.2	106.3	91.7[a]	58.3[b]	33.3	16.7
Irradiation +TCE	A	35	66.7	33.2	74.7	33.3	16.7	0	0
	B	41	95.2	38.6	103.2	58.3	33.3	16.7	0

[a] = $p < 0.0003$; [b] = $p < 0.03$ (when compared with SPS)

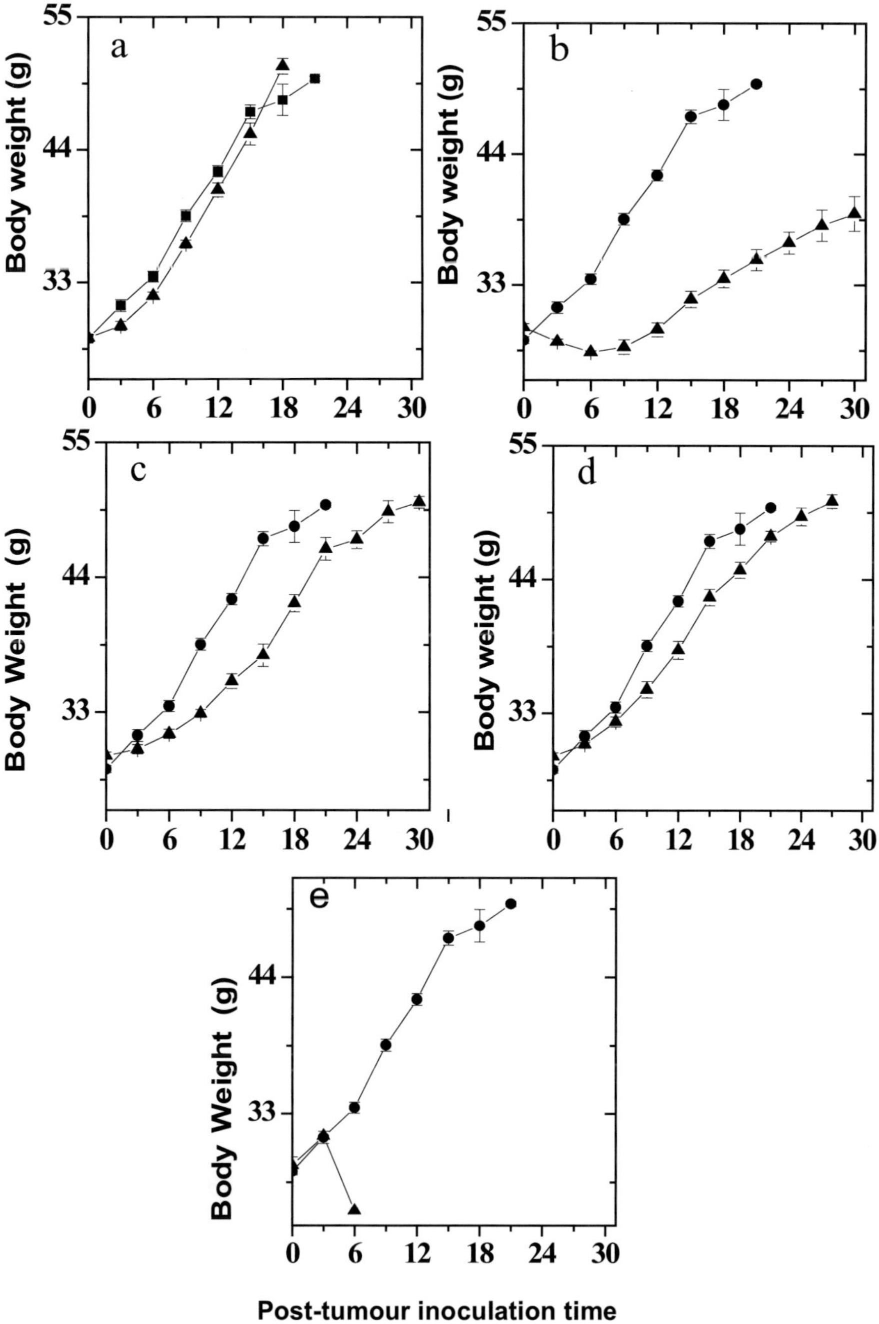

Fig. 21.1. Alterations in body weight in the Ehrlich ascites carcinoma-bearing mice treated with various doses (0, 25, 30, 50 and 100mg/kg b. wt.) of TCE when exposed to 6Gy of gamma radiation. SPS + irradiation (closed circles) and TCE + irradiation (triangles). a, 25, b, 30, c, 40, d, 50 and e, 100mg/kg b. wt. TCE.

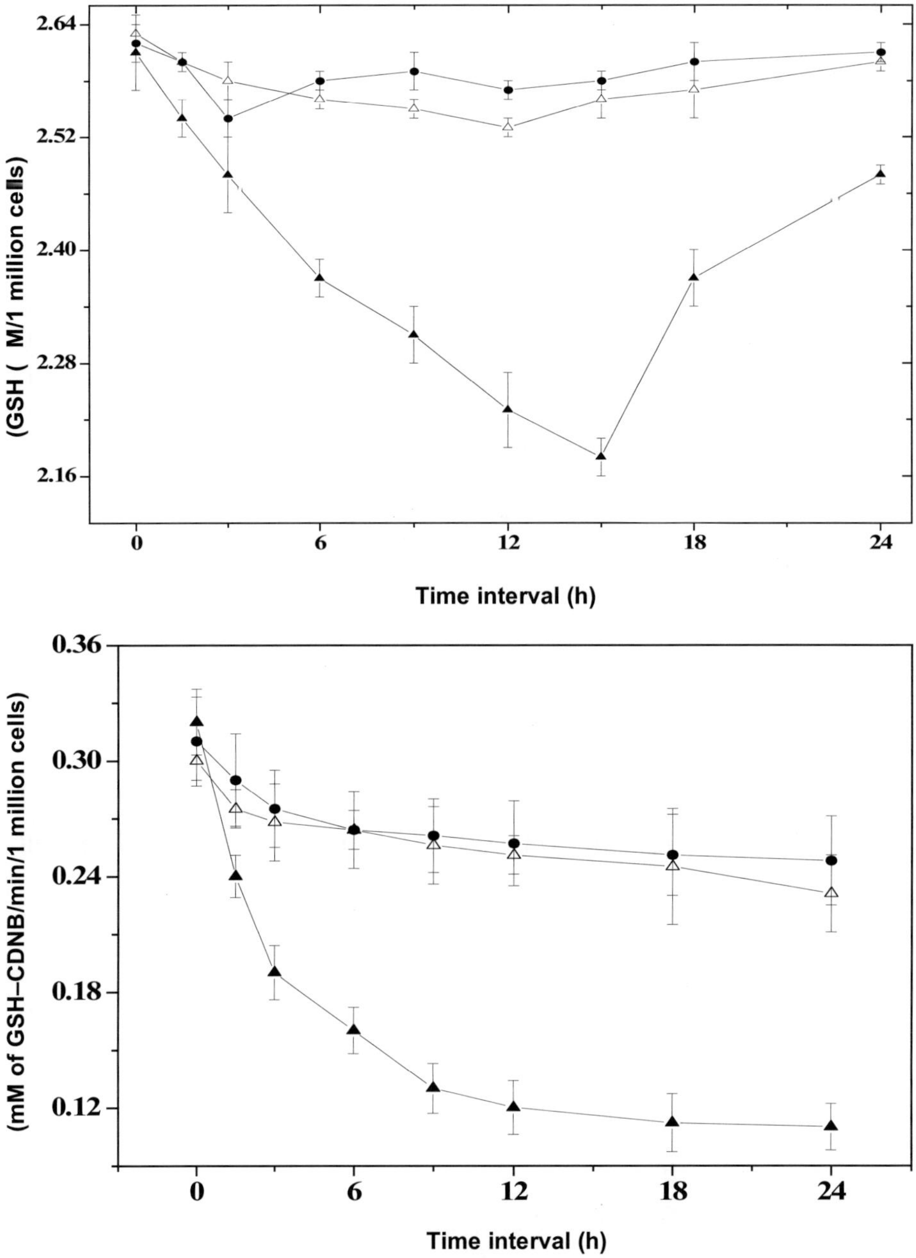

Fig. 21.2. Effect of the treatment of 30mg/kg b. wt. of dichloromethane extract of *Tinospora cordifolia* (TCE) before exposure to 6Gy of gamma radiation on the alteration in a) Glutathione and b) Glutathione-S-transferase content in mice bearing Ehrlich ascites carcinoma. Open triangles, TCE alone; Solid circles, SPS + irradiation; Solid triangles, TCE + irradiation.

This trend of decline in GST continued up to 24h post-irradiation, where it was approximately 3-fold lower than the initial value (Fig 21.2 b).

Lipid peroxidation (LPx)

Lipid peroxidation showed a slight fluctuation with time in EAC-bearing mice treated with SPS and then exposed to 6Gy radiation. Treatment of EAC-bearing mice with TCE alone showed a significant increase in LPx in EAC cells, and a peak was attained at 6h, which declined steadily thereafter up to 24h post-irradiation, where it reached an almost baseline level. Treatment of EAC-bearing mice with 30mg/kg b. wt. TCE before exposure to 6Gy hemi-body radiation increased the lipid peroxidation in Ehrlich ascites cells steadily and a maximum level of LPx was observed at 12h post-irradiation. Thereafter, the lipid peroxidation declined steadily up to 24h, where it almost reached the base-line level (Fig. 21.3).

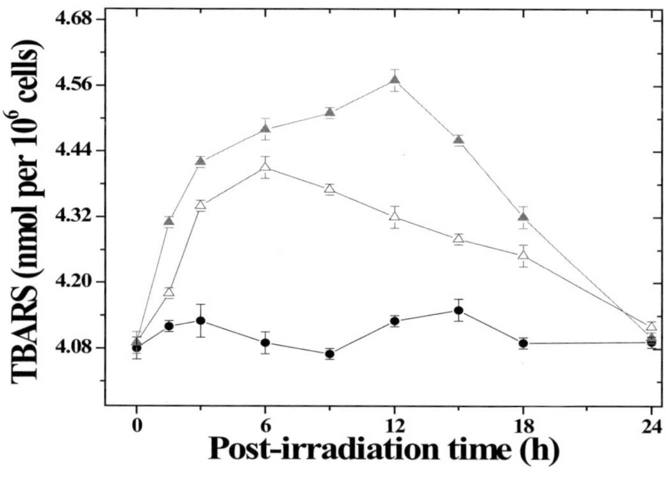

Fig. 21.3. Effect of the treatment of 30mg/kg b. wt. of dichloromethane extract of *Tinospora cordifolia* (TCE) before exposure to 6Gy of gamma radiation on the alteration in the lipid peroxidation in mice bearing Ehrlich ascites carcinoma. Open triangles, TCE alone; Solid Circles, SPS + irradiation; Solid triangles, TCE + irradiation.

Discussion

Herbal medicines appear to be better suited to alleviate the side effects of radiotherapy and chemotherapy than the exotic pharmacological agents of synthetic origin. Herbal medicines may also increase the effect of radiation on neoplastic cells, while protecting normal tissues from radiation-induced damage. Another important aspect of herbal medicines is that their use may increase the immune surveillance of normal tissues, which is adversely affected during progression of neoplasia. Radiotherapy and chemotherapy have multiple negative effects on the immune system, as do surgery, anaesthetic, antibiotics, injections, blood transfusions and all drug and chemical exposure that may form part of cancer treatment (Coleman and Mitchell, 1999). More and more people who have been diagnosed with cancer or related illnesses embrace alternative medicines, and seek out holistic practitioners who use vitamins, herbs, nutraceuticals, homeopathy and acupuncture to treat the disease. The inclusion of herbal medicine in radiotherapy regimens can improve the therapeutic index by killing neoplastic cells and reducing radiation toxicity to normal tissues (Tannock, 1996). Our earlier study had shown that dichloromethane extract of *Tinospora cordifolia* (TCE) killed neoplastic cells *in vitro* (Jagetia *et al.*, 1998; Jagetia and Rao, 2006a) and *in vivo* (Jagetia and Rao, 2006b) and reports regarding the radiosensitizing action of methylene chloride extract of *T. cordifolia in vivo* are lacking. It was, therefore, decided to screen the radiosensitizing activity of TCE in Swiss albino mice transplanted with Ehrlich ascites carcinoma.

Administration of various concentrations of TCE in combination with 6Gy hemi-body radiation caused a dose-dependent retardation in the tumour development, as was evident by the inhibition in body weight gain and increase in the life span (MST and AST). Another Indian medicinal plant, *Alstonia scholaris* has been reported to possess antineoplastic and radiosensitizing activity in different tumour models *in vitro* and *in vivo* (Jagetia and Baliga, 2003). Of all the TCE concentrations tested, highest radiomodifying effect was observed for 30mg/kg TCE, where approximately 92% long-term survivors could be observed up to 60 days ($p < 0.0001$) [50 % until 90 days ($p < 0.05$) and 17% beyond 120 days post-irradiation].

A variety of approaches have been developed for diminishing the effects of radiation on normal tissues and enhancing tumour cell kill by ionizing radiations. Different strategies may provide a therapeutic gain in radiation therapy. Results obtained with model systems do not always apply to more complicated biological systems. Many of these compounds have multiple pharmacological actions, which are sometimes antagonistic. Attempts to develop clinically relevant radiosensitizers have traditionally used an empirical approach combining radiation with standard cytotoxic chemotherapeutic agents. Although often effective in experimental models, the results obtained when these combinations are applied in a clinical setting have generally been less than expected. When a potential benefit is given for improved tumour control, quality of life and survival by a combination of radiation with herbal radiation modifiers, it is worthwhile to consider the complexity of combined modality treatment (Tannock, 1996). In view of advances made towards a thorough understanding of the molecular mechanisms of radioresponse, alteration in drug treatment regimen might increase tumour radiosensitivity.

We have developed an altered treatment modality of TCE in combination with radiation. Our results show that treatment of animals with 30mg/kg b. wt. TCE one hour before exposure to 6Gy of hemi-body gamma irradiation and further treatment of TCE once daily for another six consecutive days post-irradiation increases the life span of EAC-bearing mice, as is evident by a greater number of LTS as well as survivors beyond 120 days, when compared with the group which was irradiated first and then treated within 5 minutes with TCE for one day and another consecutive six days after irradiation. The latter showed an increase in the MST, as well as AST, when compared to the former but EAC-bearing mice failed to survive beyond 120 days. Similar types of reports regarding altered regimens in experimental system are lacking, however, other drugs such as cisplatin, 5-fluorouracil, bleomycin, methotrexate and mitomycin C (Cachin *et al.*, 1977; Crissman *et al.*, 1987; Gollin *et al.*, 1972; Gupta *et al.*, 1987; Lo *et al.*, 1976; Marcial *et al.*, 1990; Weissberg *et al.*, 1989) have been tried in clinical situations with altered regimen.

Tumour cell glutathione concentration may be among the determinants of cytotoxicity of many chemotherapeutic agents and of radiation, and an increase in glutathione concentration in cancer cells appears to be at least one of the mechanisms of acquired drug resistance to chemotherapy. Rapid glutathione synthesis in tumour cells is associated with high rates of cellular proliferation and depletion of cancer cell glutathione *in vivo* decreases the rate of cellular proliferation and inhibits cancer growth. GSH has also been shown to modulate the cytotoxicity of a variety of chemotherapeutic agents and radiation (Kennedy *et al.*, 1995). The agents that deplete GSH like N-ethylmaleamide, allyl alcohol, buthionine sulfoximine and diethyl maleate increase the radiation effect (Bridges *et al.*, 1969; Brown, 1977; Comparti *et al.*, 1991; Ozols *et al.*, 1988). Administration of TCE before 6Gy irradiation showed a steady decline in GSH concentration that reached a nadir at 15h post-irradiation. This is corroborated by a steady decline in GST activity up to 12h post-irradiation when TCE was combined with 6Gy irradiation and this trend continued up to 24h post-irradiation where it was approximately 3-fold lower than the initial value. A drug that can selectively deplete the cancer cells of their glutathione, while increasing or at least maintaining the levels of glutathione in healthy cells, may prove useful in combination treatment. This is what TCE appears to do. Selective depletion of tumour cell glutathione may in fact render cancer cells more vulnerable to the action of chemotherapy and eventually protect normal tissue against the deleterious effects of chemotherapy (Kennedy *et al.*, 1995).

The mechanisms involved in the increased radioresponse by TCE *in vivo* may not be due to a single mechanism, but due to operation of several mechanisms. Our study indicates that reduction in tumour cell glutathione, accompanied by an increased lipid peroxidation by TCE may have played an important role in tumour regression and increase in tumour free survival. The increase and reduction in LPx and GSH respectively by TCE, in combination with irradiation, may have damaged the DNA of EAC cells, thereby bringing about effective tumour cell kill. TCE has been reported to cause DNA damage in HeLa cells (Jagetia *et al.*, 1998). GSH and lipid peroxidation are inversely related and the administration of GSH-depleting agents cause lipid peroxidation and liver necrosis in the recipient mice only after the GSH level reaches a nadir (Comparti *et al.*, 1991). Further investigations are still needed to better understand the mechanisms involved in cell or tissue-dependent radiosensitization.

Natural medicines have gained popularity over synthetic drugs in recent years with the belief that they are much safer and have led to the tremendous growth of phytopharmaceutical usage. While it is known that plant extracts are active against cancer, the standard approach has been to isolate, synthesize and administer the single chemical compound responsible for this effect. However, different components in a botanical may have synergistic activities and there is also some evidence that the presence of multiple compounds in a botanical extract can buffer the toxic effects of a single constituent. Our results are in corroboration with the concept that a whole or partially purified extract of a plant offers advantages over a single isolated ingredient, and also underpins the philosophy of herbal medicine (Cragg *et al.*, 1993; Sharma and Arora, 2006).

Conclusion

The radiosensitizing effect of TCE could be attributed to depletion of glutathione and glutathione-S-transferase, accompanied by elevated levels of lipid peroxidation and DNA damage of tumour cells. Since *Tinospora cordifolia* is being used in India for treatment of various ailments, it may offer an alternative treatment strategy for cancer in combination with gamma radiation.

References

Abramowitz, M. and Stegun, I.A. (1972) *Handbook of Mathematical Functions*. Dover Publications, Inc. New York, p. 925.

Beuege, J.A. and Aust, S.D. (1978) Microsomal lipid peroxidation. *Methods in Enzymology* 30, 302–310.

Bridges, B.A., Ashwood-Smith, M.J. and Munson, R.J. (1969). Susceptibility of mild thermal and ionizing radiation damage to the same recovery mechanisms in *Escherichia coli. Biochemical and Biophysical Research Communications* 29(35), 193–196.

Brown, J.M. (1977) Cytotoxic effects of the hypoxic cell radiosensitizer R0-07-0582 to tumour cells *in vivo. Radiation Research* 72, 469–486.

Cachin, Y., Jortay, A. and Sancho, H. (1977) Preliminary results of a randomized E.O.R.T.C. study comparing radiotherapy and concomitant bleomycin to radiotherapy alone in epidermoid carcinomas of the oropharynx. *European Journal of Cancer* 13(12), 1389–1395.

Carde, P. and Laval, F. (1995) Effect of cis-dichlorodiammine platinum II and X-rays on mammalian cell survival. *International Journal of Radiation Oncology, Biology and Physics* 17, 929–933.

Coleman, C.N. and Mitchell, J.B. (1999) Clinical radiosensitization, why it does and does not work. *Journal of Clinical Oncology* 17(1), 1–3.

Comparti, M., Macllaro, E., DelBello, B. and Casini, A.F. (1991) Gutathione depletion, its effects on other oxidant systems and hepatocellular damage. *Xenobiotica* 21, 1067–1076.

Cragg, G.M., Boyd, M.R., Cardellina, J.H., Newman, D.J., Snader, K.M. and McCloud, T.G. (1994) Ethnobotany and Drug Discovery: The experience of the US National Cancer Institute. *Ciba Foundation Symposium* 185, 178-190.

Creane, M., Seymour, C.B., Colucci, S. and Mothersill, C. (1999) Radiobiological effects of docetaxel (Taxotere), a potential radiation sensitizer. *International Journal of Radiation Biology* 75(6), 731–737.

Crissman, J.D., Pajak, T.F., Zarbo, R.J., Marcial, V.A. and Al–Sarraf, M. (1987) Improved response and survival to combined cisplatin and radiation in non–keratinizing squamous cell carcinomas of the head and neck. An RTOG study of 114 advanced stage tumours. *Cancer* 59(8), 1391–1397.

Dunton, C.J., King, S.A. and Neufeld, J. (2002) Phase I study of topotecan and radiation therapy in advanced cervical cancer. *Gynecologic Oncology* 85(1), 185–187.

Geran, R.I., Greenberg, N.H., MacDonald, M., Schumacher, A.M. and Abbott, B.J. (1972) Protocols for screening chemical agents and natural products against animal tumours and other biological systems. *Cancer Chemotherapy Reports* Part III 2(3), 25–27.

Giocanti, N., Hennequin, C., Balosso, J., Mahler, M. and Favaudon, V. (1993) DNA repair and cell cycle interactions in radiation sensitization by the topoisomerase II poison etoposide. *Cancer Research* 53(9), 2105–2111.

Goel, H.C., Arora, R., Sinha, A.K., Dogra, R., Prasad, J., Singh, S., Sagar, R., Kumar, I.P., Agarwala, P.K., Sharma, A.K., Samanta, N., Gupta, D., Prakash, H., Salin, C.A., Kumar, S.S., Mittal, A. and Pathania, V. (2001) A process for preparation of a radioprotective herbal extract from *Tinospora*. Indian Patents Office, Delhi, India (Patent No. 195062).

Gollin, F.F., Ansfield, F.J., Brandenburg, J.H., Ramirez, G. and Vermund, H. (1972) Combined therapy in advanced head and neck cancer, a randomized study. *American Journal of Roentgenology, Radium Therapy and Nuclear Medicine* 114(1), 83–88.

Goodhead, D.T. (1994) Initial events in the cellular effects of ionizing radiations, clustered damage in DNA. *International Journal of Radiation Biology* 65, 7–17.

Gupta, N.K., Pointon, R.C. and Wilkinson, P.M. (1987) A randomised clinical trial to contrast radiotherapy with radiotherapy and methotrexate given synchronously in head and neck cancer. *Clinical Radiology* 38(6), 575–581.

Habig, W.H., Pabst, M.J. and Jakoby, W.B. (1974) Glutathione S-transferases. The first enzymatic step in mercapturic acid formation. *Journal of Biological Chemistry* 249(22), 7130–7139.

Iliakis, G. (1991) The role of DNA double strand breaks in ionizing radiation-induced killing of eukaryotic cells. *Bioessays* 13, 641–648.

Jagetia, G.C. and Baliga, M.S. (2003) Treatment with *Alstonia scholaris* enhances radiosensitivity *in vitro* and *in vivo*. *Cancer Biotherapy and Radiopharmaceuticals* 18(6), 917–929.

Jagetia, G.C. and Baliga, M.S. (2004) Effect of *Alstonia scholaris* in enhancing the anticancer activity of berberine in the Ehrlich ascites carcinoma-bearing mice. *Journal of Medicinal Food* 7(2), 235–244.

Jagetia, G.C. and Baliga, M.S. (2005) The effect of seasonal variation on the antineoplastic activity of *Alstonia scholaris*. *Journal of Ethnopharmacology* 96(1-2), 37–42.

Jagetia, G.C. and Rao, S.K. (2006a) Evaluation of cytotoxic effects of dichloromethane extract of Guduchi (*Tinospora cordifolia* Miers ex Hook F & Thoms) on cultured HeLa Cells. *Evidence Based Complementary and Alternative Medicine* 3(2), 267–272.

Jagetia, G.C. and Rao, S.K. (2006b) Evaluation of the antineoplastic activity of guduchi (*Tinospora cordifolia*) in Ehrlich ascites carcinoma bearing mice. *Biological and Pharmaceutical Bulletin* 29(3), 460–466.

Jagetia, G.C., Nayak, V. and Vidyasagar, M.S. (1998) Evaluation of the antineoplastic activity of guduchi (*Tinospora cordifolia*) in cultured HeLa cells. *Cancer Letters* 127(1–2), 71–82.

Jagetia, G.C., Nayak, V. and Vidyasagar, M.S. (2002) Enhancement of radiation effect by Guduchi (*Tinospora cordifolia*) in HeLa cells. *Pharmaceutical Biology* 40(3), 179–188.

Kennedy, R.S., Konok, G.P., Bounous, G., Baruchel, S. and Lee, T.D. (1995) The use of a whey protein concentrate in the treatment of patients with metastatic carcinoma, a phase I-II clinical study. *Anticancer Research* 15(6B), 2643–2649.

Lo, T.C., Wiley, A.L., Jr. and Ansfield, F.J. (1976) Combined radiation therapy and 5-fluorouracil for advanced squamous cell carcinoma of the oral cavity and oropharynx, a randomized study. *American Journal of Roentgenology* 126(2), 229–235.

Marcial, V.A., Pajak, T.F. and Mohiuddin, M. (1990) Concomitant cisplatin chemotherapy and radiotherapy in advanced mucosal squamous cell carcinoma of the head and neck. Long-term results of the Radiation Therapy Oncology Group study 81–17. *Cancer* 66(9), 1861–1868.

Masunaga, S.I., Ono, K. and Suzuki, M. (2001) Radiosensitization effect by combination with paclitaxel in vivo, including the effect on intratumour quiescent cells. *International Journal of Radiation Oncology, Biology and Physics* 50(4), 1063–1072.

Matsuura, M., Hasegawa, M. and Hayakawa, K. (2000) Experimental study of the effects on apoptosis of docetaxel alone and in combination with irradiation. *Oncology Reports* 7(2), 289–293.

McGinn, C.J., Lawrence, T.S. and Zalupski, M., (2002) On the development of gemcitabine-based chemoradiotherapy regimens in pancreatic cancer. *Cancer* 95(4 Suppl), 933–940.

Meister, A. and Anderson, M.E. (1983) Glutathione. *Annual Review of Biochemistry* 52, 711–760.

Milas, L., Milas, M. and Mason, K.A. (1999) Combination of taxanes with radiation, preclinical studies. *Seminars in Radiation Oncology* 9(2 Suppl 1), 12–26.

Moron, M.S., Depierre, J.W. and Mannervik, B. (1979) Levels of glutathione, glutathione reductase and glutathione-s-transferase activities in rat lung and liver. *Biochimica et Biophysica Acta* 582, 67–78.

Murata, R., Overgaard, J. and Horsman, M.R. (2001) Combretastatin A-4 disodium phosphate, a vascular targeting agent that improves the anti-tumour effects of hyperthermia, radiation, and mild thermoradiotherapy. *International Journal of Radiation Oncology, Biology and Physics* 51(4), 1018–1024.

Nias, A.H.W. (1990) Radiation Biology In: Sikora, K. and Halnan, K.E. (eds) *Treatment of Cancer*, Chapman and Hall Medical, London, pp. 53–75.

Owen, D.G., McNamee, J.P., Raaphorst, G.P. and Ng, C.E. (2002) Potentiation of cell killing by low-dose-rate radiation by camptothecin is related to an increase in the level of DNA double-strand breaks. *Radiation Research* 157(2), 149–157.

Ozols, R.F., Hamilton, T.C. and Young, R.C. (1988) Cellular glutathione levels and sensitivity to radiation and antineoplastic agents in human ovarian cancer. *Progress in Clinical and Biological Research* 276, 287–293.

Rauth, A.M., Mohindra, J.K. and Tannock, I.F. (1983) Activity of mitomycin C for aerobic and hypoxic cells *in vitro* and *in vivo*. *Cancer Research* 43, 4154–4158.

Rockwell, S. and Kennedy, K.A. (1979) Combination therapy with radiation and mitomycin C, Preliminary results with EMT6 tumour cells *in vitro* and *in vivo*. *International Journal of Radiation Oncology, Biology and Physics* 5, 1673–1676.

Sarma, D.N.K. and Khosa, R.L. (1993) Chemistry and pharmacology of *Tinospora cordifolia* Miers. *Indian Drugs* 30(11), 549–554.

Sharma, R.K. and Arora, R. (2006) *Herbal Drugs: A Twenty First Century Perspective*, Jaypee Brothers Medical Publishers Private Ltd., New Delhi, India, p. 688.

Sukh Dev (2006) *A Selection of Prime Ayurvedic Plant Drugs*. Ancient-Modern Concordance. Anamaya Publishers, New Delhi, India, p. 501.

Tannock, I.F. (1996) Treatment of cancer with radiation and drugs. *Journal of Clinical Oncology* 14(12), 3156–3174.

Teicher, B.A., Forler, P., Menon, K., Phares V., Amsrud, T. and Shih, C. (2000) Cryptophycin 52 and cryptophycin 55 in sequential and simultaneous combination treatment regimens in human tumour xenografts. *In vivo* 14(4), 471–480.

Tishler, R.B., Schiff, P.B., Geard, C.R. and Hall, E.J. (1992) Taxol, a novel radiation sensitizer. *International Journal of Radiation Oncology, Biology and Physics* 22, 613–6177.

Turner, A. and Pearcey, R.G. (1996) Effect of cisplatin on the clinically relevant radiosensitivity of human carcinoma cell lines. *International Journal of Radiation Oncology, Biology and Physics* 34, 367–374.

Van Belle, S., Fortan, L., De Smet, M., De Neve, W., Van der Elst J. and Storme, G. (1994) Interaction between vinblastine and ionizing radiation in the mouse MO4 fibrosarcoma *in vivo*. *Anticancer Research* 14(3A), 1043–1048.

Vickers, A. (2002) Botanical medicines for the treatment of cancer, rationale, overview of current data, and methodological considerations for phase I and II trials. *Cancer Investigation* 20(7–8), 1069–1079.

Vokes, E.E., Becket, M., Karrison, T. and Weichselbaum, R.R., (1992) The interaction of 5-fluorouracil, hydroxyurea, and radiation in two human head and neck cancer cell lines. *Oncology* 49, 454–460.

Weissberg, J.B., Son, Y.H. and Papac, R.J. (1989) Randomized clinical trial of mitomycin C as an adjunct to radiotherapy in head and neck cancer. *International Journal of Radiation Oncology, Biology and Physics* 17(1), 3–9.

SECTION VI

ANTIOXIDANTS AND RADIOTHERAPY: THE ENIGMATIC CONNECTION

Chapter 22

Do Antioxidants Reduce the Efficacy of Radiotherapy?

Ralph Moss

Introduction

In September 2005, CA: A Cancer Journal for Clinicians (a publication of the American Cancer Society) published a sweeping attack by Gabriella D'Andrea, MD, on the concurrent use of antioxidants with radiotherapy and chemotherapy (D'Andrea, 2005). That article reinforced a belief already widely held among oncologists that concurrently administered antioxidants were likely to undermine the effectiveness of cytotoxic treatments, and that patients should be strongly advised to avoid taking them during treatment. It received abundant coverage in the mainstream media, including a feature story in the Wall Street Journal.

My book, "Antioxidants Against Cancer" (Moss, 2000), reviewed the literature and found a preponderance of evidence in support of concurrent use. This analysis was later updated in a journal article (Moss, 2006).

Other researchers have also contested this broad attack on antioxidants. For instance, Kenneth Conklin, MD, PhD, of the University of California, Los Angeles (UCLA) Medical Center, has questioned the theoretical basis for prohibiting concurrent use of antioxidants during cancer treatment. He has stated that although radiation kills cells by generating high levels of free radicals, this does not necessarily imply that antioxidants are contraindicated in all cases.

Conklin pointed to the fact that radiotherapy is most effective in well-oxygenated tissues, whereas tumours are often hypoxic at their center, diminishing the effectiveness of the treatment (Conklin, 2000). Antioxidants improve blood flow and promote normal oxygenation in tissues (Cameron and Cotter, 1999). When they do this within and around tumours, they may thereby play a beneficial role by rendering tumours more not less-susceptible to the cytotoxic effects of radiation. Since free radical generation is proportional to the oxygen tension in the tissue, antioxidants given in amounts that improve blood flow, but not in amounts that quench free radicals, may also result in an improved antineoplastic effect (Conklin, 2004).

The most recent comprehensive surveys of the concurrent use controversy are: Block *et al.*, 2007; Simone *et al.*, 2007. Simone and co-workers (2007) surveyed the peer-reviewed literature from 1996 through November 2003 on the use of vitamins and antioxidants given along with chemotherapy and radiation therapy. They identified a total of 280 peer-reviewed articles on the topic, of which 50 were clinical trials, involving a total of 8,521 patients. Of these patients, 5,081 received supplemental nutrients such as beta-carotene; vitamins A, C and E; selenium; cysteine; B vitamins; vitamin D3; vitamin K3; and glutathione, as single agents or in combination.

Simone concluded that these studies have "consistently shown that non-prescription antioxidants and other nutrients do not interfere with therapeutic modalities for cancer" (Simone *et al.,* 2007). In fact, he maintains that the use of antioxidants enhanced the effectiveness of standard therapeutic modalities (including radiotherapy) for cancer, while decreasing adverse effects and protecting normal tissue. In 15 of the studies, 3,738 patients who took non-prescription antioxidants and other nutrients had increased survival, contrary to the dominant paradigm (Simone *et al.,* 2007).

Block and co-workers (2007) focused on the interaction of antioxidants primarily with chemotherapy, but their review has implications for radiotherapy as well. They surveyed 845 articles and found 19 clinical trials that met strict inclusion criteria. These studies evaluated glutathione, melatonin, vitamin A, an antioxidant mixture, vitamin C, N-acetylcysteine, vitamin E and ellagic acid. Participants in most of these studies had advanced or recurrent disease. The authors concluded: "None of the trials reported evidence of significant decreases in efficacy from antioxidant supplementation during chemotherapy" (Block *et al.,* 2007).

Instead, many of the studies indicated that antioxidant supplementation was associated with increased survival times, increased tumour responses, or both, as well as fewer toxicities than controls. They concede: "A lack of adequate statistical power was a consistent limitation" (Block *et al.,* 2007). Block and colleagues believe that "large, well-designed studies of antioxidant supplementation concurrent with chemotherapy are warranted".

However, such studies are not likely to happen soon. A search of the National Institute of Health's (NIH) website, www.clinicaltrials.gov, using the terms "antioxidants" and "radiotherapy", returns but a single entry, "Safety of Oral Antioxidants and Intravenous Vitamin C During Gyn Cancer Care." The principal investigator on this study is Jeanne Drisko, MD, a Professor of Complementary and Alternative Medicine (CAM) at the University of Kansas Medical Center, USA. Drisko's is a pilot study, involving fewer than 50 patients, and so even a positive outcome is unlikely to change the minds of many oncologists (Drisko, 2003).

As Conklin, Block, Simone and others have indicated, there have already been a number of clinical trials on this topic. The majority of these studies (as well as the preponderance of *in vitro* experiments) actually show a beneficial interaction between these two modalities. There is to my knowledge only a single clinical trial that contains some negative conclusions (Bairati, 2005a,b). Negative statements about the interaction of antioxidants and cancer treatment typically rely on theoretical arguments, and on a one-sided interpretation of results from a few exceptional studies, rather than a dispassionate review of the data showing mainly positive interactions.

Looking at the broad picture, the weight of evidence – from both the laboratory and the clinic – shows that antioxidants do not interfere with the beneficial results of radiotherapy. In fact, it is possible that they may actually enhance therapeutic results. They also seem able, post-therapy, to reverse some of the adverse effects of radiation.

In addition to the many clinical trials that have been conducted in an attempt to clarify the role of antioxidants in cancer treatment, there have also been many *in vitro* experiments on the topic. These have generally supported the safety and efficacy of concurrent use. However, because of the more immediate relevance of clinical trial data, this discussion will focus on the evidence garnered from clinical, rather than *in vitro*, studies.

If antioxidants can indeed be safely administered in various clinical settings, without apparently interfering with cytotoxic treatment, this contradicts the rationale for prohibiting the concurrent use of antioxidants during cancer therapy.

Synthetic Antioxidants

The oncology profession does not avoid all antioxidants. In fact, the US Food and Drug Administration (FDA) has approved Amifostine (Ethyol® or WR-2721), a powerful synthetic antioxidant which scavenges three types of free radicals– superoxide, hydroxyl and lipoperoxyl (Marzatico *et al.,* 2000; Food and Drug Administration Oncology Division Advisory Committee). It is an antioxidant analog of cysteamine (a phosphorylated aminothiol pro-drug), exerting its effect as a selective cytoprotective agent for normal tissues against the toxicities of chemotherapy and/or radiation (Cappizi, 1993; Food and Drug Administration Oncology Division Advisory Committee).

Developed at Walter Reed Army Institute of Research, USA, to protect soldiers from radioactive fallout, it was the first antioxidant agent to be approved by the FDA and international health agencies (Capizzi, 1993, 1999). Amifostine was evaluated in a multicenter, multinational phase III clinical trial that enrolled women with stage III/IV ovarian cancer. This, and additional clinical trials, showed that amifostine can protect normal tissues from the toxic effects of alkylating agents, organoplatinums, anthracyclines, taxanes and radiation (Capizzi, 1999).

In a randomized controlled trial of 242 patients with advanced ovarian cancer, amifostine significantly reduced the cumulative kidney damage associated with cisplatin, without any sign that it undermined that drug's anticancer effects (Kemp *et al.,* 1996). If anything, response rates and survival were better in the group receiving amifostine alongside cisplatin. And while 24 percent of the cisplatin-only patients had to discontinue treatment because of toxicity, only 9 percent of the amifostine-added patients did so (Kemp *et al.*, 1996). The FDA approved amifostine because objective response rates, time to progression, and the duration of survival were similar in the amifostine and control study groups (FDA).

Amifostine was originally approved for the management of post-irradiation xerostomia (1999), and the reduction of toxicity in advanced ovarian cancer (1995) and non-small cell lung cancer (1996). There are presently several Pub Med articles that reference amifostine and radiotherapy. In one of the most recent, researchers at Thomas Jefferson Hospital, Philadelphia, USA, found that in head and neck cancer patients receiving radiotherapy the incidence of acute xerostomia was lower than reported previously with no amifostine in a controlled study (Anne *et al.,* 2007).

The general consensus is that amifostine prevents or reduces toxicity without compromising the anticancer efficacy of standard treatments, including radiotherapy. In fact, in at least one clinical trial, not only did amifostine protect against mucositis and dysphagia, the clinical response rates were better. There were complete responses in 90.9 percent of patients in the amifostine-treated group vs. 78.3 percent in the control group. Cytoprotection with amifostine did not affect treatment outcome (Antonadou *et al.*, 2002).

Another synthetic antioxidant, dexrazoxane (Zinecard, ICRF-187 or razoxane), is a derivative of the chelating agent, EDTA. Its concurrent use with radiotherapy has been extensively reported. Dexrazoxane, a cardioprotectant, is sometimes given to patients who are also receiving the cytotoxic drug doxorubicin (Adriamycin), which carries a significant cardiotoxicity risk. The FDA approved dexrazoxane for this use in 2004.

A randomized trial of adjuvant dexrazoxane in patients with soft tissue sarcoma concluded that the treatment with radiotherapy and [dex]razoxane led to an increased response rate compared to photon irradiation alone (74 vs. 49 percent) (Rhomberg *et al.,* 1996). The local control rate was likewise improved. Radiotherapy combined with [dex] razoxane improved the local control in inoperable, residual or recurrent STS (soft tissue sarcoma) compared to radiotherapy alone (Rhomberg *et al.,* 1996).

A 2007 randomized controlled trial (RCT) from the University of Montréal, Canada, similarly concluded that dexrazoxane did not have a significant impact on the 5-year EFS (event-free survival) of high-risk patients, and there was no significant difference in outcome. The authors concluded that dexrazoxane does not interfere with the antileukemic effect of doxorubicin (Moghrabi *et al.,* 2007).

If these powerful antioxidants truly interfered with radiotherapy, this fact would surely have emerged in the past decade or more of standard use around the globe. However, the general picture that emerges from over 2,000 scientific articles on amifostine and dexrazoxane is one of safe concurrent use with conventional treatment. Opponents of the concurrent use of natural antioxidants have a difficult time explaining why nutritional supplements such as vitamins A, C and E might interfere with radiotherapy, while amifostine and dexrazoxane do not.

Vitamin E for Radiation-induced Mucositis

There are indeed clinical trials that support concurrent use of dietary antioxidants. In a 2004 randomized controlled trial, 54 patients with cancer of the oral cavity and oropharynx were randomly assigned either to rinse their mouths with a solution containing the antioxidant vitamin E, or to use a placebo mouthwash before every dose of radiation, and again 8 to 12 hours later throughout 5 to 7 weeks of radiotherapy (Ferreira *et al.,* 2004). Among the patients given vitamin E, there was a 21.6 percent incidence of radiation-induced mucositis vs. 33.5 percent among the placebo group. Vitamin E was thus associated with a 36 percent reduction in the risk of this difficult side effect of treatment. It was also associated with a reduction in WHO grades 2 and 3 pain during radiation treatment (53.8 percent in the placebo group vs. 10.7 percent in the vitamin E group, a five-fold reduction in incidence).

The authors, however, observed no significant influence on survival (Ferreira *et al.,* 2004). They concluded that alpha-tocopherol (vitamin E) decreased the incidence of symptomatic oral radiotherapy-induced mucositis in patients with cancer of the oropharynx and oral cavity. If antioxidants interfered with radiotherapy one would expect to see this reflected in this otherwise highly promising study.

Pentoxifylline and Radiotherapy

Pentoxifylline (PTX) is a xanthine-derived substance normally used to improve blood flow and to reduce aching, cramping, and tiredness in the hands and feet. It decreases the viscosity of blood, allowing it to flow more easily. PTX also has antioxidant properties (Horvath *et al.*, 2002). There have been several studies of the concurrent use of PTX, as well as vitamins, with radiotherapy: 64 papers are listed on this topic in PubMed, 15 of them reporting clinical trials. Although some of these clinical trials involve the use of PTX in the post-radiation period, and therefore do not directly address the issue of concurrent use, they nonetheless bear mentioning. For instance, PTX has been used with alpha-tocopherol (vitamin E) in the successful treatment of radiation-induced fibrosis (Delanian *et al.*, 2005a, Delanian and Lefaix, 2007) as well as to treat osteoradionecrosis (Delanian *et al.*, 2005b).

With vitamin E, PTX was also found to improve uterine function in women who had received irradiation to the pelvis, sometimes decades before. In young women who want to bear children, the combination of pentoxifylline and vitamin E can reduce fibroatrophic uterine lesions after childhood irradiation (Letur-Konirsch *et al.*, 2002). PTX also exerted a modest therapeutic effect in patients who had radiation-induced trismus (Chua *et al.*, 2001). Lee and co-workers of the University of Minnesota, USA, found that PTX, along with the B vitamin nicotinamide, increased the radio-response of tumours by improving tumour oxygenation (Lee *et al.*, 1992). PTX has also been found to convey a significant protective effect for both early and late lung radiotoxicity (Ozturk *et al.*, 2004).

In 2000, Kwon and co-workers from South Korea carried out a randomized controlled trial in which they demonstrated improved outcomes, including increased survival, in patients who received PTX concurrently with conventional radiotherapy (see Table 22.1). Forty-seven patients with non-small cell lung cancer (NSCLC) were randomly divided into a radiotherapy-alone (20 patients) vs. a radiotherapy + PTX group (27 patients). Each patient received a total tumour dose of 65–70Gy, and PTX was given to some patients at a daily dose of 1200mg during radiotherapy.

Table 22.1. Effects of radiotherapy + PTX vs. radiotherapy alone in NSCLC.

	Complete response	Partial response	Stable disease	Time to relapse	Median survival	1-yr survival	2-yr survival
RT alone	15%	65%	20%	9 mo.	7 mo.	35%	12%
RT + PTX	11%	48%	11%	11 mo.	18 mo.	60%	18%

The complete response (CR), partial response (PR) and stable disease rates in the two groups were comparable. However, patients who received the supplemental antioxidant had a median overall survival 18 months vs. just 7 months for those who received radiotherapy alone. That was an increase of 260 percent (Kwon *et al.*, 2000). Needless to say, this finding is not consistent with the notion that antioxidants as a class interfere with radiotherapy. Exactly the opposite seems to be the case. The authors concluded that PTX was a modestly effective radiation response modifier and provides benefit in the treatment of non-small cell lung cancer (Kwon *et al.*, 2000).

Another RCT carried out in 2006 studied the effect of PTX and alpha-tocopherol (vitamin E) on the clinical outcome of 66 patients with stage IIIB non-small cell lung cancer (Misirlioglu *et al.*, 2006). All patients received 46Gy of external beam radiotherapy to their primary tumours and to regional lymph nodes, followed by an additional 14Gy to the primary tumour. Half the patients also received PTX (400mg, three times daily) and vitamin E (alpha-tocopherol, 300mg, twice daily) concurrently during radiotherapy. This was followed by 400mg of PTX and 300mg of alpha-tocopherol daily for 3 months following completion of radiotherapy. A control group of 33 patients received radiotherapy alone.

After a mean follow-up of 12 months, a total of 18 patients (27.3 percent) remained alive. During this follow-up period, there were local recurrences in 14 patients and distant metastases in 18 patients. In patients who received PXT and alpha-tocopherol, 1- and 2-year overall survival rates were 55 percent and 30 percent, respectively, and median survival was 18 months. In control patients, 1- and 2-year overall survival rates were 40 percent and 14 percent, respectively, with a median survival of 10 months. These differences were statistically significant (Table 22. 2).

Table 22.2. RCT on radiotherapy with and without PTX and Vitamin E.

	1 Year Overall Survival	2 Year Overall Survival	Median Survival
PTX + Vit E Group	55%	30%	18 months
Radiation Group	40%	14%	10 months

In patients who received PTX and alpha-tocopherol, progression-free survival rates for 1 and 2 years were 48 percent and 23 percent respectively; median survival was 12 months. In the control group, the corresponding rates were 24 percent and 18 percent; median survival in this group was 8 months (Table 22.3) (Misirlioglu *et al.*, 2006).

Table 22.3. Progression-free survival (P.F.S.) with PTX and alpha-tocopherol vs. control.

	1 Yr P.F.S.	2 Yr P.F.S.	Median Survival
PTX + alpha-tocopherol	48%	23%	12 months
Control Group	24%	18%	8 months

If the standard paradigm were correct, one would expect to see decreased response rates and survival curves when antioxidants such as PTX and alpha-tocopherol were administered concurrently with radiotherapy. Yet, once again, in clinical practice the opposite was the case.

Melatonin and Radiotherapy

Among its many endocrine functions, the pituitary gland hormone melatonin also acts as an antioxidant in the body. Dr. Paolo Lissoni, Chair of Radiation Oncology at San Gerardo Hospital, Monza, Italy, and colleagues have published a large series of articles on the concurrent use of melatonin and conventional treatment. Most of these articles concern melatonin used as an adjuvant to chemotherapy, but some also involve the use of radiotherapy.

In a study of stage IV glioblastoma multiforme, 30 patients were randomized to receive radiotherapy alone (60Gy) or radiotherapy plus 20mg/daily oral doses of melatonin, until disease progression. Lissoni and co-workers reported that both the survival curve and the percent of survival at 1 year were significantly higher in patients treated with RT plus MLT, than in those receiving RT alone (Lissoni *et al.*, 1996). Six out of 14 melatonin-treated patients were alive at one year vs. just 1 out of 16 in the radiotherapy group (Lissoni *et al.*, 1996).

Moreover, the authors reported that radiotherapy- or steroid therapy-related toxicities were lower in patients concomitantly treated with melatonin (Lissoni *et al.*, 1996). The addition of this antioxidant prolonged the survival time and improved the quality of life of patients affected by glioblastoma multiforme.

Vitamin A and Radiation Proctopathy

The simple dietary antioxidant vitamin A might be able to prevent or reverse some of the worst adverse effects of radiotherapy, without interfering in its efficacy. In 2005, E.D. Ehrenpreis and co-workers of the University of Chicago Medical Center, USA, tested the use of oral retinol palmitate (vitamin A) for reducing the symptoms of radiation proctopathy. This is a category that includes diarrhoea, urgency, rectal pain, tenesmus and faecal incontinence. Their study was designed to see if vitamin A (an antioxidant noted for its ability to accelerate healing) could reduce these troublesome side effects. The study was a double-blind RCT of 10,000 IU of vitamin A by mouth for 90 days vs. placebo. Patients, all of whom had significant symptoms, were recruited more than six months after pelvic irradiation. Most of them had been irradiated for prostate disease (Ehrenpreis *et al.*, 2005).

Nineteen patients were randomized: ten to receive oral doses of retinol palmitate and nine to placebo. Five of the placebo non-responders were eventually crossed over to the retinol palmitate arm for another 90 days. Seven of 10 retinol palmitate patients responded, whereas only 2 of 9 responded to placebo.

Additionally, all 5 placebo non-responders who were crossed over to treatment with retinol palmitate then responded to treatment. The authors concluded that vitamin A "significantly reduced rectal symptoms of radiation proctopathy, perhaps because of wound-healing effects". These results can serve as the foundation for future trials examining retinol palmitate in the multi-institutional setting. Douglas K. Rex, MD, opines that if substantiated by additional studies, this discovery would dramatically advance the treatment of a difficult clinical problem (Rex, 2005).

Finnish Clinical Trial

Kaarlo Jaakkola and co-workers at the University of Jyvaskyla, Finland, compared the treatment of patients receiving radiotherapy (and chemotherapy) for small cell lung cancer (SCLC) with or without antioxidant vitamins and minerals.

Antioxidant treatment, in combination with chemotherapy and irradiation, prolonged the survival time of patients with small cell lung cancer compared to most published combination treatment regimens alone. These workers also noticed that the patients receiving antioxidants were able to tolerate chemotherapy and radiation treatment well. Surviving patients started antioxidant treatment in general earlier than those who succumbed (Jaakkola, 1992).

Bairati Study

One study frequently cited as providing evidence against the concurrent use of antioxidants with radiotherapy is that carried out by Isabelle Bairati and co-workers (2005) at the Université Laval, Québec, Canada.

On the one hand, the Bairati study concluded that supplementation with high doses of alpha-tocopherol (vitamin E) and/or beta-carotene significantly mitigated the adverse effects of radiation in patients who were undergoing treatment for head and neck cancer. There was in fact a 62 percent reduction in severe adverse effects to the larynx and other sites in patients who were randomized to receive both of these antioxidants.

On the other hand, there was also a non-significant trend towards developing second primary cancers during the supplementation period, which began during radiotherapy and extended for three years afterwards (Bairati *et al.,* 2005a).

The first (positive) finding was all but ignored in discussions of this study, while the second (negative) finding generated a great deal of adverse publicity for the concurrent use of antioxidants.

The Bairati paper was published in both the Journal of the National Cancer Institute (JNCI) and the Journal of Clincal Oncology (JCO – the official publication of the American Society for Clinical Oncology). Much was made of the fact that patients receiving supplementation, in addition to radiation, had a higher rate of second primaries while receiving the vitamins. But, as a close reading of the article makes clear, they also had a lower rate of such second primaries once the supplementation was discontinued.

In fact, by the completion of the study (eight years after start of radiotherapy) there were fewer second primaries or recurrences in the supplementation group compared to those receiving a placebo (113 vs. 119 participants, respectively) (Bairati *et al.,* 2005a). In addition, a majority of supplemented patients were spared the worst adverse effects of treatment. This pointed to a potential strategy for both minimizing adverse effects (enabling patients to better tolerate the therapeutic dose), while not increasing the long-term recurrence rate. But these mitigating facts were generally downplayed or ignored in a storm of negative publicity that was generated around the Bairati trial, especially after publication of the (second) JCO article (Bairati *et al.,* 2005b).

The Québec authors called for further trials to explore the various effects of antioxidants with radiotherapy. They opined that given the current true uncertainty surrounding these issues among patients, their treating physicians, and in the medical community, randomized controlled trials should be conducted to provide clear scientific evidence regarding the efficacy and safety of antioxidant use as adjuvant therapies for cancer (Bairati *et al.*, 2005b).

One question that needs answering was whether, even temporarily, these two dietary antioxidants actually quenched the powerful free radicals generated by radiotherapy. In an accompanying JCO editorial, Kevin A. Camphausen and co-workers of the National Cancer Institute (NCI), USA, and several colleagues cast doubt on the widespread notion that supplemental antioxidants could possibly be taken by patients in sufficient quantity or strength to interfere with the primary and secondary free radical species produced by radiation therapy (Camphausen *et al.*, 2005).

They speculated that, instead, antioxidants might suppress continued free radical production that arises from an inflammatory response following radiation therapy. This could perhaps impede anti-tumour activity, although it is not known if this inflammatory response does actually occur in tumour tissue (Block, 2005). It is all highly speculative.

Camphausen and co-workers (2005) concede that most phytochemical antioxidants, far from being simple scavengers of free radicals, also trigger complicated signal transduction pathways, which may ultimately result in tumour cell death. A few of these pathways, however, may also lead to tumour cell survival. These authors conclude that while patients should avoid "unnecessary supplementation" during and after radiotherapy, using antioxidants to improve the therapeutic index of radiotherapy is a reasonable and commendable goal. Further investigations should be conducted in cancers in which there is an effective salvage therapy, in case second primaries or recurrences do occur (Camphausen *et al.*, 2005). This was a wise recommendation, which would have left the door open for further research into this complex interaction. Unfortunately, however, for many physicians the takeaway message of the Bairati article was simply to avoid the concurrent use of all supplemental antioxidants during all forms of radiotherapy and chemotherapy.

Radiation Decreases Antioxidant Levels

Finally, the use of antioxidants during conventional treatment is sometimes presented as an irrational intervention on the part of self-willed patients attempting to meddle in their own treatment against the better judgement of their oncologists (Parker-Pope, 2005). However, an alternative view is that it is the oncologists who often fail to monitor the nutritional and biochemical status of their patients before, during and after radiotherapy. If they did, they themselves might see the need to adjust antioxidant intake in conformity with test results.

Numerous studies show that conventional treatments do indeed decrease plasma antioxidant levels, which may reflect a failure of the antioxidant defence mechanism against oxidative damage induced by commonly used anticancer drugs as well as radiation (Weijl *et al.*, 1998; Sangeetha *et al.*, 1990; Durken *et al.*, 1995; Lauterburg *et al.*, 1994). It is increasingly recognized that optimal levels of antioxidants (both dietary and endogenous) are a precondition for good health. Radiotherapy, by design, generates high levels of free radicals, but these powerful cancer-killing molecules may also upset the body's oxidant-antioxidant equilibrium, causing both immediate and long-term damage that can result, among other things, in second cancers (Koh *et al.*, 2007). This can exacerbate comorbid conditions that either predate the cancer or emerge as a result of the treatment itself.

"It's important to keep comorbid conditions in mind for the sake of the entire patient and not just focus on the cancer," said Ellen F. Manzullo, an Associate Professor at The University of Texas M. D. Anderson Cancer Center, Houston, USA, "because the patient can do extremely well as far as their cancer is concerned but subsequently die of coronary artery disease or stroke" (Galloway, 2004).

There is a new class of integrative physicians who recognize that the generation of free radicals may be a necessary tool in the effort to destroy cancer cells. However, these physicians are also cognizant of the fact that exposure to radiation diminishes antioxidant levels, as the body attempts to counteract free radical damage. While many oncologists vociferously oppose patients' self-administration of antioxidants during radiation and chemotherapy, they ignore the fact that standard therapeutic methods cause a major decline in patients' levels of micronutrients as well as intrinsic antioxidant systems.

Radiation damage to antioxidant systems has particularly been studied for its role in conditioning patients for bone marrow transplantation. In one such study, Clemens and co-workers reported on vitamin E and beta-carotene levels in the blood of 19 patients, as well as their levels of lipid hydroperoxides (i.e., free radicals), before and after total-body irradiation (TBI). The lipid hydroperoxides were found to increase significantly in the group of patients with additional TBI, whereas the other group, receiving no additional TBI, showed no significant change (Clemens *et al.,* 1989). As a result, the authors suggested adding high-dose supplementation of essential antioxidants for patients undergoing BMT (Clemens *et al.,* 1989).

At Emory University, Atlanta, researchers similarly reported that chemotherapy and radiation therapy result in increased free radical formation and depletion of tissue antioxidants. They studied the plasma antioxidant status of 24 BMT patients and reported that plasma glutathione (GSH) and alpha- and gamma-tocopherol concentrations decreased and the GSH redox state became more oxidized after conditioning treatment. They concluded that a significant decline in GSH-glutathione disulfide, cysteine-cystine and vitamin E status occurs after chemotherapy and BMT (Jonas *et al.,* 2000).

Lin (2002) also studied antioxidant status in patients undergoing BMT. Ten out of 19 patients were randomized to receive vitamin C (300mg/d) and vitamin E (600mg/d) consecutively for 15 days before BMT. The control group received BMT with this prior vitamin administration. Four measurements of antioxidant levels were carried out before and after BMT. The authors concluded that exogenous supplementation of antioxidant vitamins before BMT may improve the antioxidant capacity and reduce lipid peroxidation in patients with BMT, effectively alleviating their peroxide stress induced by high-dose chemo/radiotherapy (Lin, 2002).

Many oncologists concede that antioxidant levels are decreased by toxic treatment, but believe that these levels spring back quickly. However, this may not necessarily be the case. At Leiden University Medical Center, Holland, researchers found that the levels of the intrinsic antioxidants bilirubin, albumin and uric acid all remained low for quite a while after irradiation, as did the ratio of vitamin E to cholesterol and triglycerides. The researchers called this a failure of the antioxidant defense mechanism against oxidative damage, caused by commonly used toxic treatments (Weijl *et al.,* 1998).

In a Turkish study, Elango *et al.* (2006) collected blood samples from 63 stage III oral cancer patients before initiating radiotherapy. Twenty-seven (27) of these patients were given radiotherapy alone, while the remaining 36 were given selenium (400μg/day for 6 months). Both groups were then followed for 6 months. These authors evaluated a broad array of antioxidants, including the plasma selenium concentration, non-enzymatic systems including GSH, vitamins A, C and E, ceruloplasmin and enzymatic antioxidant systems including superoxide dismutase, catalase, glutathione peroxidase, glutathione reductase, glucose-6-phosphate dehydrogenase (Elango *et al.*, 2006). They found that the concentrations of all of these were significantly lowered in oral cancer patients compared to persons without cancer. But a "similar decrease in the concentration of selenium and antioxidants status was observed in the radiotherapy group. By contrast, in the selenium-supplemented group there was a marked increase in the concentrations of selenium as well as the general antioxidant status at six months compared to the radiation group." The authors concluded: "The observed result represents the antioxidant property of selenium through the improvement of antioxidant defense system. Selenium supplementation could be of great interest in protecting cells against oxidative stress" (Elango *et al.*, 2006).

Conclusions

To summarize, both radiotherapy and chemotherapy can diminish antioxidant levels. Patients' use of antioxidants can be seen as an attempt – admittedly sometimes a poorly informed attempt – to correct this situation by the self-administration of dietary antioxidants. Oncologists have a professional responsibility to inform themselves of the clinical and physiological effect of cytotoxic treatment on antioxidants and to remediate any deficiencies. If they fail to do so, they can hardly be surprised if patients take it upon themselves – however inexpertly – to do so.

If oncologists indiscriminately oppose the ingestion of supplemental or even dietary antioxidants during treatment, are they not then actually embracing the notion of treatment-induced diminution of endogenous antioxidants as an integral part of their proposed treatment strategy? Is hypovitaminosis then to be regarded as a component of standard treatment? By what medical theory can such a thing be proposed?

But if nutritional depletion is not part of standard treatment, then aren't oncologists obligated to at least replace antioxidants and other nutrients that are lost as a result of aggressive treatment?

References

Anne, P.R., Machtay, M., Rosenthal, D.I., Brizel, D.M., Morrison, W.H., Irwin, D.H., Chouqule, P.B., Estopinal, N.C., Berson, A. and Curran, W.J. Jr. (2007) A Phase II trial of subcutaneous amifostine and radiation therapy in patients with head-and-neck cancer. *International Journal of Radiation Oncology, Biology, and Physics* 67, 445–452.

Antonadou, D., Pepelassi, M., Synodinou, M., Puglisi, M. and Throuvalas, N. (2002) Prophylactic use of amifostine to prevent radiochemotherapy-induced mucositis and xerostomia in head-and-neck cancer. *International Journal of Radiation Oncology, Biology, and Physics* 52, 739–747.

Bairati, I., Meyer, F., Gelinas, M., Fortin, A., Nabid, A., Brochet, F., Mercier, J.P., Tetu, B., Harel, F., Masse, B., Vigneault, E., Vass, S., del Vecchio, P. and Roy, J. (2005a) Randomized trial of antioxidant vitamins to prevent acute adverse effects of radiation therapy in head and neck cancer patients. *Journal of Clinical Oncology* 23, 5805–5813.

Bairati, I., Meyer, F., Gelinas, M., Fortin, A., Nabid, A., Brochet, F., Mercier, J.P., Tetu, B., Harel, F., Masse, B., Vigneault, E., Vass, S., del Vecchio, P. and Roy, J. (2005b) A randomized trial of antioxidant vitamins to prevent second primary cancers in head and neck cancer patients. *Journal of National Cancer Institute* 97, 481–488.

Block, K. (2005) Antioxidants in the news. *Integrative Cancer Therapies* 4, 271–273.

Block, K.I., Koch, A.C., Mead, M.N., Tothy, P.K., Newman, R.A. and Gyllenhaal, C. (2007) Impact of antioxidant supplementation on chemotherapeutic efficacy: A systematic review of the evidence from randomized controlled trials. *Cancer Treatment Reviews Mar* 14 (Epub ahead of print).

Cameron, N.E. and Cotter, M.A. (1999) Effects of antioxidants on nerve and vascular dysfunction in experimental diabetes. *Diabetes Research and Clinical Practice* 45, 137–146.

Camphausen, K., Citrin, D., Krishna, M.C. and Mitchell, J.B. (2005) Implications for tumour control during protection of normal tissues with antioxidants. *Journal of Clinical Oncology* 23, 5455–5477.

Capizzi, R.L. (1993) Clinical status and optimal use of Amifostine. *Oncology* 13, 47–59.

Capizzi, R.L. (1999) Amifostine reduces the incidence of cumulative nephrotoxicity from cisplatin: laboratory and clinical aspects. *Seminars in Oncology* 2(7), 72–81.

Chua, D., Lo, C., Yuen, J. and Foo, Y.C. (2001) A pilot study of pentoxifylline in the treatment of radiation-induced trismus. *American Journal of Clinical Oncology* 24, 366–369.

Clemens, M.R., Ladner, C., Schmidt, H., Ehninger, G., Einsele, H., Buhler, G., Waller, H.D. and Gey, K.F. (1989) Decreased essential antioxidants and increased lipid hydroperoxides following high-dose radiochemotherapy. *Free Radical Research Communication* 7, 227–232.

Conklin, K.A. (2000) Dietary antioxidants during cancer chemotherapy: impact on chemotherapeutic effectiveness and development of side effects. *Nutrition and Cancer* 37, 1–18.

Conklin, K.A. (2004) Cancer chemotherapy and antioxidants. *Journal of Nutrition* 34, 3201S–3204S.

D'Andrea, G.M. (2005) Use of antioxidants during chemotherapy and radiotherapy should be avoided. *CA: A Cancer Journal for Clinicians* 55, 319–321.

Delanian, S. and Lefaix, J.L. (2007) Current management for late normal tissue injury: radiation-induced fibrosis and necrosis. *Seminars in Radiation Oncology* 17, 99–107.

Delanian, S., Depondt, J. and Lefaix, J.L. (2005b) Major healing of refractory mandible osteoradionecrosis after treatment combining pentoxifylline and tocopherol: a phase II trial. *Head Neck* 27, 114–123.

Delanian, S., Porcher, R., Rudant, J. and Lefaix, J.L. (2005a) Kinetics of response to long-term treatment combining pentoxifylline and tocopherol in patients with superficial radiation-induced fibrosis. *Journal of Clinical Oncology* 23, 8570–8579.

Drisko, J.A., Chapman, J. and Hunter, V.J. (2003) The use of antioxidant therapies during chemotherapy. *Gynecologic Oncology* 88, 434–439.

Durken, M., Agbenu, J., Finckh, B., Hubner, C., Pichlmeier, U., Zeller, W., Winkler, K., Zander, A. and Kohlschutter, A. (1995) Deteriorating free radical-trapping capacity and antioxidant status in plasma during bone marrow transplantation. *Bone Marrow Transplantation* 15, 757–762.

Ehrenpreis, E.D., Jani, A., Levitsky, J., Ahn, J. and Hong, J. (2005) A prospective, randomized, double-blind, placebo-controlled trial of retinol palmitate (vitamin A) for symptomatic chronic radiation proctopathy. *Diseases of the Colon and Rectum* 48, 1–8.

Elango, N., Samuel, S. and Chinnakkannu, P. (2006) Enzymatic and non-enzymatic antioxidant status in stage (III) human oral squamous cell carcinoma and treated with radical radiotherapy: influence of selenium supplementation. *Clinica Chimica Acta* 373, 92–98.

Ferreira, P.R., Fleck, J.F., Diehl, A., Barletta, D., Braga-Filho, A., Barletta, A. and Ilha, L. (2004) Protective effect of alpha tocopherol in head and neck cancer radiation-induced mucositis: a double–blind randomized trial. *Head Neck* 26, 313–321.

Food and Drug Administration (FDA). Oncology Division Advisory Committee (ODAC). Fulfillment of the accelerated approval requirements for the non-small cell lung cancer indication: ethyol (amifostine) reduces the cumulative renal toxicity associated with repeated administration of cisplatin in patients with advanced Non–small cell lung cancer. (Available at: http://www.fda.gov/ohrms/dockets/ac/03/briefing/3936B1_05_MedImmune–Eythol.htm).

Galloway, D. (2004) Treating patients with cancer requires looking beyond the tumour. *Oncology* 49, 7–8.

Horvath, B., Marton, Z., Halmosi, R., Alwxy, T., Szapary, L., Vekasi, J., Biro, Z., Habon. T., Kesmarky, G. and Toth, K. (2002) *In vitro* antioxidant properties of pentoxifylline, piracetam, and vinpocetine. *Clinical Neuropharmacology* 25(1), 37–42.

Jaakkola, K., Lahteenmaki, P., Laakso, J., Harju, E., Tykka, H. and Mahlberg, K. (1992) Treatment with antioxidant and other nutrients in combination with chemotherapy and irradiation in patients with small-cell lung cancer. *Anticancer Research* 12, 599–606.

Jonas, C.R., Puckett, A.B., Jones, D.P., Griffith, D.P., Szeszycki, E.E., Bergman, G.F., Furr, G.E., Tyre, C., Carlson, J.L., Galloway, J.R., Blumberg, J.B. and Ziegler, T.R. (2000) Plasma antioxidant status after high-dose chemotherapy: a randomized trial of parenteral nutrition in bone marrow transplantation patients. *American Journal of Clinical Nutrition* 72, 181–189.

Kemp, G., Rose, P., Lurain, J., Berman, M., Manetta, A., Roullet, B., Homesley, H., Belpomme, D. and Glick, J. (1996) Amifostine pretreatment for protection against cyclophosphamide-induced and cisplatin–induced toxicities: results of a randomized control trial in patients with advanced ovarian cancer. *Journal of Clinical Oncology* 14, 2101–2112.

Koh, E.-S., Tran, T.H., Heydarian, M., Sachs, R.K., Tsang, R.W., Brenner, D.J., Pintilie, M., Xu, T., Chung, J., Paul, N. and Hodgson, D.C. (2007) A comparison of mantle versus involved-field radiotherapy for Hodgkin's lymphoma: reduction in normal tissue dose and second cancer risk. *Radiation Oncology* 2, 13 [Epub ahead of print].

Kwon, H.C., Kim, S.K., Chung, W.K., Cho, M.J., Kim, J.S., Kim, J.S., Moon, S.R., Park, W.Y., Ahn, S.J., Oh, Y.K., Yun, H.G., Nah, B.S. (2000) Effect of pentoxifylline on radiation response of non-small cell lung cancer: a phase III randomized multicenter trial. *Radiotherapy and Oncology* 56, 175–179.

Lauterburg, B.H., Nguyen, T., Hartmann, B., Junker, E., Kupfer, A., Cerny, T. (1994) Depletion of total cysteine, glutathione, and homocysteine in plasma by ifosfamide/mesna therapy. *Cancer Chemotherapy and Pharmacology* 35, 132–136.

Lee, I., Kim, J.H., Levitt, S.H. and Song, C.W. (1992) Increases in tumour response by pentoxifylline alone or in combination with nicotinamide. *International Journal of Radiation Oncology, Biology, and Physics* 22, 425–429.

Letur-Konirsch, H., Guis, F. and Delanian, S. (2002) Uterine restoration by radiation sequelae regression with combined pentoxifylline-tocopherol: a phase II study. *Fertility and Sterility* 77, 1219–1226.

Lin, J.Q. (2002) Effect of nutrition intervention on antioxidant capacity and lipid peroxide in patients with bone marrow transplantation. *Di Yi Jun Yi Da Xue Xue Bao* 22, 530–532.

Lissoni, P., Meregalli, S., Nosetto, L., Barni, S., Tancini, G., Fossati, V. and Maestroni, G. (1996) Increased survival time in brain glioblastomas by a radioneuroendocrine strategy with

radiotherapy plus melatonin compared to radiotherapy alone. *Oncology* 53, 43–46.
Marzatico, F., Porta, C., Moroni, M., Bertorelli, L., Borasio, E., Finotti, N., Pansarasa, O. and Castagna, L. (2000) *In vitro* antioxidant properties of amifostine (WR-2721, Ethyol). *Cancer Chemotherapy and Pharmacology* 45, 172–176.
Misirlioglu, C.H., Erkal, H., Elgın, Y., Ugur, I. and Altundag, K. (2006) Effect of concomitant use of pentoxifylline and alpha-tocopherol with radiotherapy on the clinical outcome of patients with stage IIIB non-small cell lung cancer: a randomized prospective clinical trial. *Medical Oncology* 23, 185–189.
Moghrabi, A., Levy, D.E., Asselin, B., Barr, R., Clavell, L., Hurwitz, C., Samson, Y., Schorin, M., Dalton, V.K., Lipshultz, S.E., Neuberg, D.S., Gelber, R.D, Cohen, H.J., Sallan, S.E. and Silverman, L.B. (2007) Results of the Dana-Farber Cancer Institute ALL Consortium Protocol 95–01 for children with acute lymphoblastic leukemia. *Blood* 109, 896–904.
Moss, R.W. (2000) *Antioxidants Against Cancer*, Equinox Press, Brooklyn, New York, USA.
Moss, R.W. (2006) Should patients undergoing chemotherapy and radiotherapy be prescribed antioxidants? *Integrative Cancer Therapies* 5, 63–82.
Ozturk, B., Egehan, I., Atavci, S. and Kitapci, M. (2004) Pentoxifylline in prevention of radiation–induced lung toxicity in patients with breast and lung cancer: a double-blind randomized trial. *International Journal of Radiation Oncology, Biology, and Physics* 58, 213–219.
Parker–Pope, Tara. (2005) Cancer and Vitamins: Patients urged to avoid supplements during treatment. *The Wall Street Journal* Sept 20, Sect. D:1.
Rex Douglas, K. (2005) Vitamin A reduces non–bleeding symptoms of chronic radiation proctopathy. *Journal Watch Gastroenterology* April, 12 (Available at: http://gastroenterology.jwatch.org/cgi/content/full/2005/412/1).
Rhomberg, W., Hassenstein, E.O. and Gefeller, D. (1996) Radiotherapy vs. radiotherapy and razoxane in the treatment of soft tissue sarcomas: final results of a randomized study. *International Journal of Radiation Oncology and Biology* 36, 1077–1084.
Sangeetha, P., Das, U.N., Koratkar, R. and Suryaprabha, P. (1990) Increase in free radical generation and lipid peroxidation following chemotherapy in patients with cancer. *Free Radical Biology and Medicine* 8, 15–19.
Simone, C.B. II, Simone, N.L., Simone, V. and Simone, C.B. (2007) Antioxidants and other nutrients do not interfere with chemotherapy or radiation therapy and can increase kill and increase survival. Part 1. *Alternative Therapies in Health and Medicine* 13, 22–28.
Weijl, N.I., Cleton, F.J. and Osanto, S. (1998) Free radicals and antioxidants in chemotherapy–induced toxicity. *Cancer Treatment Reviews* 23, 209–40.

INDEX